Lecture Notes in Artificial Intelligence 11607

Subseries of Lecture Notes in Computer Science

Series Editors

Randy Goebel
  University of Alberta, Edmonton, Canada
Yuzuru Tanaka
  Hokkaido University, Sapporo, Japan
Wolfgang Wahlster
  DFKI and Saarland University, Saarbrücken, Germany

Founding Editor

Jörg Siekmann
  DFKI and Saarland University, Saarbrücken, Germany

More information about this series at http://www.springer.com/series/1244

Leong Hou U. · Hady W. Lauw (Eds.)

Trends and Applications in Knowledge Discovery and Data Mining

PAKDD 2019 Workshops
BDM, DLKT, LDRC, PAISI, WeL
Macau, China, April 14–17, 2019
Revised Selected Papers

 Springer

Editors
Leong Hou U.
University of Macau
Macao, China

Hady W. Lauw
Singapore Management University
Singapore, Singapore

ISSN 0302-9743   ISSN 1611-3349 (electronic)
Lecture Notes in Artificial Intelligence
ISBN 978-3-030-26141-2   ISBN 978-3-030-26142-9 (eBook)
https://doi.org/10.1007/978-3-030-26142-9

LNCS Sublibrary: SL7 – Artificial Intelligence

This Springer imprint is published by the registered company Springer Nature Switzerland AG
The registered company address is: Gewerbestrasse 11, 6330 Cham, Switzerland

Preface

The Pacific-Asia Conference on Knowledge Discovery and Data Mining (PAKDD) is one of the longest established and leading international conferences in the areas of data mining and knowledge discovery. It provides an international forum for researchers and industry practitioners to share their new ideas, original research results, and practical development experiences from all KDD-related areas, including data mining, data warehousing, machine learning, artificial intelligence, databases, statistics, knowledge engineering, visualization, decision-making systems, and the emerging applications. As the first joint event, PAKDD 2019 was held in Macau SAR, China, during April 14–17, 2019.

Along with the main conference, PAKDD workshops intend to provide an international forum for researchers to discuss and share research results. After reviewing the workshop proposals, we were able to accept five workshops that covered topics in intelligence and security informatics, weakly supervised learning, data representation for clustering, biologically inspired techniques, and deep learning for knowledge transfer. The diversity of topics in these workshops contributed to the main themes of the APWeb-WAIM conference. These workshops were able to accept 30 full papers that were carefully reviewed from 52 submissions. The five workshops were as follows:

- The 14th Pacific Asia Workshop on Intelligence and Security Informatics (PAISI 2019)
- PAKDD 2019 Workshop on Weakly Supervised Learning: Progress and Future (WeL 2019),
- Learning Data Representation for Clustering (LDRC 2019),
- The 8th Workshop on Biologically Inspired Techniques for Knowledge Discovery and Data Mining (BDM 2019)
- The First Pacific Asia Workshop on Deep Learning for Knowledge Transfer (DLKT 2019)

July 2019

Hady W. Lauw
Leong Hou U.

Organization

Workshop Co-chairs

Hady W. Lauw — Singapore Management University, Singapore
Leong Hou U. — University of Macau, SAR China

The 14th Pacific Asia Workshop on Intelligence and Security Informatics (PAISI 2019)

Workshop Co-chairs

Michael Chau — The University of Hong Kong, SAR China
G. Alan Wang — Virginia Tech, USA
Hsinchun Chen — The University of Arizona, USA

Program Committee

Ahmed Abbasi — Virginia Tech, USA
Indranil Bose — Indian Institute of Management, Calcutta
Robert Chang — Central Police University, Taiwan, China
Uwe Glasser — Simon Fraser University, Canada
Eul Gyu Im — Hanyang University, South Korea
Da-Yu Kao — Central Police University, Taiwan, China
Siddharth Kaza — Towson University, USA
Paul W. H. Kwan — The University of New England, Australia
Mark Last — Ben-Gurion University of the Negev, Israel
Ickjai Lee — James Cook University, Australia
You-Lu Liao — Central Police University, Taiwan, China
Hongyan Liu — Tsinghua University, China
Hsin-min Lu — National Taiwan University, Taiwan, China
Xin Robert Luo — University of Minnesota, USA
Anirban Majumdar — SAP Research, Germany
Byron Marshall — Oregon State University, USA
Dorbin Ng — The Chinese University of Hong Kong, SAR China
Shaojie Qiao — Southwest Jiaotong University, China
Jialun Qin — The University of Massachusetts Lowell, USA
Shrisha Rao — IIIT-Bangalore, India
Srinath Srinivasa — IIIT-Bangalore, India
Aixin Sun — Nanyang Technological University, Singapore
Paul Thompson — Dartmouth College, USA
Jau-Hwang Wang — Central Police University, Taiwan, China
Shiuh-Jeng Wang — Central Police University, Taiwan, China
Jennifer Xu — Bentley University, USA

| Yilu Zhou | Fordham University, USA |
| William Zhu | University of Electronic Science and Technology of China, China |

PAKDD 2019 Workshop on Weakly Supervised Learning: Progress and Future (WeL 2019)

Organizing Committee

| Yu-Feng Li | Nanjing University, China |
| Sheng-Jun Huang | Nanjing University of Aeronautics and Astronautics, China |

Learning Data Representation for Clustering (LDRC 2019)

General Chair

| Mohamed Nadif | Paris University Descartes, France |

Program Co-chair

| Lazhar Labiod | Paris University Descartes, France |

Workshop Organizers

Lazhar Labiod	Paris University Descartes, France
Mohamed Nadif	Paris University Descartes, France
Allou Same	l'Institut français des sciences et technologies des transports, de l'aménagement et des reseaux, France

Program Committee

Severine Affeldt	Paris University Descartes, France
Salah Aghiles	Singapore Management University, Singapore
Melissa Ailem	USC Michelson Center, USA
Daniel Berrar	Tokyo Institute of Technology, Japan
Patrice Bertrand	CREMADE, Paris Dauphine University, France
Marc Boullé	Orange Labs, France
Frabcusci de Assis Tenorio de Carvalho	Universidade Federal de Pernambuco, Brazil
Eruc Gaussier	University Grenoble Alps, France
Salvatore Ingrassia	University of Catania, Italy
Lazhar Labiod	Paris University Descartes, France
Vincent Lemaire	Orange Labs, France
Ahmed Moussa	ENSAT, Abdelmalek Essaadi University, Morocco
Mohamed Nadif	LIPADE, Paris University Descartes, France
François Role	Paris University Descartes, France
Fabrice Rossi	Université Paris 1 Panthéon-Sorbonne, France
Allou Same	IFSTTAR, France

The 8th Workshop on Biologically-Inspired Techniques for Knowledge Discovery and Data Mining (BDM 2019)

Program Chairs

Shafiq Alam	University of Auckland, New Zealand
Gillian Dobbie	University of Auckland, New Zealand

Program Committee

Patricia Riddle	University of Auckland, New Zealand
Kamran Shafi	University of New South Wales, Australia
Stephen Chen	York University, Canada
Mengjie Zhang	Victoria University, New Zealand
Kouroush Neshatian	Victoria University, New Zealand
Ganesh Kumar Venayagamoorthy	Missouri University of Science and Technology, USA
Yanjun Yan	Western Carolina University, USA
Mark Johnston	Victoria University, New Zealand
Asifullah Khan	Pakistan Institute of Engineering and Applied Sciences, Pakistan
Bo Yuan	Tsinghua University, China
Richi Nayak	Queensland University of Technology, Australia
Faiz Rasol	University of Auckland, New Zealand

The First Pacific Asia Workshop on Deep Learning for Knowledge Transfer (DLKT 2019)

Organizers

Fuzhen Zhuang	Institute of Computing Technology, Chinese Academy of Sciences, China
Deqing Wang	Beihang University, China
Pengpeng Zhao	Soochow University, China

Contents

The 14th Pacific Asia Workshop on Intelligence and Security Informatics (PAISI 2019)

A Supporting Tool for IT System Security Specification Evaluation Based on ISO/IEC 15408 and ISO/IEC 18045

Da Bao[✉], Yuichi Goto, and Jingde Cheng

Department of Information and Computer Sciences,
Saitama University, Saitama, Japan
{baoda,gotoh,cheng}@aise.ics.saitama-u.ac.jp

Abstract. In evaluation and certification framework based on ISO/IEC 15408 and ISO/IEC 18045, a Security Target, which contains the specifications of all security functions of the target system, is the most important document. Evaluation on Security Targets must be performed as the first step of the whole evaluation process. However, evaluation on Security Targets based on ISO/IEC 15408 and ISO/IEC 18045 is very complex. Evaluation process involves of many tasks and costs lots of time when evaluation works are performed by human. Besides, it is also difficult to ensure that evaluation is fair and no subjective mistakes. These issues not only may result in consuming a lot of time, but also may affect the correctness, accuracy, and fairness of evaluation results. Thus, it is necessary to provide a supporting tools that supports all tasks related to the evaluation process automatically to improve the quality of evaluation results at the same time reduce the complexity of all evaluator and certifiers' work. However, there is no such supporting tool existing until now. This paper proposes a supporting tool, called Security Target Evaluator, that provides comprehensive facilities to support the whole process of evaluation on Security Targets based on ISO/IEC 15408 and ISO/IEC 18045.

Keywords: IT security evaluation · ISO/IEC 15408 ·
ISO/IEC 18045 · Security target

1 Introduction

ISO/IEC 15408 [1–3] and ISO/IEC 18045 [4] (also known as Common Criteria and Common Evaluation Methodology) are a pair of international competitive standards for security evaluation on IT systems. They are used to evaluate the security capability of different target information systems. The results of evaluation can be used for comparisons among different IT systems, such that customers can easily choose the systems. Evaluation based on ISO/IEC 15408 and ISO/IEC 18045 now is widely used, there are already 28 countries have taken

© Springer Nature Switzerland AG 2019
L. H. U and H. W. Lauw (Eds.): PAKDD 2019 Workshops, LNAI 11607, pp. 3–14, 2019.
https://doi.org/10.1007/978-3-030-26142-9_1

these standards as their national standards to measure security of IT systems [9].

A Security Target (ST), which contains the specifications of all security functions of the target system, is the most important document in evaluation based on ISO/IEC 15408 and ISO/IEC 18045. Mistakes in ST can have repercussions throughout the whole evaluation process based on ISO/IEC 15408. Consequently, it is necessary to perform a strict evaluation on ST as the first step of the evaluation.

However, evaluation on Security Targets based on ISO/IEC 15408 and ISO/IEC 18045 is very complex. The evaluation process involves of many tasks and costs lots of time when evaluation works are performed by human [5,6]. Besides, it is also difficult to ensure that evaluation is fair and no subjective mistakes. These issues not only may result in consuming a lot of time, but also may affect the correctness, accuracy, and fairness of evaluation results. Thus, it is necessary to provide a supporting tool that supports all tasks related to the evaluation process automatically to improve the quality of evaluation results at the same time reduce the complexity of all evaluator and certifiers' work. However, there is no such supporting tool existing until now.

This paper proposes a supporting tool, called Security Target Evaluator, that provides comprehensive facilities to support the whole process of evaluation on STs based on ISO/IEC 15408 and ISO/IEC 18045. This paper analyzed the evaluation process of STs based on ISO/IEC 18045 and clarified 168 detailed evaluation tasks. The procedures of performing those tasks are also proposed. Then the paper provides corresponding methods to support evaluation on STs. With these methods, we can implement a comprehensive supporting tool for evaluation on STs.

2 Security Target and Its Evaluation Process

2.1 Security Target with ISO/IEC 15408

ISO/IEC 15408 gives a unified vocabulary to describe security characteristics of the target systems. The standard is composed of 3 parts: Part 1 is an introduction to the framework of ISO/IEC 15408; Part 2 is the security functional components; Part 3 is the security assurance components. The abstract security requirements are provided in Part 2 and Part 3 of ISO/IEC 15408.

An ST is the document that describes security specifications of the target system and clarifies the scope of it in the unified vocabulary provided by ISO/IEC 15408. The basic contents of an ST are organized as the Fig. 1. The figure shows the structural outline of an ST. ST introduction describes the target system, which is called Target of Evaluation (TOE) in ISO/IEC 15408, on different levels of abstraction. Conformance claims shows that the ST claims conformance to ISO/IEC 15408 Part 2, ISO/IEC 15408 Part 3, etc. Security problem definition (SPD) shows which threat, organizational security policies (OSP) and assumption that must be countered, enforced and upheld by the TOE; Security objectives shows the solution (TOE itself and operational environment) to the security problems for the TOE; Extended components definition defines new components if the components are not included in ISO/IEC 15408 Part 2 or CC Part3. Security requirements defines the SFRs and SARs; TOE summary specification shows how the SFRs are implemented in the TOE.

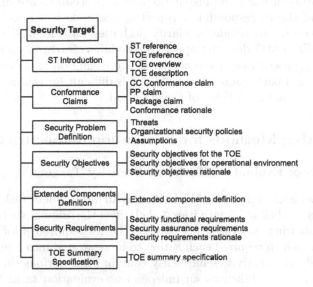

Fig. 1. The structure of security target

2.2 Evaluation Process Based on ISO/IEC 18045

ISO/IEC 18045 is a companion standard of ISO/IEC 15408, and provides a set of activities that can be followed to conduct an ISO/IEC 15408 evaluation on the target system. These activities describe the minimum actions to be performed in the evaluation. What to evaluate STs is also provided in ISO/IEC 18045, which consists of 76 activities.

For example, INT.1-10 is an activity of ISO/IEC 18045 for evaluating STs. That is to examine the TOE description to determine that it describes the logical scope of the TOE. According to INT.1-10, there are 3 actions in the activity that should be performed. The first action is to confirm whether the logical scope description of the target system is in the TOE description section. The second action is to confirm whether the security functions are described in the logical scope description. The third action is to confirm whether the security functions are described clearly and not possible to lead a misunderstanding. When INT.1-10 is followed to examine the ST, all of the 3 actions must be performed properly.

2.3 Issues of the Evaluation Process

Although, ISO/IEC 18045 has given a set of instructions to guide the evaluation activities, but some are not clear enough. Some instructions may include multiple actions about how to evaluate ST. Some activities may include implicit actions, that may cause confusion of non-experienced evaluators and result in no subjective mistakes in evaluation. The example instruction is combined with 3 targets. Those targets are the smallest unit of verification and validation. Moreover, for the tasks included in the same instruction, their procedures are different with each other, and the corresponding supporting methods will also be different. Thus, it is necessary to divide or clarify activities into detailed tasks. Additionally, ISO/IEC 18045 does not specify procedures clearly to carry out those activities. Many procedures of the activities are basing on experienced evaluators' judgment and background knowledge. It is difficult for evaluators who are lack of experience and knowledge of the standard.

3 Supporting Methods for Evaluation on Security Targets

3.1 Analysis of Evaluation Tasks on Security Targets

Considering that each action should be taken as an independent task for evaluating STs, we analyzed all of 76 activities and clarified the original set of evaluation activities by following two rules. (1) Some activities required a lot of multiple actions or steps, which required each action to be separated into a separate evaluation task. (2) Some activities implicitly contain some actions that should be specified explicitly and taken as an independent evaluation task. We clarified 168 evaluation tasks in the 76 activities. The details of each evaluation task was shown in related work, which is published in another papers [12].

Then, we clarified the procedures for each evaluation tasks. Among the procedures for 168 evaluation tasks, the procedures of some tasks are similar with each other. Therefore, we classified the 168 evaluation tasks into 6 groups. Procedures of tasks in a group have similar pattern. Table 1 shows which group the tasks belong to.

Table 1. Count of evaluation tasks in each classification

Classification of evaluation tasks	Count
Existence of content	68
Sufficiency and necessity of content	47
Correctness of outside relationship	16
Sufficiency and necessity of inside relationship	24
Correspondence of security requirements	10
Dependency among security requirements	3

The following paragraph shows the 6 groups of targets and the procedures patterns of each group. The underlined text changes according to the targets just like parameters in a function.

- **Existence of Content**
  The tasks in this group is about examining whether the contents, that should be included in the ST, is existing or not.
  procedure pattern
  1. select The task about existence of the contents
  2. examine whether **the content** that should be contained in STs according to the rule, is existing in **related section** of the ST
- **Sufficiency and Necessity of Content**
  The tasks in this group is about examining whether the contents is sufficient or necessary or possible to occur misunderstanding for this ST.
  procedure pattern
  1. select The task about the relationship between ST and other documents
  2. find out **the section** relating to The task
  3. examine the content in these sections whether the contents is sufficient or necessary or possible to occur misunderstanding for this ST according to the provided explanations and tips
- **Correctness of Outside Relationship**
  The tasks in this group is about examining whether the relationship between ST and other document (ISO/IEC 15408 Part 2, ISO/IEC 15408 Part 3, etc.) claimed in the ST is the same as actual relationship.
  procedure pattern
  1. select The task about the relationship between ST and other documents
  2. find out these **the elements** in the ST and the other document relating to target
  3. examine whether the set of elements in the ST has one different element from or is a subset of elements in the other documents by comparing the two sets
- **Sufficiency and Necessity of Inside Relationship**
  The tasks in this group is about examining whether the relationship between different parts of the ST is sufficient and necessary for this ST.
  procedure pattern
  1. select The tasks about the relationship between two parts of ST
  2. find out **these sections**, that reflect the relationship
  3. examine the sufficiency and necessity of the relationship by analysis the contents of these sections according to the provided explanations and tips
- **Correspondence of Security Requirements**
  The tasks in this group is about examining whether security requirements in this ST is an instance of the abstract security requirements (SRs) provided in ISO/IEC 15408 Part 2 and Part 3.
  procedure pattern
  1. select The task about the security requirements
  2. find out **the SRs** in the ST and the corresponding **abstract SRs** in CC Parts
  3. prove whether these SR in the ST is instance of abstract SR in CC Parts

- **Dependency among Security Requirements**
  The tasks in this group is about examining whether the dependency of security requirements in this ST satisfied the dependency that defined in ISO/IEC 15408.
  procedure pattern
  1. select The task about the security requirements
  2. find out **the SRs** in the ST
  3. examine every SR and confirm that the SRs depended by this SR is existing in the ST

The procedural order is also important parts of evaluation on STs. We divided the activities into small tasks, but all of tasks should be combined to form a complete evaluation process. For example, suppose there are two targets from group *Existence of Content* and group *Sufficiency and Necessity of Content*, the target about examining whether the content is existing must be confirmed firstly, then another target will be examined to determinate the sufficiency and necessity of existing content. Thus, we should clarify the procedural order among the targets, so that the evaluation process can be supported properly.

3.2   Supporting Methods

We proposed the supporting methods under the consideration that procedures of the evaluation tasks in the same group can be supported by the the same method. The following paragraphs show the supporting method for each group.

Existence of Content: The tasks in this group can be checked automatically if an ST is formalized in XML. For example, INT.1-2-2 is a target of this group. That is "existence of version of the ST". It is possible to examine INT.1-2-2 by defining the tags $<st\_version>$, $</st\_version>$ in the XML format of STs and checking whether a text exists between the tags $<st\_version>$, $</st\_version>$ or not.

Sufficiency and Necessity of Content: The determinations of the tasks in this group can only be made by human. Thus, the supporting method for this group is providing an environment to display only the specification of the ST and guidances or helpful explanations. It is possible to implement by tagging STs and related documents. The environment is a convenience that the evaluators can focus on making the determination and have no more need of finding out the relevant sections by themselves.

Correctness of Outside Relationship: The tasks in this group can be checked automatically by providing some functions. The functions can extract the relevant sections from the ST formatted in XML and relevant documents formatted in XML, and then compare these sections to confirm the relationship among these sections according to the targets. The extraction and comparison can be easily completed by the software that can save a lot of time for the developers.

Sufficiency and Necessity of Relationship: The determinations of the tasks in this group can only be made by human. Thus, the supporting method for this group is providing an environment in which the content of trace and the two relevant sections are displayed automatically by search the tagging ST. Some prepared explanations and tips will also be displayed in the environment to help the developers to make the determination. The environment is a convenience that the developers can focus on making the determination and have no more need of finding out the relevant sections by themselves.

Correspondence of Security Requirements: The tasks in this group can be checked by a formal method that performs the evaluation tasks based on ISO/IEC 15408 [13]. The formal method uses in Z notation to formalize the ST's security requirements and ISO/IEC15408's security requirements, and uses Z/EVES [8], a theorem proving tool, to support the tasks in this group. An environment [11] is provided to supporting the process of formalizing the STs' security requirements in Z notation [7]. The abstract security requirements in the ISO/IEC 15408 has been formalized and will be provided when the evaluation tasks is performed.

Dependency Among Security Requirements: The tasks in this group can be performed automatically if an ST is formalized in XML and the dependency is organized in structural format. To support the tasks, we built a hierarchical tree based on the dependency that is defined in the ISO/IEC 15408 Part 2. For each SFR in ST, a list, that include all necessary SFRs, will be produced according to the hierarchical tree. Every security function requirement in the list is examined to confirmed whether it is existing in the ST.

Procedural Order: Supporting for procedural order of evaluation tasks can be performed automatically according to the relationship among the tasks. To support the tasks, we built a hierarchical tree based on the procedure order. When a task is going to be examined, the relevant task will be confirmed according to the relationship. A list is produced to show all of the tasks whose examination must be performed before the selected one's. A second list is produced to show the targets whose examination can be performed after the selected one. It will provide a convenient for the developers, because there is no need to prepare the execution order for the tasks and focus on determination and checking contents.

4   Security Target Evaluator: A Supporting Tool for Evaluation on Security Targets

4.1   Requirements for Security Target Evaluator

Security Target Evaluator (STE for short) is a supporting tool which provides functionality to support the whole process of evaluation on STs based on ISO/IEC 15408 and ISO/IEC 18045. We defined 3 basic requirements for STE. Its requirements are as follows:

R1. STE must guarantee that all of evaluations tasks for STs must be performed properly. According to our analysis and classification on evaluation tasks, we need to implement components for executing each supporting method.

R2. STE must provide facilities for all evaluators to perform their tasks in a guided regular order. Evaluation based on ISO/IEC 15408 and ISO/IEC 18045 is a very complex process. The evaluation process corresponds to a set of evaluation tasks from different classification, wherefore different components must collaborate with each other to perform these evaluation task completely.

R3. STE must guarantee that each relevant document and relevant data must be managed in an appropriate format. The evaluation process of STs would refer to ISO/IEC 15408 and ISO/IEC 18045, the Security Target document, all of comments from evaluators and kinds of temporary files in evaluation tasks. All of document and data must be managed in an appropriate format, so as to be accessed and reused easily.

4.2 Components for Supporting Methods in Security Target Evaluator

To perform all evaluation tasks on STs, we implemented corresponding components to execute each supporting method. We devised the algorithm for each component according to the procedural pattern for each supporting method. We also designed each component with the capability of handling different inputs that is required by the evaluation tasks in corresponding group. Additionally, we assigned *Group ID* for each evaluation task so that each tasks can refer to the right component. The Fig. 2 shows all components of STE and relevant data.

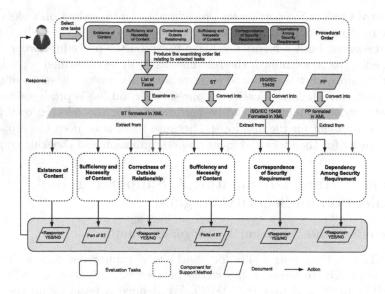

Fig. 2. Components in security target evaluator

To support the evaluators for performing the evaluation task in regular order, we assigned *Task ID* for each evaluation tasks and defined the hierarchical tree to organize the IDs. We also devised the function for produce the sequential list for each tasks. The list can be shown to the evaluators who can make their own plan according to the sequential.

We designed STE as a web application for evaluators to ubiquitously evaluate STs without depending on a particular environment. We implemented it with Java (Version 8 Update 171) and Play Framework (Version 2.2.1). Figure 3 show user interfaces for choose ST and the evaluation tasks.

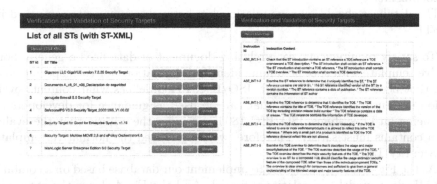

Fig. 3. Choosing ST and the evaluation tasks.

4.3   Management of Relevant Documents and Data

To manage the relevant documents and data, we prepared appropriate formats for them. Because XML has the ability of the flexible representing, we used XML to design formats for ST, ISO/IEC 15408, each kind of list and hierarchical tree. Considering that the ST formalized in XML is the basis of all the support methods, we defined XML-based format (named ST-XML) for the ST according to ISO/IEC 15408 and ISO/IEC 18045. ISO/IEC 15408 explains what an ST must contain, and defines a normative structure for STs. That can be used as guidance in the producing an ST. ISO/IEC 18045 provides the criteria that it is necessary for ST to satisfy. From another perspective, the criteria also put out information about what an ST must contain. Therefore, we adapted the normative structure in ST-XML, and defined every necessary tags that are sufficient for the production of STs. Furtherly, we defined some necessary tags relating to examining the targets of verification and validation to make the ST-XML more suitable for the verification and validation. With the help of ST-XML, relating sections can be easily found out by searching the corresponding tags. There is no need to execute the searching actions by the developers themselves, and they can focus on the judgment whether the ST satisfied the targets of verification and validation. The Fig. 2 shows all the relevant document and data, and also the relationship between components and these data.

In order to manage ISO standards and relating data by database, various characteristics of the standards and documents have to be taken into a careful consideration.

- The database should satisfy following requirements according to their characteristics.
- The database should correspond with semistructured structure of the standards.
- The database should manage all relating documents of evaluation process and be able to extract needed parts from the standards and documents.
- The database should maintain the relationship among the standards and the documents.
- The database should manage relationship between components composed in STE.

To satisfy requirements, each relating document or data need a general normative template with specific unique identify informations for each specific content that need to be searched out. ISO/IEC 15408 defines what contents evaluation evidences must contain, that can be used as guidance for developers to producing an IT systems. ISO/IEC 18045 provides the criteria that the evaluation contains must satisfy. Therefore, we summarized the normative template for Security Target.

As the Fig. 2 shows, we choose to implement our database based on a combine of XML data model and relational model. And the Fig. 4 shows the structure of our database. We chose to implement the evaluation management database by using IBM DB2 Express-C. Because data model of evaluation management database is based on a combine of XML data model and relational model. IBM DB2 Express-C is a free hybrid type database management system with strong functions to support such data models.

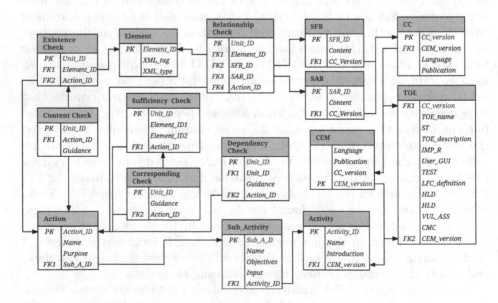

Fig. 4. Data model in security evaluation database

We have designed XML templates for evaluation relating documents, summarized all tasks of evaluation process based on Version 3.1 of ISO/IEC 15408 and ISO/IEC 18405. We implemented a prototype of database with all the tables we designed by using IBM DB2 Express-C Database Management System [10].

5   Concluding Remarks

This paper proposes a supporting tool, called Security Target Evaluator, that provides facilities to support the whole process of evaluation on STs based on ISO/IEC 15408 and ISO/IEC 18045. To implement the supporting tool, this paper analyzed the evaluation process of STs based on ISO/IEC 18045 and clarified 168 detailed evaluation tasks. The we clarified the procedures of each evaluation task and provided supporting methods for them. This paper also shown the design and implementation of STE. STE is the first tool that can support the whole process of evaluation on STs. Currently, STE can provide partial facilities for evaluating STs. The next step, we are going to implement all components and provide full facilities for evaluators to perform all evaluation tasks on STs.

Besides evaluating STs, ISO/IEC 15408 and ISO/IEC 18045 also provides instruction to guide evaluators to perform evaluation on development process, evaluation on the guidance document, evaluation on life-cycle support process, evaluation on test process, evaluation on vulnerability assessment process, and evaluation on composition process. The evaluation activities should be also supported properly. In the future, we will construct the supporting environment for IT system security evaluation and certification based on ISO/IEC 15408 and ISO/IEC 18045 that provides comprehensive facilities to support the whole evaluation and certification process. This supporting environment can no only provide evaluators with a high time efficiency, but also can provide more fairness, more correctness and more accuracy of evaluation results.

References

1. International Organization for Standardization: ISO/IEC 15408: 2009, Information Technology Security Techniques Evaluation Criteria for IT Security Part 1: Introduction and General Model (2009)
2. International Organization for Standardization: ISO/IEC 15408–2: 2008, Information Technology Security Techniques Evaluation Criteria for IT Security Part 2: Security Functional Components (2008)
3. International Organization for Standardization: ISO/IEC 15408–3: 2008, Information Technology Security Techniques Evaluation Criteria for IT Security Part 3: Security Assurance Components (2008)
4. International Organization for Standardization. ISO/IEC 18045: Information Technology Security Techniques Methodology for IT Security Evaluation (2008)
5. Herrmann, D.S.: Using the Common Criteria for IT Security Evaluation. Auerbach Publications, New York (2002)

6. Higaki, W.H.: Successful Common Criteria Evaluations: A Practical Guide for Vendors. CreateSpace, Lexington (2010)
7. Lightfoot, D.: Formal Specification Using Z, 2nd edn. Red Globe Press, London (2000)
8. ORA Canada. Z/EVES. http://oracanada.com/z-eves/welcome.html. Accessed 29 Sept 2018
9. Members of the Common Criteria Recognition Arrangement. https://www.commoncriteriaportal.org/ccra/members/. Accessed 29 Sept 2018
10. IBM DB2 Express-C. https://www.ibm.com/analytics/jp/ja/technology/db2/db2-trials.html. Accessed 20 Nov 2018
11. Yajima, K., Morimoto, S., Horie, D., Azreen, N.S., Goto, Y., Cheng, J.: FORVEST: a support tool for formal verification of security specifications with ISO/IEC 15408. In: Proceedings of the 4th International Conference on Availability, Reliability and Security (ARES 2009), Fukuoka, Japan, pp. 624–629. IEEE Computer Society Press (2009)
12. Bao, D., Miura, J., Zhang, N., Goto, Y., Cheng, J.: Supporting verification and validation of security targets with ISO/IEC 15408. In: Proceedings of 2nd International Conference on Mechatronic Sciences, Electric Engineering and Computer (MEC 2013), Shenyang, China, pp. 2621–2628. IEEE Press (2013)
13. Morimoto, S., Shigematsu, S., Goto, Y., Cheng, J.: Classification, formalization and verification of security functional requirements. In: Geffert, V., Karhumäki, J., Bertoni, A., Preneel, B., Návrat, P., Bieliková, M. (eds.) SOFSEM 2008. LNCS, vol. 4910, pp. 622–633. Springer, Heidelberg (2008). https://doi.org/10.1007/978-3-540-77566-9_54

An Investigation on Multi View Based User Behavior Towards Spam Detection in Social Networks

Darshika Koggalahewa$^{(\boxtimes)}$ and Yue Xu

Queensland University of Technology, Brisbane, Australia
darshikaniranjan.koggalahewa@hdr.qut.edu.au,
yue.xu@qut.edu.au

Abstract. Online Social Networks have become immensely vulnerable for spammers where they spread malicious contents and links. Understanding the behaviors across multiple features are essential for successful spam detection. Majority of the existing methods rely on single view of information for spam detection where diversified spam behaviors may not allow these techniques to be survived. As a result, Multiview solutions are getting emerged. Based on homophily theory, a hypothesis of spammer's behaviors should be inconsistent across multiple views compare to legitimate user behaviors is defined. We investigated the consistency of the user's content interest and popularity over multiple topics across multiple views. The results confirm the existence of notable difference of average similarity between legitimate and spam users. It proved that the legitimate user behaviors are consistent across multiple views while spammers are inconsistent. This indicates that consistency of user behavior across multiple views can be used for spam detection.

Keywords: Spam detection · Content interest · Behavior consistency

1 Introduction

The nature and the structure of a complex social system are determined by the members and their relationships of the system. Online Social Networks (OSN) enable new forms of social interaction with profound influence on our social, political and business cultures. They are inherently liable to be exploited for malicious and criminal purposes. For example, malicious users spread unsolicited or harmful information (i.e., spamming), or persuade other users to provide sensitive information (i.e. phishing) that can be exploited for further attacks [1]. The development of techniques to protect against OSN-based threats is critically important and a matter of urgency. Among all types of OSN attackers, Social spammers have shown a significant growth in OSNs by spreading phishing scams, publishing malicious contents with links and promotion campaigns [1]. The spammers frequently change their spamming strategies and try to disguise as benign users.

State of the art reveals that, mainly classification methods have been proposed for spam detection [2, 3] while they used user features and OSN features for spam

© Springer Nature Switzerland AG 2019
L. H. U and H. W. Lauw (Eds.): PAKDD 2019 Workshops, LNAI 11607, pp. 15–27, 2019.
https://doi.org/10.1007/978-3-030-26142-9_2

detection. Many efforts try to find the new features or use combination of features to improve the accuracy of the classifiers [3, 4]. Essentially, classifiers are constructed based on a labelled training dataset with selected features. However, the labelled training datasets are expensive to generate and mostly unavailable to users. Moreover, attackers' features change over time, meaning that classifiers that are constructed based on previous data cannot detect new forms of attacks. In addition to the limitations associated with classification techniques, successful classifiers developed for spam detection even suffered with the problems of data labelling, spam drift, imbalance datasets and data fabrication in terms of spam detection.

The areas of sociology and phycology have revealed that the human behaviors are consistent across different contexts. Nevertheless, they suffered with limitations of using less samples of human beings for their experiments. There is no such investigation which used the digital data to prove this nature of human behaviors. The homophily theory [7] suggests that, people with similar characteristics or interests are likely to connect. Cardoso [6] has investigated that two users who follow reciprocally have the same user interest in their shared contents. Further they highlighted the fact that "there exists a higher topical similarity between connected users in social networks". Consistency can be defined as the adherence to the same principles or agreement or compatibility among the parts of complex processes. Sharemen et al. defined that the situational similarity along with behavioural and usage similarity imply the behavioural consistency of human personalities in their day to day lives [8]. Consistency can be considered as one of the fundamental factors to determine the trust and loyalty among individuals or communities. According to above theories, benign users have similar interest over the same contents and their behaviors should be consistent across different situations. At the same time malicious users may not similar in terms of their content interest and their behaviors are not consistent across different situations.

Based on the theories of homophily and consistency we developed a general hypothesis: "The user behaviors should be consistent across multiple aspects. We considered the user's information interest and the popularity derived through the retweeting behavior as two separate views. Based on two views, we extend the general hypothesis: "Legitimate users should have a higher consistency over multiple views while spam users are less consistent or inconsistent". The intention was towards the spam detection and we conducted a set of experiments to verify the hypothesis. Three publicly available twitter data sets, Social Honeypot [9], HSpam14 [10] and "The fake project" [11] were used for the experiments. The results were supporting the hypothesis and indicated that the user behavior consistency across multiple views can be useful for spam detection. The contributions of this paper can be summarized as follows. The paper presents an in-depth empirical analysis on three real world twitter datasets and the investigation results proved that the features distribution across different views for legitimate users and spammers are different. Hence it is possible to use multiple views for spam detection. The investigation reveals that the user's behaviors across multiple views for legitimate users are consistent while spam user behaviors are inconsistent. Thus user behavior consistency across multiple views could be used for spam detection.

2 Related Work

There are numerous efforts for spam detection which used the raw twitter contents from both message and account aspects. They employ the machine learning techniques to detect spammers. Few others tried to combine features from twitter contents and apply some statistical methods to detect the spams. Connecting with third-party services for the detection has become the third approach for spam detection [12]. Majority of the spam detection techniques employed with classification techniques by utilizing different features to train the classifiers. SMFSR used the information of user activities to train the classifier while SSDM introduced a combination of text information and network information for spam detection through classification [13]. Some efforts tried to learn classifiers by only using the features extracted from user's behaviors [4, 14] by distinguishing the features for identifying spammers. Some content features used for spam detection include *post to comment similarities, noun phrase analysis, word usage such as keywords or patterns, frequency of words, bag of words* etc. It is essential to combine these text features with other features such as url or connection features to detect the spammers [4]. Some efforts have utilized the OSN features for spam detection. Ghosh et al. [15] proposed a ranking scheme by using the link farming and a SybilRank algorithm is presented in [16] by considering the social graph properties as inputs. Social account relationships are used to construct a social relationship building algorithm and they extended their algorithm to detect spammers based on account relationships [17]. The current spam detection approaches suffered with the problems of data collection and labelling [18], spam drifting, class imbalance problems [13] where many of these problems have a higher influence from the features used in these classification techniques.

In most scientific data analysis problems, data are collected from diverse domains or features are extracted from different extractors. They are classified in to different groups. These groups are named as "views" and multiple views are considered for final solution by combining them in different ways. Multi View clustering (MVC) is used to group the objects in to different clusters by considering the features derived from different views [8]. NMF can be considered as the common method for Multiview clustering. They used the joint factorization of different metrics. By applying the clustering methods, results could be obtained from each view. Shen et al. [3] have developed a Multiview based spam detection approach by considering text, url and hash tags as different views. They try to combine the multiple social information extracted from multiple views to learn a model based on NMF. Nevertheless, their intention was to test the possibility of getting accurate results by combining multiple features to train the classifiers. Homophily is the principle that a contact between similar people occurs at a higher rate than among dissimilar people. The pervasive fact of homophily means that cultural, behavioural, genetic or material information that flows through networks will tend to be localized" [7]. In our investigation we try to use the idea of user's interest should be consistent across multiple views. We also assumed that combination of different features of OSN across multiple views could be used to distinguish spammers from legitimate users in online social networks.

3 Hypothesis

Investigations conducted in psychology and sociology have revealed the nature of user behaviors and their changes in different contexts, situations and time frames. Funder [19] studied the behavioural changes and stated that the "set of user behaviors expected to be operant are tended to be more consistent across situations". Further study conducted by Furr and Funder [20] categorized the behaviors into *automatic* and *controlled* where they stated automatic behaviors are highly consistent in both similar and dissimilar situations. They found that "participants were more behaviorally consistent across similar pairs of situations than across dissimilar pairs". More importantly they observed the nature of "greater similarity is related to greater consistency". Tapper [21] studied the behavior of criminal offenders in her thesis and stated that the offender behaviors are consistent in all their criminal offences. By adhering to the investigations of "User behavior consistencies", we developed the following hypothesis to conduct our investigation. "The user behavior should be consistent across multiple views and if pair of users are consistent in one view, they must be consistent in other views also".

Hypothesis 1: Legitimate users should have a higher similarity over multiple topics while spam users have either less similarity in terms of content or the "similarity difference" between two user groups must have a significant difference.

Hypothesis 2: Legitimate users should have a higher consistency over multiple views while spam users have either less consistency or "consistency difference" between two user groups must have a significant difference in terms of multiple views.

In other terms legitimate users should be consistent across multiple views and the malicious users should not be consistent across multiple views.

4 Approach and Experimental Setup

In the investigation, we used the twitter social network which contains short text messages. The hash tags are considered as relevant topics for the twitter content. We used two important factors in twitter messages to represent users. 1. The user's information interest derived through the content similarity over different topics (hash tags) in tweets. 2. The popularity of a certain user which is calculated through average number of retweets per each topic. The investigation is conducted to validate the hypothesis by considering the idea of users' information interest should be consistent across all the topics, and if a pair of individuals or a set of users are consistent in one view, they should be consistent in other view also.

4.1 Dataset Description

Three publicly available data sets were used to conduct the investigation.

Social Honey Pot [9] contains 22,223 content polluters with 2,353,473 tweets, and 19,276 legitimate users with 3,259,693 tweets.

HSpam14 million Tweets-HSpam14 [10] contains over 10 million of labelled genuine tweets along with over three million of spam tweets with their hashtags. The tweets were collected during the year of 2015 and labelled using a mixture of human annotation and a combination of traditional classification algorithms.

The Fake Project dataset [11] is a dataset of twitter spam bots and genuine accounts released by Institute of Informatics and Telematics of Italian National Research Council (CNR). The data set contains over eight million of genuine tweets from more than 3000 user accounts and over nine million of tweets collected across more than 8000 different user accounts.

4.2   Representation of User's Content Interest for the Investigation

Topics: (hash tags): consist of all hash tags used by users in U, U is the set of users in the social networks. A topic is a text content given by users to describe the content in the posts. We assume that the tags given by users in twitter posts have some relationship/relevance or meaning to the post content. Topics are denoted as $T = \{t_1, t_2, \ldots, t_n\}$. The topics in T are frequent topics extracted from all the tweets by users in U based on the percentage of tweets which contain each topic. The topics are used to highlight the content of a post and there must be some correlation between the topic and the content. We can assume that, posts can be clustered in to groups by using the given hash tags if we consider the semantic meaning of the topic.

Posts (Twitter tweets): Consist of all the posts with tagged topics by all the users in U. Let $P = \{p_1, p_2, \ldots, p_N\}$ denote a set of posts each of which contains some topics in T.

Users: Consist of all the users in the Online Social network who used hash tags to label their posts (tweets). The set of users is denoted as $U = \{u_1, u_2, \ldots, u_m\}$. The investigation has focused on two types of users in terms of their content usage or other behaviors in social network. In the datasets used in this investigation, Legitimate users and spam users have been labelled as separate groups, denoted as U_L and U_S, respectively.

Words: From P using tf-idf, we extract the frequent terms or words. Let $P_i \subseteq P$ contain all the tweets posted by user $u_i \in U$, P_i is considered as one document concatenating all tweets posted by u_i, $\{P_1, \ldots, P_n\}$ is a collection of documents for users in U. For the words in P_i for user u_i, based on their tf-idf values on the collection $\{P_1, \ldots P_n\}$, the top frequent words, denoted as W_i, are selected for user u_i. Overall, $W = \bigcup_{u_i \in U} W_i$ contains all the frequent words from all users.

Simple User Profile Based on Tweet Content ($\mathcal{TC}$)**:** The three-dimensional relationship with hash tags, users and the content (words) are defined as $\mathcal{TC} : U \times T \times W \rightarrow \{0, 1\}$ If user u_i published posts with tag t_j that contain word w_k, then $\mathcal{TC}(u_i, t_j, w_k) = 1$, otherwise $\mathcal{TC}(u_i, t_j, w_k) = 0$. A three order tenser representing the users' content was constructed to represent the users' content interest for the investigation. The three dimensions of the tensor represents the User, Topic and the Word. The values were word availability (1 or 0) over certain topic for a given user. Another matrix is generated to represent the popularity of the user by using the User, Topic and the retweet percentage. For a user $u_i \in U$ and a topic $t_j \in T$, $\mathcal{TC}(u_i, t_j) = \langle w_1, \ldots, w_{|W|} \rangle$ is the user's profile under topic t_j. Using two individual user profile vectors, the similarity between the two users' content over the topic is calculated using the "jaccard similarity" index using the following formula. The Eq. (1) calculates the jaccard similarity between two users:

$$Sim_j(u_i, u_k) = \frac{TC(u_i, t_j) \bigcap TC(u_k, t_j)}{TC(u_i, t_j) \bigcup TC(u_k, t_j)} \tag{1}$$

4.3   Representation of User's Popularity for the Investigation

The popularity of a user in a social network can be determined using the number of friends and number of follower/followee relationships. In general, a user who is having more friends or follower/followee relationships can be considered as a popular user in the community and otherwise the user is unpopular. In addition to that if a user's posts over different topics are being shared (retweeted) for many times by the others, we could assume that the user is popular among the members of the community. We may need to consider the number of posts and the average of retweets for those posts to determine the popularity of that user. If the average retweet over all the topics is high we considered the user as a popular user. We assume that a user will share some post since they are interested in that post and the owner of the post is reliable. Generally there exists some level of trustworthiness between post owner and the person who retweeted. In general people share someone else's content, if they accept the content or if the content is reliable and trustworthy. Spam users also have the habit of sharing spam content over multiple user groups, nevertheless, we expect that the popularity of a spam user should not be consistent over multiple topics or content. For each post we calculated the number of retweets and we represented the retweet as an average across each topic to normalize the differences in posting behaviors. In order to derive users' popularity, we defined the following tensor and matrix.

Retweet Tensor: $RT \in \mathbb{N}^{|U| \times |T| \times |P|}$, $RT(u_i, t_j, p_k)$ is the number of times (i.e., count) that user u_i's tweet p_k was retweeted by other users. The tweet p_k is in topic t_j. Average retweet count: $AR \in \mathbb{R}^{|U| \times |T|}$, $AR(u_i, t_j)$ is the average retweet count of user u_i's tweets in topic t_j, i.e., $AR(u_i, t_j) = \frac{1}{|P_{ij}|}\sum_{p \in P_{ij}} RT(u_i, t_j, p)$, P_{ij} is a set of tweets that user u_i posted in topic t_j. For example, if a user u_i has posted 5 tweets in topic t_j, each of the tweets has been retweeted by other users 5, 10, 0, 12, 20 times (i.e., counts), respectively. The average retweet count for the user in this topic is 9.4, i.e., $AR(u_i, t_j) = 9.4$. By considering all topics, each user can be represented by a vector containing the average retweet counts of all topics, i.e., $\langle AR(u_i, t_1), \ldots, AR(u_i, t_{|T|}) \rangle$. Over the all users' average retweet vectors, an average vector can be derived, which is called the centroid average retweet vector, denoted as $\langle CA(t_1), \ldots, CA(t_{|T|}) \rangle$, where $CA(t_j) = \frac{1}{|U|}\sum_{u \in U} AR(u, t_j)$ is the average retweet count for topic t_j over all users. For some users their retweet count for several tweets were comparably high while some users had less number of retweets. To normalize this behavior, for each user we compare the retweet count of each post in each topic with the centroid average retweet count, and use the percentage of posts whose retweet count is above the average retweet count to represent the user' popularity. Let $\mathcal{UP}(u_i, t_j)$ represent user u_i's popularity in topic t_j, $\mathcal{UP}(u_i, t_j)$ is calculated as follows:

$$\mathcal{UP}(u_i, t_j) = \frac{|HR_{ij}|}{|P_{ij}|}$$
$$HR_{ij} = \{p | p \in P_{ij}, \mathcal{AR}(u_i, t_j)\rangle\mathcal{CA}(t_j)\}$$

(2)

4.4 Experimental Methods for User Behavior Analysis

Experiment 1: User's Content Interest Over Multiple Topics

The investigation used the derived user similarities to analyze the interest similarity between pairs of users over each topic. The investigation was conducted for afore-mentioned three data sets for both legitimate and spam users separately and together. For each pair of users in each user set, we calculate the content similarity of the two users under each topic. Then based on the pairwise similarities, the average similarity for each user over all the other users is calculated for each topic. Figures 1, 2 and 3 show the average similarity distribution for all datasets for both legitimate and spam users. For legitimate users there exists a higher average similarity while spammers have a comparably lower similarity. The highest similarity of spammers and the lowest of the legitimate users had at least a difference of 0.1 in all datasets. It is noticed that for some spam users there is no similarity at all.

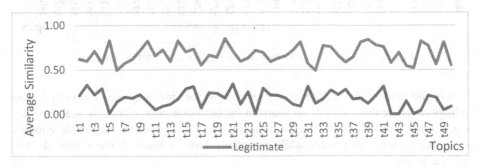

Fig. 1. Average similarity distribution of "HSpam14 data set"

The investigation revealed that the content usage for these spammers for a certain topic were highly diversified and it reduced the similarity in this user group. The analysis of datasets indicated that spammers include the hash tags for their content without having any correlation with the content. The nature of malicious users targeting the trending topics confirms the findings from the previous efforts [14, 22]. We further analyzed the spam user groups in all data sets as a separate collection and discovered that some of the frequent hash tags are never or rarely used by the benign user tweets. This indicates that the malicious users frequently embed some hash tags to their tweets to promote their contents. We have investigated the nature of average similarity variation over number of topics by increasing the number of topics. With increased topics, the average similarity for both user groups gone down and the average difference also getting closer to each other. This happens due to a smaller number of users who shared those topics.

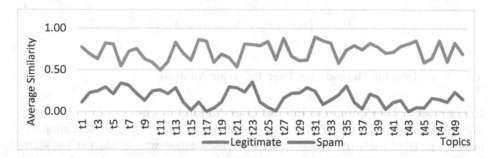

Fig. 2. Average similarity distribution of "The fake Project data set"

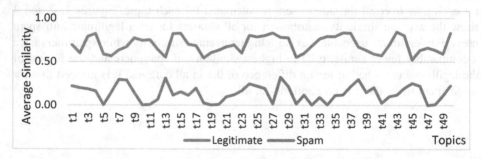

Fig. 3. Average similarity distribution of "Social Honeypot data set"

It is also evident that for the spam user groups, the similarity stays closer to 0. The findings conclude the similarity comparison for both user groups, there is a relationship between user's post/contents and the hash tag. The user's interest over multiple topics are higher in legitimate users while spammers have a lesser content interest over multiple topics.

Experiment 2: Investigating the User's Popularity Over Multiple Topics
User's popularity of the social network is essential to determine the user's nature in a social network. We assumed that their posts were retweeted due to their popularity in the network. To prepare the popularity view, we determine the popularity of a user for each topic. In there, we get number of retweet for each post under each topic and compare that with the centroid retweet value of the topic. If it is above the centroid, the post is popular for that topic and otherwise unpopular. We calculate the percentage of posts above the centroid and defined that as the popularity of that user over the topic. If the popularity of a user is above the threshold (0.5) for a certain topic, we considered

that the user is popular within that topic. By following that, we calculated the percentage of users who are popular for each topic. It is evident from the literature that the spammers also have the retweeting behavior for different posts [11, 15]. We assume that a spam tweet will not get a higher percentage of retweet compare to the legitimate users. Figure 4 depicts the behavior of retweets distribution for the two user groups across multiple topics. It is evident from the results that there is a clear separation between the two user groups for the top frequent topics. In comparison the difference of minimum and maximum values and the averages are larger than the results of experiment 1.

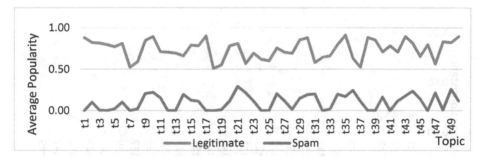

Fig. 4. Average Popularity distribution of "The fake project dataset"

There were very few outliers in the spammers group with very high retweets, nevertheless, their percentage over a certain topic for multiple posts were not significant. Analysis support our hypothesis by confirming that there exists a very low popularity for spam users while legitimate users have a higher popularity rate.

Experiment 3: User Behavior Consistency Across Multiple Views
It is noticed that the similarity of legitimate users and spammers have considerable variation in terms of content and the popularity. In Experiment 3, we investigated the consistency of users' behavior across the two views, i.e., content interest and popularity. In general, for the malicious user detection, the definition for the consistency would be varied on the nature of the views used. We observed that both user interest similarity and popularity are high for the legitimate users while they are low for the spam users. To determine the values, a threshold is maintained. If we consider two values, {High, Low}, for each view, for all the combinations, we defined the consistency of user behaviors across the two views as follows. High in both views are considered as consistent while Low in both and H/L or L/H are considered as inconsistent. For the investigation, we expect that most of the legitimate users have consistent behavior, while spammers don't. In the investigation we calculated the Content interest and popularity for each user for each topic. For a given topic ti, user ui 's content interest compare to all the users is calculated using a predefined threshold and it can be either High or Low compare to the threshold. The popularity of that user for topic ti can be calculated using another threshold and it can also considered as high or low. To determine whether user u i is consistent for a given topic ti, we used the above-

mentioned consistency definition. Figure 5 depicts the consistency of content interest and popularity for top 50 topics. The results indicated that the consistency between two views has a notable difference for both user groups. It is indicated that the majority of the users who are similar in terms of content has a higher popularity in terms of their posts. For legitimate users their consistency ranged from (0.52–0.89) while spammer shows a distribution range of 0.0 to 0.35 across multiple views. For spam users their consistency has gone down for multiple views which indicated a positive direction towards our hypothesis. The legitimate user's consistency variation is also expected to increase by some notable margin nevertheless the increase was only 1.5%. We believe that this rate should be increased with a usage of better user profile or representation.

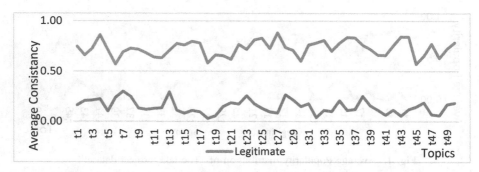

Fig. 5. Consistency of user's similarity and popularity over multiple topics

5 Discussion

The homophily effect for user relationships in OSNs have been investigated in different forms. They explored the similarities of features such as user contents, connections etc. They all have emphasized the importance of using these features for spam detection. Homophily effect for user behavior analysis is discussed in few efforts where they mainly focused on similarities of different features in social networks. Nevertheless, majority of them did not focus on considering multiple features together. This paper aimed to build a hypothesis based on the homophily theory. The investigation used the findings of behavior consistencies in the fields of psychology and sociology. Based on the general hypothesis, an extended hypothesis was proposed to classify the spam and legitimate users based on consistency between content interest and the popularity. This paper used frequent terms and frequent topics (hash tags) to represent user's interest. The analysis confirmed that the spammers are using trending topics in their tweets. We successfully revealed two important findings: 1. The user's interest similarity over multiple topics has a considerable difference between spam users and legitimate users. 2. Most importantly the fact of consistency between a user's content interest with the popularity of a user is determined through the investigation. In there if user's interest similarity is high, they are popular in most of the occasions. It is evident that the legitimate users are consistent among the multiple views and the spammers are inconsistent or poorly consistent across the multiple views. The current features used in the literature for spam detection are easily forged by the attackers. Nevertheless, maintaining the consistency across multiple aspects would be hard to forged. The investigation conducted on user interest similarity depicts the maximum similarity ranges we can use to distinguish the spammers from legitimate users. he strengths of the investigation can be summarized as follows.

1. Most important finding of the investigation is the applicability of user behavior consistencies across multiple views to identify the spam users. It is evident from the investigation that we can use the consistency of user's content interest and popularity across multiple views to separate the spammers from non-spammers. Even though we only focused on user's content interest and popularity, consistency of other user features across multiple views should depict the same pattern for many other features.
2. The investigation considered the user's content interest through engagement of user activities in social networks and it is difficult for a spammer to manipulate this feature since many spammers have a non-focus interest.
3. The popularity as a feature to classify the spammers would be independent from an individual user where the user's popularity over a certain content is generated through the interaction with other users in the social network.

Finally, we need to discuss the limitations of this investigation too. The representation of user interest used a 3-order tensor with frequent word as features. A better user representation would have a provision to improve the results in our investigation. We didn't use the community connection data such as friends, groups that users belong to. These communities may have some effect on user's behaviors. We didn't consider the focused and non-focused spammers where focused spammers may have some different behaviors.

6   Conclusion

The investigation was conducted to analyse the user behaviors across multiple aspects in online social networks. We operated on two main hypotheses, the content interest of legitimate spammers should be higher than the spammers and the user's content interest and the popularity should be consistent across multiple views. In there we assumed that the legitimate users should be consistent and the spammers should be inconsistent across multiple views. We used the hash tag oriented user content and retweet percentages for our investigation and used three publicly available datasets Social Honeypot, HSpam14 and "The fake project data set" for our investigation. We measured the user's interest similarity over multiple topics for both user groups and observed that it is comparably high for legitimate users and low for spammers in both data sets. The similarity variations across multiple topics for both user groups were closer within each group. The popularity over multiple topics for both user groups were also depicted a similar behavior where legitimate users have a higher popularity and the spammers have very low popularity over multiple topics. We analyzed the consistency of the content and popularity across multiple views and noted that the legitimate users are consistent across both views while spammers were depicted a very low consistency across both views.

Findings of the investigation are encouraging to develop a novel approach of multi view based spam detection framework based on user behavior consistencies. In future directions we planned to improve the user profile which may improve the results. The investigation has shown the provision of trust based spam detection framework using the user behavior consistencies.

References

1. Workman, M.: Gaining access with social engineering: an empirical study of the threat. Inf. Syst. Secur. **16**, 315–331 (2007)
2. Lee, K., Caverlee, J., Webb, S.: The social honeypot project. In: Proceedings of the 19th International Conference on World Wide Web - WWW 2010 (2010)
3. Shen, H., Ma, F., Zhang, X., Zong, L., Liu, X., Liang, W.: Discovering social spammers from multiple views. Neurocomputing **225**, 49–57 (2017)
4. Adewole, K., Anuar, N., Kamsin, A., Varathan, K., Razak, S.: Malicious accounts: dark of the social networks. J. Netw. Comput. Appl. **79**, 41–67 (2017)
5. Hua, W., Zhang, Y.: Threshold and associative based classification for social spam profile detection on Twitter. In: Ninth International Conference on Semantics, Knowledge and Grids (2013)
6. Cardoso, F.M.: Topical homophily in online social systems (2017)
7. McPherson, M., Smith-Lovin, L., Cook, J.: Birds of a feather: homophily in social networks. Ann. Rev. Sociol. **27**(1), 415–444 (2001)
8. Sherman, R., Nave, C., Funder, D.: Situational similarity and personality predict behavioral consistency. J. Pers. Soc. Psychol. **99**, 330–343 (2010)
9. Lee, K., Caverlee, J., Webb, S.: Uncovering social spammers. In: Proceeding of the 33rd International ACM SIGIR Conference on Research and Development in Information Retrieval - SIGIR 2010 (2010)
10. Sedhai, S., Sun, A.: HSpam14. In: Proceedings of the 38th International ACM SIGIR Conference on Research and Development in Information Retrieval - SIGIR 2015 (2015)
11. Cresci, S., Di Pietro, R., Petrocchi, M., Spognardi, A., Tesconi, M.: The paradigm-shift of social spambots. In: Proceedings of the 26th International Conference on World Wide Web Companion - WWW 2017 Companion (2017)
12. Heymann, P., Koutrika, G., Garcia-Molina, H.: Fighting spam on social web sites: a survey of approaches and future challenges. IEEE Internet Comput. **11**, 36–45 (2007)
13. Chen, C., et al.: Investigating the deceptive information in Twitter spam. Future Gener. Comput. Syst. **72**, 319–326 (2017)
14. Martinez-Romo, J., Araujo, L.: Detecting malicious tweets in trending topics using a statistical analysis of language. Expert Syst. Appl. **40**(8), 2992–3000 (2013)
15. Ghosh, S., et al.: Understanding and combating link farming in the Twitter social network. In: Proceedings of the 21st International Conference on World Wide Web - WWW 2012 (2012)
16. Cao, C., Caverlee, J.: Behavioral detection of spam URL sharing: posting patterns versus click patterns. In: International Conference on Advances in Social Networks Analysis and Mining (ASONAM), pp. 138–141 (2014)
17. Cui, P., Wang, F., Liu, S., Ou, M., Yang, S., Sun, L.: Who should share what? Item-level social influence prediction for users and posts ranking. In: Proceedings of the 34th International ACM SIGIR Conference on Research and Development in Information Retrieval, pp. 185–194. ACM (2011)

18. Wu, T., Wen, S., Xiang, Y., Zhou, W.: Twitter spam detection: Survey of new approaches and comparative study. Comput. Secur. **76**, 265–284 (2018)
19. Funder, D.: Global traits: a neo-allportian approach to personality. Psychol. Sci. **2**(1), 31–39 (1991)
20. Furr, R., Funder, D.: Situational similarity and behavioral consistency: subjective, objective, variable-centered, and person-centered approaches. J. Res. Pers. **38**(5), 421–447 (2004)
21. Tapper, S.: Testing the assumption of behavioural consistency in a New Zealand sample of serial rapists. Ph.D. dissertation, Victoria University of Wellington (2008)
22. McCord, M., Chuah, M.: Spam detection on Twitter using traditional classifiers. In: Calero, J.M.A., et al. (eds.) ATC 2011. LNCS, vol. 6906, pp. 175–186. Springer, Heidelberg (2011). https://doi.org/10.1007/978-3-642-23496-5_13

A Cluster Ensemble Strategy for Asian Handicap Betting

Yue Chen[1,2(✉)] and Jian Shi[1,2(✉)]

[1] Academy of Mathematics and Systems Science, Chinese Academy of Sciences,
Beijing 100190, China
jshi@iss.ac.cn
[2] School of Mathematical Sciences, University of Chinese Academy of Sciences,
Beijing 100049, China
chenyue14@mails.ucas.ac.cn

Abstract. Football betting has grown rapidly in the past two decades, among which fixed odds betting and Asian handicap betting are the most popular mechanisms. Much previous research work mainly focus on fixed odds betting, however, it is lack of studying on Asian handicap betting. In this paper, we focus on Asian handicap betting and aim to propose an intelligent decision system that can make betting strategy. To achieve this, a cluster ensemble model is presented, which is based on the fact that matches with similar pattern of expected goal trend series may have the same actual outcome. Firstly, we set up the component cluster which classifies matches by the pattern of expected goal trend series and then makes the same betting decision for matches that belong to the same group. Furthermore, we adopt plurality voting approach to integrate component clusters and then determine the final betting strategy. Using this strategy on the big five European football leagues data, it yields a positive return.

Keywords: Asian handicap · Component cluster · Cluster ensemble · Betting strategy

1 Introduction

Association Football (hereafter referred to simply as football) is the most popular sport internationally [1] and attracts a significant share of the total sports betting pie, particularly after its introduction online [2].

Asian handicap is a popular form of spread betting on football which creates a more level betting environment between two mismatched competing teams by giving a "handicap" (expressed in goals) to the teams before kick-off. In Asian handicap, a goal deficit is given to the team which is more likely to win (the favourite) and a head start is given to the team which is less favoured to win (the underdog) [3].

Football is one of the few sports in the world where a tie is a fairly common outcome. However, in Asian handicap, the chance for a tie is eliminated by use of

© Springer Nature Switzerland AG 2019
L. H. U and H. W. Lauw (Eds.): PAKDD 2019 Workshops, LNAI 11607, pp. 28–37, 2019.
https://doi.org/10.1007/978-3-030-26142-9_3

a handicap that forces a winner. This simplification delivers two betting options that each have a near 50% chance of success [4]. The bookmakers' aim is to create a handicap that will make the chance of either team winning (considering the handicap) as close to 50% as possible.

The handicap often changes from the time the first bet is placed to the time the last bet is placed. The betting handicap at the close of the book on a match is called the closing handicap. The closing handicap incorporates all available information and any biases of the market participants, which make it difficult, if not impossible, to find profitable betting strategy at the close of book that beat the vigorish of the sportsbook. The football outcome and bookmaker's odds in Asian handicap between two very popular clubs Chelsea and Manchester United are shown in Table 1.

This paper aims to propose an intelligent decision system that can make bets at the close of Asian handicap. We propose a cluster ensemble model to determine betting strategy. First, we set up the component cluster model which makes the same decision for matches in the same group. Furthermore, we adopt the plurality voting method to integrate component clusters and then determine the final betting strategy. We evaluate our proposed model on a real-world dataset for the big five European football leagues.

Table 1. A typical example of football outcome and bookmaker's odds

Favourite team	Underdog team	Match date	Result	Date	Favourite odds	Handicap	Underdog odds
Chelsea	Man united	2017-03-14 03:45	1:0	2017-02-27 21:59	2.12	−0.50	1.81
				2017-03-02 04:37	2.18	−0.50	1.77
				2017-03-04 21:20	2.12	0.50	1.81
				2017-03-05 04:06	2.06	−0.50	1.86
				2017-03-07 23:58	1.86	−0.50	2.06
				2017-03-09 22:56	1.83	−0.50	2.06
				2017-03-10 22:23	1.93	−0.50	1.99
				2017-03-11 06:42	1.96	−0.50	1.96
				2017-03-12 00:47	1.93	−0.50	1.99
				2017-03-13 19:11	2.08	−0.75	1.85
				2017-03-13 24:29	1.80	−0.50	2.08
				2017-03-14 01:14	1.90	−0.50	2.02
				2017-03-14 01:44	1.92	−0.50	2.00
				2017-03-14 03:35	1.92	−0.50	2.00
				2017-03-14 03:39	1.96	−0.50	1.96
				2017-03-14 03:44	2.00	−0.50	1.92

2   Related Works

There exists a large empirical literature analyzing the existence of weak form efficiency in the football betting market [9]. In the market based on fixed-odds betting, there are some mixed results. Pope and Peel [5] found that although there was some evidence of inefficiency, there did not appear to be profitable betting strategies that could have abnormal returns. But Dixon and Coles [6] proposed a time-dependent Poisson parametric model, which yielded positive profits against published market odds. Constantinou et al. [2] presented a Bayesian network model that was used to generate forecasts about the English Premier League matches during season 2010/2011, by considering both objective and subjective information for prediction.

To the extent that such inefficiency exists, a natural question is whether it can be systematically exploited. Poisson and Negative Binomial models have been used to estimate the number of goals scored by Cain et al. [7]. They drew the conclusion that even though the fixed odds offered against particular score outcomes did seem to offer profitable betting opportunities in some cases, these were few in number. Goddard and Asimakopoulos [8] proposed an ordered probit regression model to forecast the English Premier League match results in an attempt to test the weak-form efficiency of prices in the fixed-odds betting market. Even though they reported a loss of 10.5% for overall performance, the model appeared to be profitable (on a pre-tax gross basis) at the start and at the end of every season. Up to now, the study on fixed-odds betting is much more, but it is lack of studying on Asian handicap betting, especially on the betting decision which can make a profit. This motivates us to give a study on the Asian handicap betting.

3   Methodology

3.1   Expected Goal Difference Trend

Over-under bet is also a popular type of bets. An over-under bet is a wager in which a bookmaker will predict a number for the combined goal of the two teams (total) in a given game, and bettors can wager that the actual number in the game will be either higher or lower than that number [13].

Focusing on the published odds, in Asian handicap, the handicap, the odds for favourite win and undergo win are denoted as δ, d_1 and d_2, respectively. In over-under, the total, the odds for over win and under win are denoted as $\tilde{\delta}, \tilde{d}_1$ and $\tilde{d}_2$, respectively. Based on bookmakers' odds, the implied probabilities for favourite win and over win are $p_1 = d_2/(d_1 + d_2)$ and $p_2 = \tilde{d}_2/(\tilde{d}_1 + \tilde{d}_2)$.

Based on the Dixon and Coles' model [6], the implied probability functions for favourite win and over win are $f_1(\delta, \tilde{\delta}, \rho, s_1, s_2)$ and $f_2(\delta, \tilde{\delta}, \rho, s_1, s_2)$, respectively, where s_1, s_2 are favourite and undergo expected goals, respectively, ρ is a preset correlation coefficient.

We propose an inversion algorithm to calculate expected goals derived from bookmakers' odds. The core principle is that implied probability by Dixon and Coles' model and implied probability by bookmakers' odds are assumed to be equal, namely,

$$\begin{cases} f_1(\delta, \tilde{\delta}, \rho, s_1, s_2) = p_1, \\ f_2(\delta, \tilde{\delta}, \rho, s_1, s_2) = p_2. \end{cases} \tag{1}$$

The solutions of Eq. (1) are favourite expected goal and undergo expected goal, denoted by s_1^*, s_2^*, respectively. The expected goal difference between the favourite and the underdog team is:

$$z = s_1^* - s_2^*. \tag{2}$$

Odds vary from the opening to the closing of the book. From the close of book forward, the expected goal difference at j-th odds change point is z_j based on Eq. (2), and the corresponding expected goal difference trend x_j between the favourite and the underdog teams is defined as:

$$x_j = \begin{cases} 1, & z_j \geq z_{j+1}, \\ -1, & z_j < z_{j+1}. \end{cases} \tag{3}$$

We group the x_j into feature vector $\boldsymbol{x}$, and denote $\boldsymbol{x}_i$ as the feature vector for match i.

Assuming we wager 1 pound stake for each possible outcome at the close of Asian handicap, based on the actual outcome and dividend rules [3], return can be obtained. For match i, the return of wagering on the favourite and the return of wagering on the undergo are denoted by a_i and u_i, respectively.

3.2 Model Design

The basic thought of modeling is that matches with similar pattern of the expected goal difference trend series may have the same actual outcome. Hence, classifying matches by the pattern of expected goal difference trend series and making the same betting decision for matches in the same category can expect to yield a positive return.

The k-means clustering algorithm [11] is exactly the method of grouping a set of unlabelled instances in such a way that instances in the same group (called a cluster) are more similar to each other than to those in other clusters and each cluster corresponds to a potential category. The disadvantage of k-means algorithm is that an inappropriate choice of initial centers may yield poor results [14].

The clustering ensemble is a powerful tool to improve the performance of k-means algorithm, but this task is not trivial because there is no guarantee that the clusters discovered by one component cluster can correspond to the clusters discovered by the other component cluster [10].

Aiming at the particularity of data, a new ensemble model for bet decision-making is presented. The proposed model runs the k-means algorithm t times to obtain t different component clusters and then adopts plurality voting approach to integrate those component clusters.

3.3 Component Cluster

The k-means clustering is a centroid-based method and cluster is represented by a central vector [11]. Here, k-means clustering partitions the n expected goal difference trend series $D = \{x_1, x_2, \ldots, x_n\}$ into k clusters $\mathcal{C} = \{C_1, C_2, \ldots, C_k\}$ so as to minimize the within-cluster sum of distance. That is, the objective is to find:

$$\arg\min_{\mathcal{C}} \sum_{i=1}^{k} \sum_{x \in C_i} \text{dist}(x, \mu_i), \tag{4}$$

where

$$\mu_i = \frac{1}{|C_i|} \sum_{x \in C_i} x, \tag{5}$$

$$\text{dist}(x, \mu_i) = \frac{x^T \mu_i}{\| x \| \| \mu_i \|}, \tag{6}$$

μ_i is the mean of instances in C_i, $C_{l'} \cap C_l = \varnothing$, $D = \bigcup_{l=1}^{k} C_l$, and $\| \cdot \|$ is the Euclidean norm.

The particularity of data we focus on is that the return of wagering on the favourite a_i and the return of on the undergo u_i are both available for matches in the training set. We introduce and define the decision label λ_j for the cluster C_j as follows:

$$\lambda_j = \begin{cases} 1, & \frac{1}{|C_j|} \sum_{x_i \in C_j} a_i \geq \max\left(\tau, \frac{1}{|C_j|} \sum_{x_i \in C_j} u_i\right), \\ -1, & \frac{1}{|C_j|} \sum_{x_i \in C_j} u_i \geq \max\left(\tau, \frac{1}{|C_j|} \sum_{x_i \in C_j} a_i\right), \\ 0, & \text{otherwise}, \end{cases} \tag{7}$$

where $|C_j|$ is the number of matches in cluster C_j, and τ is the return threshold. Betting on the favourite, on the underdog and no betting are denoted by $\lambda_j = 1, -1, 0$, respectively.

The clustering that divides D into k clusters could be regarded as a betting decision label vector $\lambda = (\lambda_1, \ldots, \lambda_k)^T$. Formula (7) shows that matches in the same cluster will make the same betting decision. The decision-making rule is that betting on the outcome with the highest average return, if and only if the highest average return is greater than return threshold.

For match l in the prediction set, we apply nearest centroid classifier to classify the instance x_l into the existing cluster , and then make betting decision θ_l based on decision label of the cluster to which x_l belongs.

$$\hat{l} = \arg\min_{i \in \{1,2,\ldots,k\}} \text{dist}(x_l, \mu_i), \\ \theta_l = \lambda_{\hat{l}}, \tag{8}$$

where $\theta_l = 1, -1, 0$ express betting on the favourite, on the underdog and no betting on the match l, respectively.

The procedure corresponding to component cluster is shown in Fig. 1. It illustrates how to train model and make a prediction.

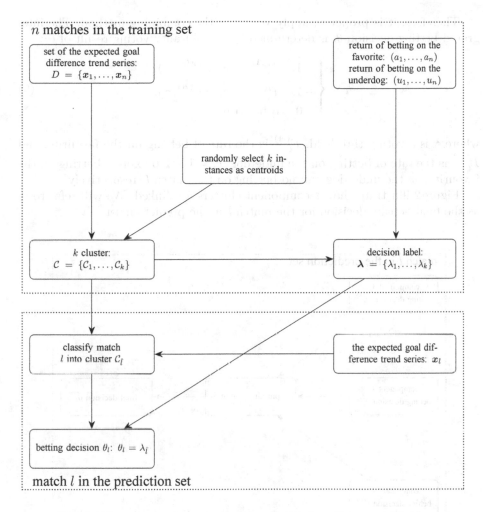

Fig. 1. Component cluster: training model and making a prediction

3.4   Cluster Ensemble

We run the k-means algorithm t times with different initial points to obtain t component clusters based on the method in Subsect. 3.3. For the match l in the prediction set, we obtain t betting decisions based on those component clusters, denoted by $\boldsymbol{\theta}_l = (\theta_l^{(1)}, \theta_l^{(2)}, \ldots, \theta_l^{(t)})^T$.

A critical problem in cluster ensemble is how to combine multiple clusters to yield a superior result. But it is easy to deal with this problem here, because $\theta_l^{(i)}$ made by component cluster i corresponds to $\theta_l^{(j)}$ made by component cluster j.

Therefore, it is proper to apply voting as the combining method. The integrated betting decision $\tilde{\theta}_l$ is determined by the plurality voting result of $\boldsymbol{\theta}_l$:

$$\tilde{\theta}_l = \begin{cases} 1, & P_l^{(A)} \geq \max(P_l^{(U)}, \gamma), \\ -1, & P_l^{(U)} \geq \max(P_l^{(A)}, \gamma), \\ 0, & \text{otherwise}, \end{cases} \tag{9}$$

where γ is a voting threshold, $P_l^{(A)}$ is the rate of betting on the favourite, and $P_l^{(U)}$ is the rate of betting on the underdog. $\tilde{\theta}_l = 1, -1, 0$ express betting on the favourite, on the underdog and no betting on the match l, respectively.

Figure 2 illustrates how t component clusters are linked. We will refer to $\tilde{\theta}_l$ as the final betting decision for the match l in the prediction set.

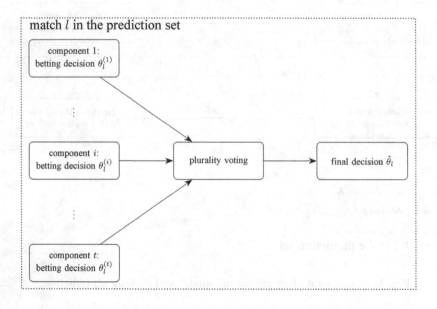

Fig. 2. How components are linked

4 Case Studies

4.1 Data

The basic dataset comprises the records of 5580 matches in the big five European football leagues during seasons 13/14–15/16. The records include the actual outcome, published market odds in Asian handicap and in over-under provided by the leading UK bookmaker Bet365. We split the dataset into a training set spanning the 13/14–14/15 seasons of which include 3720 matches and a prediction set over the 15/16 season of which includes 1860 matches. The data used to model are the returns for betting on each outcome based on the closing handicap and the goal difference trend series. These information are all available according to the method in Subsect. 3.1.

4.2 Evaluation Metrics

There are various ways in which the quality of a model can be assessed. In particular, we can consider accuracy (how close the forecasts are to the actual results) and profitability (how useful the forecasts are when used as a betting strategy). For assessing the accuracy of the proposed model we use the win rate which is the rate of forecasts are same as the actual outcome. We measure the profitability on the basis of return rate relative to stakes. Wing and Yi have already concluded that there is only a weak relationship between commonly used measures of accuracy and profitability, and that a combination of the two might be the best [12]. Hence we use assessments of both accuracy and profitability in order to get a more informative picture about the performance of the proposed model in Sect. 3.

4.3 Model Parameters Selection

Given that the training data include a large number of matches, we adopt cross validation method for model parameters selection. The performance of the model proposed in Sect. 3 depends on the number of clusters k, the number of components t, the return threshold τ and the voting threshold γ. If the number of clusters k is lower than the actual number of clusters, the within-cluster sum of distance would rise rapidly. Thus we use the dichotomy method to cluster for many times and compare the clustering results to determine the best value of k. To select the best parameters, we perform a grid search under the ranges $t \in \{500, 1000\}, \tau \in \{0.01, 0.02\}$ and $\gamma \in \{0.55, 0.60, 0.65\}$.

The model with parameters $k = 12, t = 500, \tau = 0.02, \gamma = 0.65$ performs the best. Therefore we opt for such parameters in the remainder of this paper.

4.4 Out of Sample Prediction

In real environment, the data should be regularly updated as new match data becomes available. Hence, for testing predictive capability of the selected model in previous subsection, we adopt a more realistic fixed size rolling window evaluation scheme that performs model update and discards the oldest data.

Figure 3 demonstrates the cumulative profit/loss generated after each subsequent match, assuming a 1-pound stake when the betting condition is met. The model generates a profit and almost never leads into a negative cumulative loss even allowing for the in-built bookmakers' profit margin. Table 2 shows that the proposed model wins 55.93% of the bets, with return rate at 2.35%.

The betting results for various ranges of handicap (the sign term omitted) are shown in Table 3. It demonstrates that the majority of matches have a low

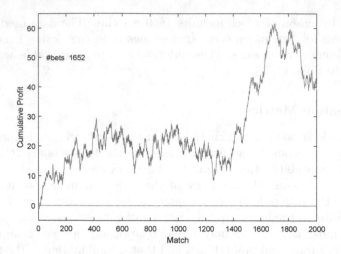

Fig. 3. Cumulative profit/loss observed

Table 2. Betting statistics

No. of bets	No. of bets win	Win rate (%)	Return rate (%)	Min. P/L balance observed (£)	Max. P/L balance observed (£)	Final balance observed (£)
1652	924	55.93	2.35	−1	61.55	38.89

handicap. The higher the handicap, the greater the win rate of the proposed model. This suggests that the more equally matched two teams are, the harder it is to predict the outcome. What is the most interesting is that the favourite team and the underdog team really approximately win 50% of all the matches respectively, which is consistent with the design intention of Asian handicap.

Table 3. Betting statistics for different handicap ranges

Handicap range	No. of favourite win	No. of push[a]	No. of underdog win	Favorite win rate	Underdog win rate	No. of bets	No. of bets win	Win rate
[3, +∞)	1	2	1	50.00	50.00	3	2	66.67
[2, 3)	35	10	35	50.00	50.00	73	43	58.90
[1, 2)	146	35	124	54.07	45.93	265	162	61.13
[0, 1)	737	138	736	50.03	49.97	1311	717	55.09
[0, +∞)	919	185	896	50.63	49.37	1652	924	55.93

[a]Push means stake refund.

5   Conclusions

In this paper, we studied Asian handicap to find a profitable betting decision and presented a novel cluster ensemble method to determine the betting strategy. The basic thought of our model is that matches with similar pattern of the expected goal difference trend series may have the same actual outcome. The innovation of the proposed model mainly embodies in the following two aspects. Firstly, we set up the component cluster which classifies matches by the pattern of the expected goal difference trend series and make the same betting decision for matches in the same group. Next, we use the plurality voting method to integrate component clusters and determine final betting strategy. The application of our model to the real-world dataset shows that it is profitable. Our model wins approximately 56% of the bets, with the return rate at approximately 2.35%.

There are three main problems that need to be solved to further advance our model in future research: (1) to consider more feature information of Asian handicap; (2) to adopt more different ways of cluster methods as component clusters; and (3) to pay more attention to betting amounts, not just concentrating on the side to bet.

Acknowledgements. This work is partially supported by the Beijing StausWin Lottery Operations Technology Ltd. And we thank Yiran Gao and Jiang Yu for providing the data and odds.

References

1. Karen, D.: Sport matters: sociological studies of sport, violence, and civilization by eric dunning. Br. J. Sociol. **20**(4), 756–757 (2000)
2. Constantinou, A.C., Fenton, N.E., Neil, M.: Profiting from an inefficient association football gambling market: prediction, risk and uncertainty using Bayesian networks. Knowl.-Based Syst. **50**(3), 60–86 (2013)
3. Asian handicap. https://en.wikipedia.org/wiki/Asian_handicap. Accessed 21 Dec 2018
4. Gandar, J., Zuber, R., O'Brien, T., Russo, B.: Testing rationality in the point spread betting market. J. Finan. **43**(4), 995–1008 (1988)
5. Pope, P.F., Peel, D.A.: Information, prices and efficiency in a fixed-odds betting market. Economica **56**(223), 323–341 (1989)
6. Dixon, M.J., Coles, S.G.: Modelling association football scores and inefficiencies in the football betting market. J. Roy. Stat. Soc.: Ser. C (Appl. Stat.) **46**(2), 265–280 (1997)
7. Cain, M., Law, D., Peel, D.: The favourite-longshot bias and market efficiency in UK football betting. Scott. J. Polit. Econ. **47**(1), 25–36 (2000)
8. Goddard, J., Asimakopoulos, I.: Forecasting football results and the efficiency of fixed-odds betting. J. Forecast. **23**(1), 51–66 (2004)
9. Forrest, D., Goddard, J., Simmons, R.: Odds-setters as forecasters: the case of English football. Int. J. Forecast. **21**(2), 551–564 (2005)
10. Zhou, Z.H., Tang, W.: Clusterer ensemble. Knowl.-Based Syst. **19**(1), 77–83 (2006)
11. Hartigan, J.A., Wong, M.A.: Algorithm AS 136: a K-means clustering algorithm. J. R. Stat. Soc. Ser. C (Appl. Stat.) **28**(1), 100–108 (1979)
12. Wing, C.K., Yi, K.L.: The use of profits as opposed to conventional forecast evaluation criteria to determine the quality of economic forecasts. Differ. Uravn. **18**(7), 1164–1170 (2007)
13. Williams, L.V.: Weak form information efficiency in betting markets. Leighton Vaughan Williams **51**(1), 1–30 (2005)
14. Hamerly, G., Elkan, C.: Alternatives to the k-means algorithm that find better clusterings. In: 11th International conference on Information and knowledge management, pp. 1–2. ACM Press, Vancsouver (2002)

Designing an Integrated Intelligence Center: New Taipei City Police Department as an Example

Chun-Young Chang[1], Lin-Chien Chien[2(✉)], and Yuh-Shyan Hwang[1]

[1] Department of Electronic Engineering,
National Taipei University of Technology, Taipei, Taiwan
[2] Department of Information Management, Central Police University,
Taoyuan City, Taiwan
lynn@mail.cpu.edu.tw

Abstract. The rapid advancement and prevalence of Internet technology was the biggest development of the 20th century. Because criminals use the Internet to commit crimes, police work has needed to shift from traditional methods to modern technology. As investigation strategies have evolved, sharable databases and integrated intelligence have become increasingly important. By maintaining data, information professionals play a crucial role in police departments, and an integrated, sharable system is essential for supporting police work.

New Taipei City is one of six main cities in Taiwan and has the largest population. Social order and traffic are the city's greatest issues. The New Taipei City Police Department (NTPD) uses integrated resources and new technology to help first-line police and investigators focus on these issues. To solve complicated problems, an integrated intelligence center (IIC) was designed to provide needed data, help team members analyze information, and guide users in searching information systems.

The IIC's services have successfully supported social maintenance tasks such as covering celebration events and city elections. The support of IIC team members has been recognized with Outstanding Government Employee and Best Contribution awards from the Taiwanese government. This paper shows the IIC's structure and how it is designed with innovation approach how it functions.

Keywords: New Taipei City · New Taipei City Police Department (NTPD) · Integrated intelligence center (IIC) · Intelligence-Led Policing · Design thinking

1 Introduction

The mission of the New Taipei City Police Department (NTPD) is to improve criminal investigation capability and integrate criminal investigation information so that the most productive and efficient methods for countering criminal activity and apprehending criminals are available. End users and stakeholders benefit when actionable information is disseminated and shared in a timely manner. First-line police officers and investigators require prompt access to accurate information about suspects so that they

L. H. U and H. W. Lauw (Eds.): PAKDD 2019 Workshops, LNAI 11607, pp. 38–44, 2019.
https://doi.org/10.1007/978-3-030-26142-9_4

can, for example, identify the model of a car to determine whether it has been stolen or recognize a face to know whether the person is suspected of a crime. As the data might be stored on different systems, an integrated intelligence center (IIC) is needed.

The information unit shoulders the responsibility of carrying out these tasks. The IIC consolidates information and distributes needed information to users. A well-designed structure makes the IIC reliable and convenient, while its operational standards makes it user-friendly and efficient.

This research used design thinking to develop the IIC's services. End users' needs were collected and used in designing the center's structure. The input resulted in a prototype that offered six services: Queries, crime information analysis, search assistance, image analysis, real-time image transmission, and Internet information collection.

2 Literature Review

Building a cross-system platform and providing accurate and up-to-date information for law enforcement use requires an integrated intelligence system [1]. Adapting a system from another country or industry might not meet the needs of local end users and stakeholders due to inherent stereotypes or cultural differences. Design thinking can be used to prototype a system that meets the needs of first-line police officers and investigators.

Design thinking is generally defined as an analytic and creative process that engages a person in opportunities to experiment, create and prototype a model [4]. Good design combines technology, cognitive science and human needs, and successful service models have been developed using design thinking [5]. Design thinking was successfully used to develop an IIC for the NTPD.

3 Method

Design thinking is a methodology that provides a solution-based approach to tackling problems. Many variants of design thinking are being used in the 21st century. This research involved three main steps: discovery (learn the end users' needs), understanding (define the problem), and implementation (find a solution). The steps are as following:

1. Discovery
(Interviews, surveys, etc.)
↓
2. Understanding
(Synthesize everything that has been learned, define the problem)
↓
3. Implementation
(Construct prototypes, focus on problem, solve the problems)

3.1 Discovery

The needs of the police officers on patrol, detectives in the field, analysts, and senior leadership were determined:

1. Suspects' personal information
2. Suspects' vehicle information (model, etc.)
3. Suspects' background information
4. Suspects' criminal record
5. Background information on local environment
6. Evidence collection

3.2 Understanding

Based on the users' needs, a mapping table was used to match the needs with an information source (Fig. 1):

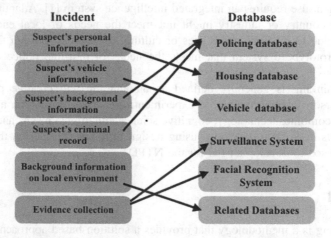

Fig. 1. Relation between Events and Supported Database System.

3.3 Implementation

Based on the incident and database matches, an integrated structure with six services was designed. The services are illustrated as Fig. 2, which are composed of Query Service, Crime Info Analysis, Applied System Q&A, Image Analysis Service, Real-Time Image Transmission, and Internet Information Collection.

Fig. 2. Prototype of an Integrated Intelligence center.

3.3.1 Query Service

The service gathers policing information by searching across 44 database platforms in the National Police Agency, the Crime Investigation Agency and the NTPD (Fig. 3). All of the information from databases should be automatically updated to the latest data once a day at last, in order to provide the law enforcement officer with the criminal information without any loss of immediacy. Furthermore, the security diagnostic of databases must be conducted regularly. Due to the fact that most of information are full of privacy and confidentiality, the implementation of information security measure and remote backup of databases should be concerned.

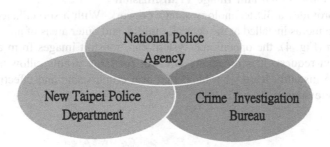

Fig. 3. Information System Services.

3.3.2 Crime Information Analysis

The service analyzes criminal cases and operational work. For example, it can analyze family member data as well as accomplices and provide information on families with a high risk of criminal involvement to police officers and investigators. The analysis can make use of international investigate analysis softwares, such as I2 Analyst's Notebook and QDA Miner, which combine with artificial intelligence (AI) as well as function of visual analysis, relation analysis and text mining, for the sake of rapidly transforming criminal-related information into intelligence. With the function of visual analysis, the law enforcement officers can keep abreast of criminal intelligence and effectively generate the strategies to solve the crime case, moreover to achieve the crime prevention.

3.3.3 Applied System Q&A
The service provides assistance in searching criminal files on systems at the National Police Agency, the Criminal Investigation Bureau, and New Taipei Police Department. With the view to making every single user has a good command of the system, the IIC educates the end-users, as well as hold training courses at regular intervals for the office units. Furthermore, the organizer conduct the questionnaire survey of users' request to add the new function to the system and make well maintenances of system periodically.

3.3.4 Image Analysis Service
The service provides facial recognition, vehicle information (model, vehicle license plate, etc.), and suspect information. Firstly, the facial recognition can be implemented to the all kinds of civil demonstration and assembly parade. As soon as the fugitives get into the meeting place, the law enforcement officers can follow theirs trace in instance and get them busted. Secondly, the image analysis service of vehicle can be combined with the surveillance system equipped with intelligent video condensing and recognition technology. With the car license plate numbers of the suspects and accomplices, the system can generate the trajectory of the target vehicles in a short time. Integrating crime information analysis with the image analysis service, the law enforcement officers can effectively locate the criminal and bring them to justice. The last but not the least, when the surveillance system capture the image of suspect, image analysis in suspect information can play its full role to find out the information of the suspect.

3.3.5 Real-Time Video and Image Transmission
The service provides a 3D technology safety network. With a surveillance system of high-quality cameras installed at the central, middle, and outer areas of an event or civil demonstration (Fig. 4), the operational system can transmit images from anywhere at the event upon request. It is especially efficient for large events, allowing the event commander to instantly learn what happened at a crime scene and effectively collect images of any evidence.

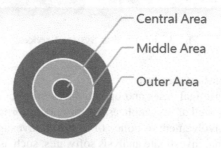

Fig. 4. Set-up of Surveillance System and Cameras.

3.3.6   Internet Information Collection

The service provides other information upon request, presented as Fig. 5. The IIC regularly holds training classes and serves as a consultant to other information units within its jurisdiction. Information cultivation is one of its missions.

Fig. 5. Integrated Intelligence Center (IIC) Working Model. (NPA: National Police Agency; CIB: Criminal Investigation Bureau; NTC: New Taipei City; NTPD: New Taipei City Police Department)

4   Discussion and Future Work

With an eye to making the systems well-functional and meeting all the end-users' request, there are some problems and challenges as following to be solved. To begin with, since the surveillance system equipped with 4G Internet technology could be disconnected with uncontrollable factor occasionally, especially in the crowded area. The law enforcement officers have to stay by the side to keep the system working normally. If the terrorist launch the hostile attack in the assembly parade, and surveillance system coincidentally get disconnected at the same time, the law enforcement officers can't receive the firsthand video and information of the scene. The situation can lead to the irrecoverable results. Furthermore, 44 database platforms from various departments should be updated to the latest information to keep the law enforcement officers up to date on criminal information. The time intervals between the update time should be short as possible, accordingly, the servers in the computer facilities should be maintained as perfect condition so as to provide the operation service smoothly and effectively.

5   Conclusion

The NTPD's IIC has been successfully used to perform case analysis for investigations involving homicides, defraud of the wealth, drug crimes, sexual assault and human trafficking. The IIC has also evaluated the risk of criminals recommitting crimes and determined locations that have a higher risk of criminal activity. The procedures involve big-data techniques. The IIC provides information, data analysis (Fig. 5) and user education. The multifunctional IIC solves complex data problems and meets the informational needs of New Taipei City law enforcement.

References

1. Guan, C.-C.: Functions and development of fusion center. In: 7th International Proceedings on Terrorism and Homeland Security, pp. 1–20 (2000)
2. Peterson, M.: Intelligence-led policing: the new intelligence architecture. Department of Justice, Bureau of Justice Assistance, Washington, DC (2005)
3. Mueller, R., Thoring, K.: Design thinking vs. lean startup: a comparison of two user-driven innovation strategies. In: International Design Management Research Conference, DMI, pp. 1–12 (2012)
4. Rim, R., Shute, V.: What is design thinking and why is it important? Rev. Educ. Res. **82**, 330–348 (2012)
5. Allio, L.: Design thinking for public service excellence. UNDP Global Centre for Public Service Excellence (2014)

Early Churn User Classification in Social Networking Service Using Attention-Based Long Short-Term Memory

Koya Sato[✉], Mizuki Oka, and Kazuhiko Kato

University of Tsukuba, Tsukuba, Japan
{koya,mizuki,kato}@osss.cs.tsukuba.ac.jp

Abstract. Social networking services (SNSs) see much early churn of new users. SNSs can provide effective interventions by identifying potential early churn users and important factors leading to early churn. The long short-term memory (LSTM) model, whose input is the user behavior event sequence binned at constant intervals, is proposed for this purpose. This model better classifies early churn users than previous machine learning models. We hypothesized that the importance of each temporal part in the event sequence is different for classifying early churn users because user behavior is known to consist of coarse and dense parts and initial behavior influences long-term behavior. To treat this, we proposed attention-based LSTM for classifying early churn users. In an experiment conducted on RoomClip, a general SNS, the proposed model achieved higher classification performance compared to baseline models, thus confirming its effectiveness. We also analyzed the importance of each temporal part and each event. We revealed that the initial temporal part and users' actions have high importance for classifying early churn users. These results should contribute to providing effective interventions for preventing early churn.

Keywords: Identify early churn users · Social networking service

1 Introduction

Various social networking services (SNSs) have been launched with the increasing number of Internet users. SNSs have changed how people share information, and they are gradually playing the role of an information source. The richness of information shared on SNSs depends on the number of users because SNSs host user-generated content. For SNS to remain attractive to users and grow, they must acquire new users and prevent user churn from the service.

SNSs see early churn of more than half of new users because of their low barrier to registration and lack of penalty for churn [3,7,10]. Furthermore, acquiring new users is well known to be more expensive than retaining existing users [11].

© Springer Nature Switzerland AG 2019
L. H. U and H. W. Lauw (Eds.): PAKDD 2019 Workshops, LNAI 11607, pp. 45–56, 2019.
https://doi.org/10.1007/978-3-030-26142-9_5

Therefore, many SNSs are using interventions such as welcome messages, user recommendations, and support for posting to prevent early churn of users [2,4]. To select effective interventions, it is important to identify potential early churn users and factors influencing early churn.

Machine learning models (e.g., random forest, logistic regression) have been widely applied for this purpose [3,10]. However, such models cannot well identify user behavior sequences that reflect early churn potential. To overcome this drawback, the long short-term memory (LSTM) model has been proposed; in this model, the input is the user behavior event sequence binned at constant intervals [6,15]. The LSTM model better identifies early churn users compared to previous machine learning models.

We hypothesize that each temporal part in the event sequence has different importance for the classification of early churn users for the following two reasons: (1) the user behavior sequence consists of coarse and dense parts, suggesting that some temporal parts include a lot of user behavior whereas others do not [1,14], and (2) previous studies have suggested that users' initial behavior after registration may affect their long-term behavior [8,9]. However, the LSTM model does not explicitly consider such information. By considering the different importance of each temporal part in the event sequence, we can expect the LSTM model to better identify early churn users.

In this study, we propose the attention-based LSTM model to identify early churn users by explicitly considering the different importance of each temporal part in the event sequence. Previous research has reported that the attention mechanism can accurately classify contents, such as text, that have different sequential importance [16]. Therefore, the proposed model should show good performance for identifying early churn users. We evaluate this model on Room-Clip, a popular Japanese SNS for sharing interior photos. We then investigate which temporal part in the event sequence is important for identifying early churn users. We also show the types of event sequences that are important for identifying early churn users and discuss effective interventions for preventing early churn on RoomClip. The main contributions of this study are as follows.

1. We proposed the attention-based LSTM model to identify early churn users.
2. We confirmed that the proposed model shows the highest identification performance in terms of identifying early churn users compared to baseline models.
3. We confirmed that initial events occurring immediately after registration have large importance for identifying early churn users.

2   Data

2.1   RoomClip

We evaluated the proposed model and conducted feature importance analysis by using anonymized data from RoomClip[1] for December 12, 2012, to May

---

[1] http://roomclip.jp/.

Table 1. Basic statistics of new users

	Early churn user (N = 29,988)				Nonearly churn user (N = 25,513)			
	Action		Reaction		Action		Reaction	
	Mean	SD	Mean	SD	Mean	SD	Mean	SD
Posts	0.122	0.943	-	-	1.42	4.85	-	-
Likes	1.43	11.9	1.24	12.2	14.7	66.3	21.9	109
Comments	0.105	1.89	0.0925	1.31	1.89	8.51	1.98	9.45
Clips	0.967	4.06	0.0264	0.409	3.30	8.07	0.464	3.19
Follows	1.40	3.79	0.203	1.29	4.64	24.7	1.74	10.8

12, 2015. RoomClip restricts posting photos to a user's room; its interface is otherwise similar to that of Instagram. Users can post photos of their room with freely chosen tags, communicate via posted photos, and follow other users. They can also comment, like, and clip (i.e., make a list of) posted photos.

We defined user's actions as *actions* and feedbacks from other users as *reactions*; actions and reactions are both a type of *event*. We considered nine types of events: post, like, liked, comment, commented, clip, clipped, follow, and followed. We obtained timestamps of every event and registration.

2.2 Definition of Early Churn Users in RoomClip

We target users who have taken at least one action (i.e., post, like, comment, clip, and follow) during O days since the registration day and define these users as new users. We also exclude newly registered users on days when the number of registered users has more than doubled in comparison with that on the previous day to ignore spam users. Ultimately, the total number of new users is $55,501$.

We defined users who did not take any actions during C days after O days since registration as early churn users and users who did take actions more than once during this period as nonearly churn users. Therefore, we treat this as a binary classification problem. We defined $O = 7$ and $C = 180$ in reference to Dror et al. [3]. $C = 180$ was considered reasonable on this data because only 5.62% of the target users have an action interval exceeding 180 days.

Table 1 shows basic statistics for the new users. The number of early and nonearly churn users is $29,988$ and $25,513$, respectively. The probability of early churn users is around 55%. We confirmed that RoomClip has an early churn rate similar to that of other SNSs [3,10]. A comparison of the mean number of each event between early and nonearly churn users showed that the former had lower mean values for all nine events than the latter.

2.3 Distribution of Event Interval of New Users

We investigate whether the event sequence of new users on RoomClip consists of coarse and dense parts, as mentioned in the introduction, by observing the

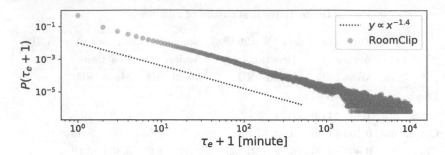

Fig. 1. Distribution of the event interval (τ_e) of new users within $O = 7$ days from the registration day. Vertical axes indicate the probability. Horizontal axes indicates $\tau_e + 1$. To plot this data on a log-log scale, we add 1 to each τ_e value. The dotted line is shown for reference.

distribution of the event interval (τ_e). When the distribution of τ_e obeys the power law, the event sequence is characterized by coarse and dense parts [1,14]. Figure 1 shows the distributions of τ_e of new users within O days since the registration day. We confirmed that the distribution of τ_e obeys the power law.

We confirmed that the event sequence of new users consists of coarse and dense parts. We considered that these coarse and dense parts contribute to the different importance of each temporal part in the event sequence for classifying early churn users. Therefore, by considering such information, we can classify early churn users more accurately.

3   Proposed Model

We describe the proposed attention-based LSTM model that explicitly considers the different importance of each temporal part in the event sequence to classify early and nonearly churn users by observing the event sequence on the first O days. Figure 2 shows the proposed model. It mainly consists of three layers: sequence encoding layer, attention layer, and classification layer.

3.1   Input

We define the event sequence as $\mathbf{e} = (e(1), ..., e(t), ..., e(T))$. t is the index of the temporal part of the event sequence. The t–th temporal part $e(t)$ consists of event occurrences summed over each b hour which takes an arbitrary value. Before inputting $e(t)$ to the sequence encoder, we apply a single-layer perceptron. Furthermore, $x(t) = \text{ReLU}\left(W_e e(t) + b_e\right)$, where $x(t)$ is the hidden representation of $e(t)$ and the actual input to the sequence encoder.

3.2   Sequence Encoding Layer

We used the LSTM to encode the sequence of $x(t)$ [6]. Here, we used LSTM with the forget gate, a basic expansion of the model proposed by Gers et al. [5]. The

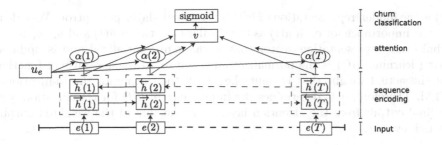

Fig. 2. Proposed model.

internal structure of LSTM with the forget gate is described as follows.

$$i(t) = \sigma_h \left(W_{xi}x(t) + W_{\tilde{h}i}\tilde{h}(t-1) + b_i \right),$$

$$f(t) = \sigma_h \left(W_{xf}x(t) + W_{\tilde{h}f}\tilde{h}(t-1) + b_f \right),$$

$$c(t) = f(t)c(t-1) + i(t)\tanh\left(W_{xc}x(t) + W_{\tilde{h}c}\tilde{h}(t-1) + b_c \right), \qquad (1)$$

$$o(t) = \sigma_h \left(W_{xo}x(t) + W_{\tilde{h}o}\tilde{h}(t-1) + b_o \right),$$

$$\tilde{h}(t) = o(t)\tanh\left(c(t) \right).$$

$i(t)$ is the input gate that determines the amount of input information to maintain. $f(t)$ is the forget gate that determines the amount of information for forgetting the previous cell state $(c(t-1))$. $o(t)$ is the output gate that determines the amount of information of the current cell state that is applied to the final output of the LSTM layer. $\tilde{h}(t)$ is the final output of the LSTM layer in each t. σ_h is the recurrent activation function, and it is a sigmoid function.

We used the bidirectional LSTM in which the forward LSTM output that feeds $x(t)$ from $t = 1$ to $t = T$ and the backward LSTM output that feeds $x(t)$ from $t = T$ to $t = 1$ are concatenated [13]. We expected that the bidirectional LSTM can encode richer information than the forward LSTM. We defined the t-th output from the forward and backward LSTM as $\overrightarrow{h}(t)$ and $\overleftarrow{h}(t)$, respectively. We defined the final concatenated output as $h(t) = [\overrightarrow{h}(t), \overleftarrow{h}(t)]$.

3.3   Attention Layer

To consider the different importance for each t, we used the attention after the sequence encoding layer. To simplify the interpretation of the attention result, we used the attention in reference to Yang et al. as follows [16].

$$u(t) = \tanh\left(W_u h(t) + b_u \right)$$

$$\alpha(t) = \frac{\exp\left(u(t)^\mathsf{T} u_e \right)}{\sum_t \exp\left(u(t)^\mathsf{T} u_e \right)} \qquad (2)$$

$$v = \sum_t \alpha(t)h(t).$$

$u(t)$ is the hidden representation of $h(t)$ by the single-layer perceptron. We calculate the importance for each $u(t)$ as the similarity between $u(t)$ and u_e. u_e is the global event representation vector that is randomly initialized and is updated during learning. $\alpha(t)$ is the normalization result of u_e by the softmax function and the actual weight that is multiplied with the output of the bidirectional LSTM. Therefore, $\alpha(t)$ determines the importance of $h(t)$ for classification. v is the final output from the attention layer; it is the sum of the weighted output through every t.

3.4 Early Churn Classification Layer

To classify early and nonearly churn users, we assign the labels $\hat{y} = 1$ to early churn users and $\hat{y} = 0$ to nonearly churn users. We used the sigmoid function as the classification layer ($y = \mathrm{sigmoid}(W_y v + b_y)$).

3.5 Learning Algorithm

We set binary cross-entropy as the loss function ($loss(y, \hat{y}) = y \log \hat{y} + (1 - y) \log (1 - \hat{y})$) and minimize this loss with back-propagation through time [12].

4   Experiment

4.1 Evaluation Criteria

To evaluate the classification performance of each model, we choose the area under the receiver operating characteristic curve (AUC). The AUC is a graphical plot created by plotting the true positive rate (TPR) as a function of the false positive rate (FPR) while changing the classification threshold. In this task, TPR and FPR are the proportion of correctly identified early churn users among actual early churn users and the proportion of incorrectly identified nonearly churn users among actual nonearly churn users, respectively. Because AUC is calculated by changing the classification threshold, AUC is independent of the threshold. Numerically, AUC ranges from 0 to 1, with values close to 1 indicating good classification performance. To check the classification performance of unlearned data during model training, we conducted 10-fold cross-validation.

4.2 Preprocessing

For RoomClip data with early churn user ratio of 55%, the number of each label is biased. The learning result is known to depend on the major label. To treat this, we applied undersampling. In undersampling, the major label is resampled to have the same number of labels as the minor label.

We select the following binning intervals: $b = 1, 2, 3, 6, 12$, and $24\,\mathrm{h}$. Smaller b means that the input data contains more detailed temporal change. As b increases, the input data contains less detailed temporal change. Because the event sequence of new users consists of coarse and dense parts, we also considered that the different importance for each t for classification becomes more remarkable.

4.3 Baseline Models

We evaluate the classification performance of the proposed model through comparisons with baseline models. In this study, we used two machine learning models, random forest and logistic regression, and LSTM without attention as the baseline models.

Logistic regression is a classical model that is often used in binary classification problems. Random forest is a type of ensemble learning approach that learns a lot of weak trees and takes a decision by majority. To implement these two machine learning models, we used scikit-learn that is provided for machine learning in the Python library[2]. In logistic regression, we applied grid search to the penalty and strength of regularization. In random forest, we set the number of weak trees as 500. Furthermore, we applied grid search to two hyperparameters: the number of features to consider in an individual tree and the minimal number of samples to stop separation. We considered the effect of the remaining hyperparameters to be low. Therefore, we used the default values of scikit-learn for the remaining hyperparameters.

To implement LSTM-based models, we used Keras that is provided for deep learning in the Python library[3]. To evaluate the effectiveness of the attention on early churn user classification from the comparison between with attention and without attention, we calculate the mean of all bidirectional LSTM outputs and feed the calculated mean to the classification layer and call it as LSTM without attention. We used the same hyperparameters between the proposed model and LSTM without attention. We did not apply any hyperparameter tuning method such as grid search because of the learning time and the large number of influential hyperparameters. We use the Z-score to normalize the input. We set the number of units in the LSTM and single perceptron as 10. We set the learning epoch and batch size as 30 and 64, respectively. We used Adam for determining the learning rate. We set the default value in Keras for the remaining hyperparameters.

4.4 Result

Figure 3 shows the mean AUC for each model with varying b and ROC for $b = 1$. For $b = 12$, the proposed model, LSTM without attention, and random forest show almost the same performance. As b decreases, the proposed model gradually outperforms other comparative models. Furthermore, we confirmed that the proposed model shows the highest AUC among these comparative models for $b = 1$.

Next, we confirm whether the proposed model classified early churn users by considering the different importance for each t. Figure 4 shows the relationship between $\alpha(t)$ for $b = 1$ and the event sequence in which the label was correctly classified. The left- and right-hand-side figures show examples of nonearly and

---

[2] https://scikit-learn.org/stable/.
[3] https://keras.io/.

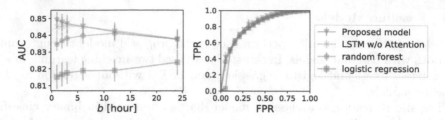

Fig. 3. Results of AUC and ROC of each model. The horizontal and vertical axis of the left-hand-side figure indicate the mean AUC and b, respectively. The horizontal and vertical axis of the right-hand-side figure indicate the TPR and FPR for $b = 1$, respectively. Each vertical bar indicates the standard deviation. The highest mean AUC among all b values of the proposed model, LSTM without attention, random forest, and logistic regression is $(Mean = 0.849, SD = 0.00499)$, $(Mean = 0.845, SD = 0.00469)$, $(Mean = 0.841, SD = 0.00390)$, and $(Mean = 0.823, SD = 0.00757)$, respectively.

early churn users, respectively. From Fig. 4, $\alpha(t)$ shows higher and lower values than the theoretical value for each t.

We confirmed that the proposed model shows the highest performance for small b by considering the difference in relative importance for each t. The event sequence captures detailed temporal changes as b decreases. However, it is also considered that the number of the less important bin increases. Because the LSTM without attention nor other baseline models were not able to properly handle it and it had a negative influence on the classification, we considered that the result worsened with decreasing value of b to a certain degree. In contrast, the proposed model seems to properly handle it, as seen in Fig. 4. These reasons explain the highest performance of the proposed model for small b.

5   Feature Importance Analysis

We conducted two feature importance analyses using the proposed model for $b = 1$ to consider effective interventions for preventing early churn in RoomClip: we clarify the importance of each temporal part in the event sequence and we clarify which event type is important for classification. We also used 10-fold cross-validation to calculate the feature importance.

5.1   Importance of Each Temporal Part in Event Sequence

We clarify the importance of each temporal part in the event sequence for the classification of early churn users because it contributes to considering effective interventions from the aspect of elapsed time since the registration. We defined $\alpha_f(t)$ as the mean $\alpha(t)$ of all test data on each fold. Then, we calculated the mean $\alpha_f(t)$ of all folds. This indicates the global importance of the t–th temporal part in the event sequence for classification. Figure 5 shows the mean $\alpha_f(t)$ of all folds.

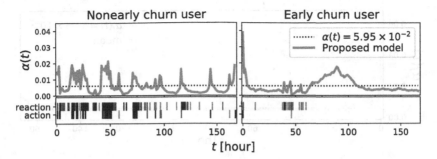

Fig. 4. Relationship between the event sequence and attention $(\alpha(t))$ for $b = 1$. Left-and right-hand-side figures show examples of nonearly and early churn users, respectively. Horizontal axes of all figures indicate t. Vertical axes in the top figures indicate $\alpha(t)$. The dotted line indicates $\alpha(t) = 5.95 \times 10^{-2}$, the theoretical value for every $\alpha(t)$ that has the same importance for $b = 1$. Vertical axes in the bottom figures indicate the type of events. The vertical gray bar indicates the occurrence of an action or reaction. The darker the line color, the more frequently the event occurs.

Figure 5 shows that the mean $\alpha_f(t)$ for $t < 30$ and $t > 155$ exceeded the theoretical value for every $\alpha_f(t)$ that has the same importance for $b = 1$. In particular, we found that the mean $\alpha_f(t)$ for $t = 0$ is remarkably large.

We confirmed that the initial temporal part from the registration has high importance for the classification of early churn users. This is because the initial user behavior of early and nonearly churn users on RoomClip is different, as confirmed in previous studies [8]. As discussed, the event sequence of new users consists of coarse and dense parts; we considered that users who took some actions immediately before the end of $O = 7$ days take an action again after $O = 7$ days. This explains the high importance of the period immediately before the end of $O = 7$ days.

5.2 Importance of Event Type

Identifying the different importance of each event type for classification helps in understanding effective interventions. To identify this, we performed classifications using only the features of each event sequence.

Figure 6 shows the classification performance of models classified by each event sequence. We found that the classification performance of all action sequences is around the same as that of the model classified from all event sequences including reaction sequences. Among action sequences, we found that the classification performance of like and follow sequences was high and that of post and comment sequences was low. By contrast, we found that the classification performance of each reaction and all reaction sequences was low.

Users who newly register on the service rarely post photos, as confirmed from Table 1. Furthermore, new users seem to have weak relations to other users. Therefore, new users have few opportunities to receive reactions from other users.

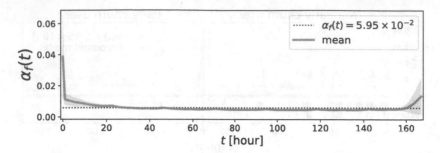

Fig. 5. Result of mean $\alpha_f(t)$. This figure is only slightly different from Fig. 4 above. The blue line indicates the mean of $\alpha_f(t)$. $\alpha_f(t)$ is the mean $\alpha(t)$ of all test data per fold in 10-fold cross-validation. The light blue shading indicates the standard deviation. (Color figure online)

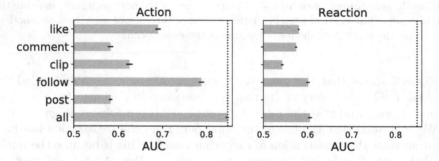

Fig. 6. Mean AUC of model classified by each event sequence. Horizontal axis indicates the AUC and vertical axis, the used event sequence. Error bars indicate the standard deviation. Vertical dotted lines indicate the classification performance of all event sequences. Left- and right-hand-side figures show the results of actions and reactions, respectively.

This leads to low importance of reactions for the classification of early churn users. Table 1 confirms that the number of comments and posts is lower than the number of likes and follows. New users seem to hesitate in performing both actions because they need to think about contents when they post and comment. This leads to the low importance of posts and comments and high importance of likes and follows for the classification of early churn users.

6   Related Research

Many studies have used machine learning models to classify early churn users in SNSs. Classification based on random forest showed high performance [3,10]. Recently, deep learning models have been proposed for this purpose. Yang et al. proposed the LSTM model to classify early churn users on Snapchat and showed that it achieves higher performance than other machine learning models [15]. Yang et al. also proposed an ensemble model by learning multiple LSTMs and combining the weighting outputs from each LSTM based on the attention. Our proposed model differs from their work in that the different importance of each temporal part in the event sequence for classification can be considered by using the attention. To the best of our knowledge, no early churn user classification model considers the different importance of each temporal part in the event sequence for classification in SNS. Because this study aims to clarify the effect of the different importance of each temporal part in the event sequence for classification, we did not consider how the classification performance changes when we modified the ensemble model of Yang et al. by replacing LSTMs with proposed models.

Previous studies have suggested that the initial behavior affects users' long-term behavior. For example, Pal et al. showed that users who will become experienced users can be predicted from the first few weeks' behavior after registration [9]. Karumur et al. showed that early churn users can be classified from their first consecutive actions after registration [8]. Previous studies focused on only certain temporal parts in the event sequence and confirmed their effect on long-term behavior. Therefore, it is unclear whether the initial behavior will have relatively high influence among all temporal parts in the event sequence. This study revealed the importance of each temporal part in the event sequence for classifying churn users.

7   Discussion and Conclusion

In SNSs, the early churn rate is typically high. It is essential to identify potential early churn users and important factors that lead to early churn in SNSs because doing so can enable effective intervention for preventing early churn. We hypothesized that the importance of each temporal part in the event sequence differs for classification. Therefore, we proposed attention-based LSTM. In the experiment conduced on RoomClip, the proposed model achieved higher classification performance compared to baseline models. This result confirmed the effectiveness of the proposed model. Because many human sequential activities consists of coarse and dense parts like SNSs (e.g., exchanging e-mails, working patterns), we believe that our approach is effective for them [1].

Furthermore, our analysis of the importance of each temporal part in the event sequence revealed that the time immediately after registration has high importance for classification. We also found that new users' actions, especially likes and follows, have high importance for classification whereas reactions from other users have little influence. These two results can be used to consider effective interventions, such as presenting popular users and photos to a newly joined user immediately after registration to encourage follows and likes.

These results can help apply interventions for potential early churn users and general interventions for preventing early churn. These results can be applied to general SNSs such as Instagram and Twitter for which temporal information is available, and we hope that they will be of use to SNS developers.

Acknowledgement. We would like to thank RoomClip Inc. for providing data.

References

1. Barabasi, A.L.: The origin of bursts and heaby tails in human dynamics. Nature **435**(7039), 207–211 (2005)
2. Choi, B., Alexander, K., Kraut, R.E., Levine, J.M.: Socialization tactics in Wikipedia and their effects. In: Proceedings of the 2010 ACM Conference on Computer Supported Cooperative Work, pp. 107–116. ACM (2010)
3. Dror, G., Pelleg, D., Rokhlenko, O., Szpektor, I.: Churn prediction in new users of Yahoo! answers. In: Proceedings of the 21st International Conference on World Wide Web, pp. 829–834. ACM (2012)
4. Freyne, J., Jacovi, M., Guy, I., Geyer, W.: Increasing engagement through early recommender intervention. In: Proceedings of the Third ACM Conference on Recommender Systems, pp. 85–92. ACM (2009)
5. Gers, F.A., Schmidhuber, J.A., Cummins, F.A.: Learning to forget: continual prediction with LSTM. Neural Comput. **12**(10), 2451–2471 (2000)
6. Hochreiter, S., Schmidhuber, J.: Long short-term memory. Neural Comput. **9**(8), 1735–1780 (1997)
7. Joyce, E., Kraut, R.E.: Predicting continued participation in newsgroups. J. Comput.-Mediat. Commun. **11**(3), 723–747 (2006)
8. Karumur, R.P., Nguyen, T.T., Konstan, J.A.: Early activity diversity: assessing newcomer retention from first-session activity. In: Proceedings of the 19th ACM Conference on Computer-Supported Cooperative Work & Social Computing, pp. 595–608. ACM (2016)
9. Pal, A., Chang, S., Konstan, J.A.: Evolution of experts in question answering communities. In: The 6th International AAAI Conference on Weblogs and Social Media, pp. 274–281. AAAI (2012)
10. Pudipeddi, J.S., Akoglu, L., Tong, H.: User churn in focused question answering sites: characterizations and prediction. In: Proceedings of the 23rd International Conference on World Wide Web, pp. 469–474. ACM (2014)
11. Reichheld, F.F., Sasser, J.W.: Zero defections: quality comes to services. Harvard Bus. Rev. **68**(5), 105–111 (1990)
12. Rumelhart, D.E., Hinton, G.E., Williams, R.J.: Learning rrepresentations by backpropagating errors. Nature **323**(6088), 526–533 (1986)
13. Schuster, M., Paliwal, K.K.: Bidirectional recurrent neural networks. IEEE Trans. Signal Process. **45**(11), 2673–2681 (1997)
14. Yan, Q., Wu, L., Zheng, L.: Social network based microblog user behavior analysis. Phys. A **392**(7), 1712–1723 (2013)
15. Yang, C., Shi, X., Jie, L., Han, J.: I know you'll be back: Interpretable new user clustering and churn prediction on a mobile social application. In: Proceedings of the 24th ACM SIGKDD International Conference on Knowledge Discovery & Data Mining, pp. 914–922. ACM (2018)
16. Yang, Z., Yang, D., Dyer, C., He, X., Smola, A., Hovy, E.: Hierarchical attention networks for document classification. In: Proceedings of the 14th Conference of the North American Chapter of the Association for Computational Linguistics: Human Language Technologies, pp. 1480–1489 (2016)

PAKDD 2019 Workshop on Weakly Supervised Learning: Progress and Future (WeL 2019)

Weakly Supervised Learning by a Confusion Matrix of Contexts

William Wu(✉)

Faculty of Engineering and Information Technology,
University of Technology Sydney, Ultimo, Australia
William.Z.Wu@student.uts.edu.au

Abstract. Context consideration can help provide more background and related information for weakly supervised learning. The inclusion of less documented historical and environmental context in researching diabetes amongst Pima Indians uncovered reasons which were more likely to explain why some Pima Indians had much higher rates of diabetes than Caucasians, primarily due to historical, environmental and social causes rather than their specific genetic patterns or ethnicity as suggested by many medical studies.

If historical and environmental factors are considered as external contexts when not included as part of a dataset for research, some forms of internal contexts may also exist inside the dataset without being declared. This paper discusses a context construction model that transforms a confusion matrix into a matrix of categorical, incremental and correlational context to emulate a kind of internal context to search for more informative patterns in order to improve weakly supervised learning from limited labeled samples for unlabeled data.

When the negative and positive labeled samples and misclassification errors are compared to "happy families" and "unhappy families", the contexts constructed by this model in the classification experiments reflected the Anna Karenina principle well - "Happy families are all alike; every unhappy family is unhappy in its own way", an encouraging sign to further explore contexts associated with harmonizing patterns and divisive causes for knowledge discovery in a world of uncertainty.

Keywords: Weakly supervised learning · Context · Confusion matrix ·
Context construction · Contextual analysis

1 Introduction

In many real-world data mining tasks, such as healthcare data investigation and financial profile classification, it is often the case that not all sample records in training sets are fully labeled due to various constraints such as cost and capability restriction, sampling and classification limitation. The weakly supervised learning approach is meant to handle these situations adaptively when only a small number of training samples are labeled by initial probing classifications, and then some forms of self-learning are engaged to revise progressively and to label the rest accordingly.

Weakly supervised learning may have emanated from the "bootstrapping" way of semi-supervised self-learning in the field of data mining and classification [1, 2] and

© Springer Nature Switzerland AG 2019
L. H. U and H. W. Lauw (Eds.): PAKDD 2019 Workshops, LNAI 11607, pp. 59–64, 2019.
https://doi.org/10.1007/978-3-030-26142-9_6

many strategies and techniques have been developed, for example, active learning by uncertainty sampling, query-by-committee and a decision-theoretic approach assisted by human experts as "oracles" [3], semi-supervised learning by an expectation–maximization approach, co-training and multi-view learning via separation of attribute sets with integration of result learning [4], and reinforcement learning by an iteration of an agent's action and reward functions in a Markov decision process [5, 6]. Collectively speaking, the more information available on the samples, the more informed decisions can be made by the experts as "oracles" and by the researchers to improve the self/co-training rulesets and the action-reward learning logic.

As an example of improved learning by considering contexts around the available data, the inclusion of less documented historical and environmental context in researching diabetes amongst Pima Indians by Schulz el al. and Phihl [7, 8], uncovered reasons which were more likely to explain why some Pima Indians "have up to ten times the rate of diabetes as Caucasians" which were not due to their specific genetic patterns or ethnicity, but because of their loss of access to water from the Gila River in the 1900s which adversely affected their agricultural supply and farm work, and their later adaption of modern Western fast food and life-style. In comparison, another group of Pima Indians from the same tribe migrated to Mexico in the 1890s and continued their traditional farm work and traditional diet, have a significant lower rate of diabetes despite these two groups sharing a common genetic background [8, 9].

A confusion matrix is a simple and effective way to summarize classification results and algorithm performance in a tabular form by tallying the category results, making result comparison easy between class label categories [10–12]. When using the term "incremental context" to describe how data samples are sorted in ascending order to emulate the value growth path of a lead attribute, and the term "correlational context" to describe the subsequent and correlated value variation of other related attributes along the value growth path of the lead attribute, then the term "categorical context" can describe the cross-category interrelation when each matrix cell hosts both "incremental context" and "correlational context" for their respective result category, which makes cross-category context comparison simple, visual and systematic, and facilitates learning in a category-by-category and side-by-side manner.

2   Context Construction with Confusion Matrix

2.1   Post-classification Analysis and Data Normalization for Easy Comparison

Context construction can start as a part of the post-classification analysis process based on the initial set of training samples which have been labeled, and the underlying classification platform can be based on any classification algorithm, e.g. decision tree, neural network or naive Bayes. The resulting confusion matrix can be evaluated first and if suitable, be transformed into a table with each of its category cells to house the incremental and correlational contexts of their respective training samples.

In order to construct incremental and correlational context for an easy and consistent comparison between attributes in a dataset, the values of different attributes measured in different units are converted into one standardized value range between 0 and 100 via a min-max normalization routine, and ten key value growth stages of each attribute can then be established by ten deciles based on these standardized values.

2.2 Construction of Categorical, Incremental and Correlational Context

Key steps in this construction process include:

1. Evaluate initial classification result and its confusion matrix
2. Define a selection list of lead attributes and their rotation/significance order
3. Select a lead attribute and redistribute its data records with standardized values into categorized subsets according to their confusion matrix result categories
4. For each categorized subset sorted by the current lead attribute
    4a. Extract the median record from each decile to simulate ten value growth stages in a vertical way within its categorized cell as incremental context
    4b. Start from the current lead attribute and for its ten growth stages, connect and plot lines across other attributes accordingly as correlational context
    4c. Evaluate these incremental and correlational contexts in the form of line and value patterns between matrix categories as categorical context
5. Select the next lead attribute from the rotation list determined in Step 2 and repeat Steps 3 & 4 to construct more contextual models for further analysis

This context construction process starts as a multi-contextual model with categorical, incremental and correlational context. It is hoped that this is only the beginning, and other factors can later be introduced to this multi-contextual model to expand the scope of data exploration and knowledge discovery for weakly supervised learning.

3   Experiments and Analysis

3.1 Context Construction for Attribute "Length" in the Page Blocks Dataset

Experiments on the labeled training sets are based on ten UCI datasets [13] but only two are highlighted in this short paper. WEKA's [14] C4.5/J48 decision tree is utilized as the initial probing classifier, but other algorithms can also be used.

When testing on the Page Blocks dataset, the attribute "length" is selected as the first lead attribute because it is ranked as the most significant attribute by the gain ratio evaluator, and its multi-contextual model is shown in Table 1. The four cells in Table 1 represent the four confusion matrix result categories based on an initial binary classification for the labeled text or non-text samples. The y-axis in each cell represents the ten key value growth stages of attribute "length", and the x-axis represents the other related attributes in the training set. The ten plots starting from "length" and connecting across the other attributes represent how the values of other attributes from the same key growth stage samples change according to the incremental growth of the "length" value in a correlational way, showing how the converging and diverging value patterns differentiate between the four confusion matrix categories and reflecting the Anna Karenina principle [15] - text type page blocks share the same regular characteristics, non-text type page blocks and classification errors differ in various ways due to numerous factors such as varying color pigment and text font-size, etc. Context models constructed for other attributes also present similar patterns.

Table 1. A confusion matrix of contexts constructed for the Page Blocks dataset

3.2 Context Construction for Attribute "plas" in the Pima Diabetes Dataset

In this test case demonstration of the Pima Indians diabetes dataset, the attribute "plas" is selected as the lead attribute and its multi-contextual model is shown in Table 2. One distinctive pattern is that the majority of "plas" values in the true negative category are bounded in the middle range between the normalized median value of 37 and 70, and as the "plas" value increases, the values of the other attributes vary rather uniformly, and most are bounded by a narrow value range. In contrast, "plas" values in the true positive category start from a higher point of 55 and scatter in a wider value range up to 97, and the values of the other related attributes, e.g. "mass", "age" and "pedigree", also fluctuate in a wider value range with higher starting points. The two error categories, false positives and false negatives, share similar "bigger and rougher" value variation patterns similar to the true positive category.

In essence, the true negative samples show more uniformity than the true positive samples and the error samples. This is a theme that matches the contrast between the benefits of harmonized patterns and the burdens of unrestrained value fluctuations signified by the Anna Karenina principle and shared by other attributes and datasets.

Table 2. Matrix of incremental, correlational & categorical context for the Pima dataset

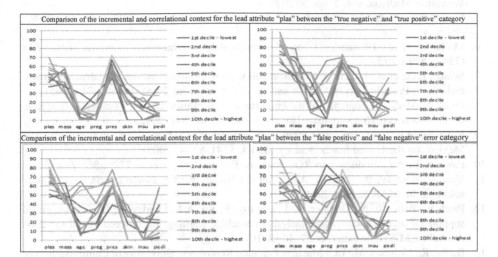

4 Potential Issues and Conclusion

Context construction by coupling incremental sampling with cross-attribute plotting under a confusion matrix structure may sound trivial, and its representation of incremental, correlational and categorical context may look inconsequential. However, the focus of this study is not about breaking and winning in complexity, it is about sorting and connecting data elements to gain more understanding about their internal relationship within context and to initiate such contextual analysis with simplicity and practicality, hence the inclusion of ROC curve and F-score will be considered next.

A more specific issue concerns the suitability and adequacy of using ten sorted deciles to represent ten key value growth stages of a lead attribute. While systematic sampling can be considered as a kind of stratification with equal sampling fractions, its value growth representation can be over-optimistic and will need improvement.

In conclusion, the experiments and examination of this multi-contextual model can indeed highlight certain converging and diverging patterns in a contextual and meaningful way, and they may potentially be applicable to unlabeled data for pattern-matching and error-estimation when learning from labeled samples. It may seem far-fetched, but supported by some interesting findings, it is worth exploring.

References

1. Nadeau, D., Sekine, S.: A survey of named entity recognition and classification. Lingvist. Investig. **30**(1), 3–26 (2007)
2. Zhou, Z.H.: A brief introduction to weakly supervised learning. Natl. Sci. Rev. **5**(1), 44–53 (2017)
3. Settles, B.: Active learning. Synth. Lect. Artif. Intell. Mach. Learn. **6**(1), 1–114 (2012)

4. Zhu, X.: Semi-supervised learning literature survey. Computer Science, University of Wisconsin-Madison, vol. 2, no. 3 (2006)
5. Sutton, R.S., Barto, A.G.: Introduction to Reinforcement Learning. MIT Press, Cambridge (1998)
6. Mnih, V., et al.: Human-level control through deep reinforcement learning. Nature **518** (7540), 529 (2015)
7. Schulz, L.O., et al.: Effects of traditional and western environments on prevalence of type 2 diabetes in Pima Indians in Mexico and the US. Diab. Care **29**(8), 1866–1871 (2006)
8. Phihl, T.: Medical Geography of the Pima Indian Reservation Diabetes Epidemic: The Role of the Gila River. Undergraduate Honors Theses, Paper 304 (2012)
9. Valencia, M.E., et al.: Impact of lifestyle on prevalence of kidney disease in Pima Indians in Mexico and the United States. Kidney Int. **68**, S141–S144 (2005)
10. Fawcett, T.: An introduction to ROC analysis. Pattern Recogn. Lett. **27**(8), 861–874 (2006)
11. Visa, S., Ramsay, B., Ralescu, A.L., Van der Knaap, E.: Confusion matrix-based feature selection. In: MAICS, pp. 120–127 (2011)
12. Patel, K., Bancroft, N., Drucker, S.M., Fogarty, J., Ko, A.J., Landay, J.: Gestalt: integrated support for implementation and analysis in machine learning. In: Proceedings of the 23nd Annual ACM Symposium, pp. 37–46 (2010)
13. Bache, K., Lichman, M.: UCI Machine Learning Repository. University of California, School of Information and Computer Science, Irvine, CA (2013). http://archive.ics.uci.edu/ml
14. Hall, M., Frank, E., Holmes, G., Pfahringer, B., Reutemann, P., Witten, I.: The WEKA data mining software: an update. SIGKDD Explor. **11**(1), 10–18 (2009)
15. Diamond, J.M.: Guns, germs and steel: a short history of everybody for the last 13,000 years. Random House (1998)

Learning a Semantic Space for Modeling Images, Tags and Feelings in Cross-Media Search

Sadaqat ur Rehman[1]([✉]), Yongfeng Huang[1], Shanshan Tu[2],
and Basharat Ahmad[1]

[1] Tsinghua National Laboratory for Information Science and Technology,
Tsinghua University, Beijing 100084, People's Republic of China
`z-sun15@mails.tsinghua.edu.cn`, `yfhuang@mail.tsinghua.edu.cn`
[2] Faculty of Information Technology, Beijing University of Technology,
Beijing 100124, China
`sstu@bjut.edu.cn`

Abstract. This paper contributes a new, real-world web image dataset for cross-media retrieval called FB5K. The proposed FB5K dataset contains the following attributes: (1) 5130 images crawled from Facebook; (2) images that are categorized according to users' feelings; (3) images independent of text and language rather than using feelings for search. Furthermore, we propose a novel approach through the use of Optical Character Recognition (OCR) and explicit incorporation of high-level semantic information. We comprehensively compute the performance of four different subspace-learning methods and three modified versions of the Correspondence Auto Encoder (Corr-AE), alongside numerous text features and similarity measurements comparing Wikipedia, Flickr30k and FB5K. To check the characteristics of FB5K, we propose a semantic-based cross-media retrieval method. To accomplish cross-media retrieval, we introduced a new similarity measurement in the embedded space, which significantly improved system performance compared with the conventional Euclidean distance. Our experimental results demonstrated the efficiency of the proposed retrieval method on three different public datasets.

Keywords: Cross-media search · Text-image-feeling embeddings · FB5K dataset

1 Introduction

The current era has seen rapid growth in Multimedia Information Retrieval (MIR). Despite constant hard work in the development and construction of new MIR techniques and datasets, the semantic gap between images and high-level concepts remains high. We need a promising model to focus on modeling high-level semantic concepts, either by image annotation or by object recognition to diminish this semantic gap. Numerous real-world methods [1,2] have been introduced for this kind of concept-based multimedia search system. Among several

© Springer Nature Switzerland AG 2019
L. H. U and H. W. Lauw (Eds.): PAKDD 2019 Workshops, LNAI 11607, pp. 65–76, 2019.
https://doi.org/10.1007/978-3-030-26142-9_7

of these methodologies, the first step is dataset selection for high-level concepts and small semantic gaps, which are relatively easy for machine understanding and training.

This paper presents a novel resource evaluation dataset for cross-media searching, called FB5K along with a benchmark learning system. Existing cross-media or multi-modal retrieval datasets have some limitations. Firstly, some datasets lack context information i.e. link relations. Such context information is quite accurate, and can provide significant evidence to ameliorate cross-media retrieval system accuracy. Similarly, the Pascal VOC 2012 dataset[1] [3] consists of only 20 categories. However, cross-media retrieval implicates numerous domains under real-world internet conditions. Cross-media retrieval systems trained on scanty domain datasets have difficulties in handling queries from anonymous domains. Secondly, popular cross-media datasets are small in size, for example Xmedia [4], IAPR TC-12 [5], and Wikipedia[2] [6]. This deficiency in appropriate data makes it difficult for retrieval systems to learn and evaluate the robustness in real-world galleries. Thirdly, datasets such as, ALIPR [7], SML [8], either just used all the image annotation keywords associated with training images, or unenforced any constraint to the annotation vocabulary for example ESP [9], LabelMe [10], and AnnoSearch [11]. Therefore, these datasets essentially neglect the differences among keywords relating to semantic gaps.

Fig. 1. Some examples of FB5K dataset used for cross-media search.

---

[1] http://host.robots.ox.ac.uk/pascal/VOC/.
[2] http://www.svcl.ucsd.edu/projects/crossmodal/.

Considering the aforementioned problems, this paper makes three major contributions. The first is the collection of a new resource evaluation cross-media retrieval dataset, named FB5K. It contains 5130 image-feeling pairs collected from Facebook[3], introduced for the first time in the cross-media retrieval research community. This dataset is differentiated from current datasets in three aspects: varied domains, high-level semantic information incorporation, and rich context information. Eventually, it should provide a more accurate standard for cross-media study. Therefore, we constructed a standard dataset, keeping in mind the research issues to focus researcher/developer efforts on cross-media retrieval algorithm development, instead of laboriously comparing methods and results. The second is that, to the best of our knowledge, this is the first effort to collect a dataset of high-level concepts with small semantic gaps based on users' semantic descriptions i.e. image-feeling relationships. Third, this approach aims to learn the cross-media embeddings of users' feelings, images, and tags/texts. We propose a novel method by using Optical Character Recognition (OCR), explicit incorporation of high-level semantic information, and a new similarity measurement in the embedded space, which significantly overcomes the conventional distance measurement methods and improve retrieval performance.

2 Proposed Dataset

This section describes a new dataset called FB5K, which comprises 5130 images collected from Facebook. The complete FB5K dataset will be made available via ABC[4].

2.1 Dataset Collection

Each step in the dataset collection is briefly explained below.

Seed User Gathering. In order to obtain the genuine emotions of users associated with an image rather than image contents, we obtained seed users by sending queries to Facebook with numerous key words, for example *happy, hungry, love*, etc.

User Candidate Generation. To generate user candidates we implement a web spider to crawl the user accounts of individuals who were following the seed users. This step was repeated a number of times until we obtained a lengthy list of user candidates.

Feelings Collection. Another web spider collected feelings as text associated with the matching images by visiting the web pages of different users present in the candidate list. Our finding suggested that about 80% of the users' feelings accompanied the images.

---

[3] facebook.com.
[4] http://www.xyz.com/.

Data Pruning. We refer to an image, tag, or feeling-text pair as a tweet. Data were pruned based on the following criteria (pruned out data were referred to as garbage data):

- Feelings without images;
- Tweets not associated with images or feelings;
- Repeated images with the same ID;
- Error images.

As a result, a total of 5130 image-tag pairs were obtained. Figure 1 presents some examples from this benchmark dataset.

2.2   Dataset Characteristics

The performance of cross-media retrieval methods is highly dependent on the nature of the dataset used for their evaluation. The FB5K dataset includes a set of images that are closely associated with user feelings. These images were crawled from Facebook along with the user-associated feelings. The FB5K dataset has the following attributes:

- First, since this dataset was collected from a social media website, it contains a broad variety of domains under single examples of feelings such as, *hungry, love, sad, thankful* etc.
- Second, the relationship between images and users' feelings is often very strong. In the examples given in Fig. 1, the images have strong ties with the associated feelings. Such is the case in a realistic scenario.
- Third, FB5K is a large-scale dataset, containing 5130 image-text pairs, which helps to avoid overfitting during system training. In other words, it helps to test the cross-media retrieval method's robustness via a wealth of data.
- Fourth, this dataset helps to reduce the semantic gap by providing more accessible visual content descriptors using high-level semantic concepts.

To our knowledge, this is the first cross-media dataset that consists of the above-mentioned characteristics. Also, we believe that FB5K is the first dataset collected from Facebook that comprises high-level concepts with minor semantic gaps between users' semantic descriptions, and a ground-truth of 70 concepts for the whole dataset.

3   Proposed Retrieval Method for the FB5K Dataset

This section briefly explains the proposed cross-media retrieval algorithm for FB5K. Numerous features are used for image representation, for example SIFT [12], color features [13], GIST [14] and HOG [15,16]. These features are useful for extracting colors and shapes of images, but not for words represented by the images. In this regards, we first propose OCR then adopt explicit incorporation of high-level semantic information and finally develop a novel similarity

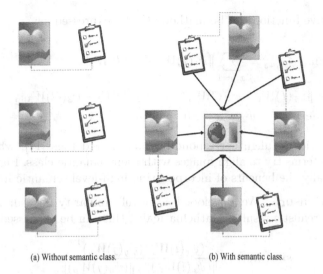

(a) Without semantic class.   (b) With semantic class.

Fig. 2. Graphical representation of high-level semantic information incorporation: (a) without semantic class and (b) with semantic class.

measurement in the embedded space to improve the retrieval performance. A detailed explanation is provided as follows:

Text Extraction. First is the extraction of words on each image using tessart[5].

Incorporation of High-Level Semantic Information. To facilitate OCR text extraction we incorporate high-level semantic information for learning a common space for image, text/tag, and semantic information (user feelings). Assume we have n training images having i_f-dimensional visual feature vectors and t_f-dimensional tag feature vectors. Where $I \in R^{n \times i_f}$ and $T \in R^{n \times t_f}$. Furthermore, we also associated each training image with a high-level semantic class, $C \in R^{n \times c}$, where c represents the number of categories. Individual images are labeled with one c class (only one specific class in each row of K is 1 and the remaining are 0).

Let i, j denote two points. We define similarity as:

$$K_x(i,j) = \psi_x(i)\psi_x(j)^T \qquad (1)$$

where K_x is a kernel function and $\psi_x(.)$ represent a function embedding the original feature vector into a nonlinear kernel space. The goal is to find matrices W_x that project the embedded vector $\psi_x(i)$ to minimize the distance between data items.

---

[5] https://github.com/tesseract-ocr/tesseract.

The objective function can be mathematically expressed as:

$$\min_{w_1,w_2,w_3} = \sum_{x,y=1}^{3} \|\psi_1(I)W_1 - \psi_2(T)W_2\|_2$$
$$+ \|\psi_1(I)W_1 - \psi_3(C)W_3\|_2 + \|\psi_2(T)W_2 - \psi_3(C)W_3\|_2$$
$$where\ w_1 = w_2 = w_3 = 0 \tag{2}$$

this equation tries to align corresponding images and tags [17], whereas, the remaining two terms try to align images with their semantic class. Figure 2 illustrates graphically the benefits of incorporating high-level semantic information.

Similarity Measure. We developed a novel similarity measurement that yielded better realistic results. Mathematically, this can be expressed as:

$$sim(x_i y_i) = \frac{(\psi_x(i)W_x)(\psi_y(j)W_y)^T}{\|(\psi_x(i)W_x)\|_2 \|(\psi_y(j)W_y)\|_2} \tag{3}$$

where x_i represents the training image and y_i represents the corresponding tweet. W_x projects the embedded vector $\psi_x(i)$ and W_y projects the embedded vector $\psi_y(j)$ to minimize the distance between image and text.

Distance in Common Subspace. In this paper, we represent the cosine distance between two different modalities in the common subspace as $Cos(Twt, Img)$, where Twt and Img represent the tweet and image. It was learned by retrieval methods such as, Correspondence Auto Encoder (Corr-AE) and subspace methods.

Ranking. Each candidate in the gallery was ranked, based on similarity distances between the queries and candidates.

4   Experimental Results and Discussion

4.1   Experimental Setup

All experiments were performed on four subspace learning methods, which were Canonical Correlation Analysis (CCA) [6], Bilinear Model (BLM) [18], Partial Least Square (PLS) [19], and Generalized Multi-view Marginal Fisher Analysis (GMMFA) [20] and three Corr-AE methods [21]: Corr-AE, cross Corr-AE and full Corr-AE.

In the case of subspace learning methods, we used the implementation from [20] to compute the linear projection matrix. For Corr-AE methods, we use the implementation of [21] to calculate the hidden vectors of the two different modalities. We employed a 1024-dimensional hidden layer. For Corr-AE, cross Corr-AE and full Corr-AE the weight factors for reconstruction errors and correlation distances were set to 0.8, 0.2 and 0.8, respectively.

Dataset Splitting. We used three datasets in each experiment: Wikipedia, Flickr30k and FB5K. We split each dataset into a training set, a testing set, and a validation set, as illustrated below:

1. *Wikipedia dataset.* In the case of subspace learning, we used 2173 and 500 image-text pairs for training and testing respectively, while for Corr-AE methods a further 193 pairs served as a validation set. We utilized all of the data in a test set as a query.
2. *Flickr30k dataset.* For subspace learning, we used 15000 image-text pairs for both training and testing while for Corr-AE methods an additional 1783 image-text pairs were added for validation. We randomly selected 2000 images and texts from the test set to function as a query.
3. *FB5K dataset.* We split the dataset into 80% and 20% image-text pairs for training and testing respectively. We used the same split for subspace learning, while for Corr-AE methods 250 additional image-text pairs served as a validation set.

Representation. All images were first resized to dimension of 224×224. Then we extracted the last fully connected (fc7) Convolution Neural Network (CNN) features using VGG16 [22] with CAFFE [23] implementation. Text representation was based on Latent Dirichlet Allocation (LDA) [24]. An LDA model was learned from all texts and used to compute the probability of each text under 50 hidden topics. We used this probability vector for text representation. A Bag-of-Word (BoW) model was used for text representation in Corr-AE methods. Initially, texts were converted into lower case, with all stopping-words removed. A unigram model was adopted to form a dictionary of the most recurrent 5000 words. Based on this dictionary, for each text we generated a 5000-dimensional BoW model.

Evaluation Parameters. We assessed the retrieval performance using Cumulative Match Characteristic (CMC) curves and mean rank. CMC is a useful approach that is used as the evaluation metric in many applications such as face recognition [25–27] and biometric systems [28,29]. For cross-media retrieval, CMC can be illustrated by a curve of average retrieval accuracy with respect to the average ranks of the correct matches for a series of queries, K, where rank is:

$$Rank = \frac{1}{|K|} \sum_{x=1}^{K} rank_x, \qquad (4)$$

$rank_x$ refers to the rank position of the correct match for the xth query.

4.2   Retrieval Methods Comparison Using Different Datasets

We tested different cross-media retrieval methods on Flickr30k, Wikipedia and FB5K datasets. Figure 3 shows the effectiveness of the different retrieval methods. We drew several conclusions from this.

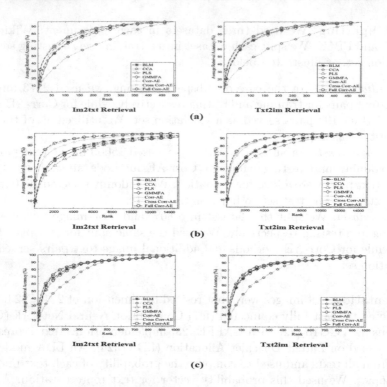

Fig. 3. CMC curves compared for different Corr-AE and subspace learning methods using different cross-media datasets: (a) Wikipedia, (b) Flickr30k, and (c) FB5K.

Corr-AE methods performed well compared to subspace learning methods with all three datasets. However, CCA showed a significant improvement in performance as the number of training samples increased using FB5K. The logic behind this is that correlation is ignored between different modalities in subspace learning when representation learning is performed. However, representation learning and correlation learning are merged into a single process in Corr-AE methods. Furthermore, Corr-AE is used to train a model by minimizing linear combinations of representation learning error and correlation learning error for individual modalities, and between hidden representations of two modalities [16]. This minimization of correlation learning error helps the model in learning hidden representations, while minimization of representation learning error makes better hidden representations to reconstruct the input of individual modalities.

The retrieval performance was highest for FB5K and lowest for Wikipedia. This shows that the tweets are highly correlated when using FB5K compared with Flickr30k and Wikipedia. The main reason that FB5K obtained the highest retrieval accuracy is twofold: first, it contained high-level concepts with small semantic gaps. Second, text and images were highly correlated in this dataset. We conclude that user descriptions on tweets are highly correlated to the scenarios.

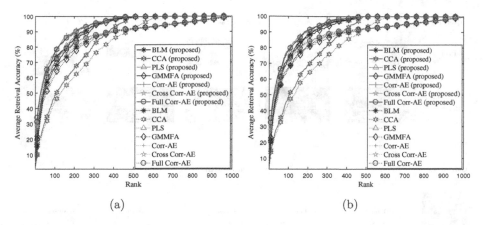

Fig. 4. CMC curves on FB5K with the proposed method and baselines. (a) Im2txt retrieval. (b) Txt2im retrieval.

4.3 Proposed Method Performance

In this section, we describe the proposed method for evaluation of FB5K. We compared its performance with the baseline methods.

Figure 4 clearly shows that using the proposed method in the baseline learning systems significantly improved their performance. In particular, using OCR, explicit incorporation of high-level semantic information, and a specially developed similarity measurement in the embedded space improved cross-media retrieval accuracy when similar retrieval methods were used. For example, in the case of Txt2im retrieval, CCA achieved 45% accuracy at rank = 110, whereas the BLM, PLS, and GMMFA methods achieved the same accuracy at ranks 20, 25 and 18, respectively. Incorporating the proposed method boosted the accuracy of CCA, BLM, PLS, and GMMFA to 6.5%, 4%, 5% and 7% respectively, at the same rank.

4.4 FB5K Retrieved Examples

This section describes the retrieval examples for FB5K using CCA and the proposed method.

Figure 5(a) shows the retrieval image results for different query tags. It shows that the proposed method was successful in learning colors, background and class information, e.g. in Fig. 5(a) we used the keyword *cold* to retrieve the images on right, which strongly indicate that the keyword information lay in the retrieved image. Moreover, incorporation of semantic class not only improved the retrieval accuracy, but also provided higher weights to more minor concepts during the formation of query tag vectors.

Figure 5(b) shows the tagging results retrieved by the proposed method on some test images. It is clear that using the proposed method with FB5K significantly outperformed the baseline methods, despite its diverse features.

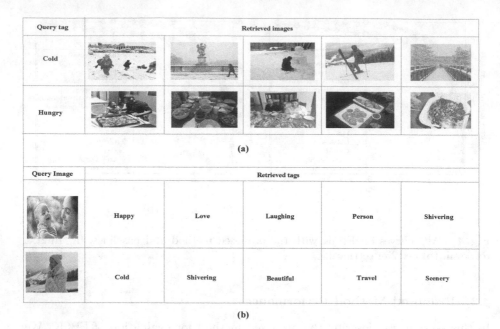

Fig. 5. Retrieval examples for FB5K using CCA and the proposed method. The first two rows represent the query tag and its corresponding top five retrieved images, whereas the last two rows show query images and their corresponding top five retrieved tags. (a) tag/txt2img retrieval. (b) img2tag/txt retrieval.

Furthermore, FB5K provides information that is more realistic to the user. It incorporates high-level semantic information by providing the class probability for individual images. For example, in Fig. 5(b), with a query image of a baby, the proposed method retrieved *happy* and *love* as high frequency words in the retrieved text. This shows that despite the sentiment of an image being hidden under high-level concepts, opinion characteristics can have an impact on multimodal retrieval.

5   Conclusion

This paper introduced a novel cross-media dataset called FB5K. We also presented a more realistic embedding approach for images, tags/texts, and their semantics. Specifically, in order to learn the cross-modal embeddings of user feelings, images and tags/texts, we developed a novel method by utilizing OCR, explicit incorporation of high-level semantic information, and a new similarity measurement in the embedded space, to improve the retrieval performance.

We believe that FB5K and the proposed cross-media retrieval method suffice as a reference guide for researchers and developers to facilitate the design and implementation of better evaluation protocols.

References

1. Lu, Y.-J., Nguyen, P.A., Zhang, H., Ngo, C.-W.: Concept-based interactive search system. In: Amsaleg, L., Guðmundsson, G.Þ., Gurrin, C., Jónsson, B.Þ., Satoh, S. (eds.) MMM 2017. LNCS, vol. 10133, pp. 463–468. Springer, Cham (2017). https://doi.org/10.1007/978-3-319-51814-5_42
2. Kambau, R.A., Hasibuan, Z.A.: Concept-based multimedia information retrieval system using ontology search in cultural heritage. In: Second International Conference on Informatics and Computing (ICIC), pp. 1–6. IEEE (2017)
3. Hwang, S.J., Grauman, K.: Reading between the lines: object localization using implicit cues from image tags. IEEE Trans. Pattern Anal. Mach. Intell. **34**(6), 1145–1158 (2012)
4. Peng, Y., Huang, X., Zhao, Y.: An overview of cross-media retrieval: concepts, methodologies, benchmarks and challenges. IEEE Trans. Circ. Syst. Video Technol. **28**, 2372–2385 (2017)
5. Grubinger, M., Clough, P., Muller, H., Deselaers, T.: The IAPR TC-12 benchmark: a new evaluation resource for visual information systems. In: International Workshop onto Image, vol. 5, p. 10 (2006)
6. Rasiwasia, N., et al.: A new approach to cross-modal multimedia retrieval. In: Proceedings of the 18th ACM International Conference on Multimedia, pp. 251–260. ACM (2010)
7. Li, J., Wang, J.Z.: Real-time computerized annotation of pictures. IEEE Trans. Pattern Anal. Mach. Intell. **30**(6), 985–1002 (2008)
8. Carnciro, G., Chan, A.B., Moreno, P.J., Vasconcelos, N.: Supervised learning of semantic classes for image annotation and retrieval. IEEE Trans. Pattern Anal. Mach. Intell. **29**(3), 394–410 (2007)
9. Von Ahn, L., Dabbish, L.: Labeling images with a computer game. In: Proceedings of the SIGCHI Conference on Human Factors in Computing Systems, pp. 319–326. ACM (2004)
10. Russell, B.C., Torralba, A., Murphy, K.P., Freeman, W.T.: LabelMe: a database and web-based tool for image annotation. Int. J. Comput. Vis. **77**(1–3), 157–173 (2008)
11. Wang, X.-J., Zhang, L., Jing, F., Ma, W.-Y.: Annosearch: Image auto-annotation by search. In: 2006 IEEE Computer Society Conference on Computer Vision and Pattern Recognition, vol. 2, pp. 1483–1490. IEEE (2006)
12. Lowe, D.G.: Distinctive image features from scale-invariant keypoints. Int. J. Comput. Vis. **60**(2), 91–110 (2004)
13. Zheng, L., Wang, S., Liu, Z., Tian, Q.: Packing and padding: coupled multi-index for accurate image retrieval. In: IEEE Conference on Computer Vision and Pattern Recognition (2014)
14. Oliva, A., Torralba, A.: Modeling the shape of the scene: a holistic representation of the spatial envelope. Int. J. Comput. Vis. **42**(3), 145–175 (2001)
15. Dalal, N., Triggs, B.: Histograms of oriented gradients for human detection. In: IEEE Computer Society Conference on Computer Vision and Pattern Recognition, CVPR 2005, vol. 1, pp. 886–893. IEEE (2005)
16. Gong, Y., Ke, Q., Isard, M., Lazebnik, S.: A multi-view embedding space for modeling internet images, tags, and their semantics. Int. J. Comput. Vis. **106**(2), 210–233 (2014)
17. Hardoon, D.R., Szedmak, S., Shawe-Taylor, J.: Canonical correlation analysis: an overview with application to learning methods. Neural Comput. **16**(12), 2639–2664 (2004)

18. Tenenbaum, J.B., Freeman, W.T.: Separating style and content. In: Advances in Neural Information Processing Systems, pp. 662–668 (1997)
19. Rosipal, R., Krämer, N.: Overview and recent advances in partial least squares. In: Saunders, C., Grobelnik, M., Gunn, S., Shawe-Taylor, J. (eds.) SLSFS 2005. LNCS, vol. 3940, pp. 34–51. Springer, Heidelberg (2006). https://doi.org/10.1007/11752790_2
20. Sharma, A., Kumar, A., Daume, H., Jacobs, D.W.: Generalized multiview analysis: a discriminative latent space. In: 2012 IEEE Conference on Computer Vision and Pattern Recognition (CVPR), pp. 2160–2167. IEEE (2012)
21. Feng, F., Wang, X., Li, R.: Cross-modal retrieval with correspondence autoencoder. In: Proceedings of the 22nd ACM International Conference on Multimedia, pp. 7–16. ACM (2014)
22. Rasiwasia, N., Mahajan, D., Mahadevan, V., Aggarwal, G.: Cluster canonical correlation analysis. In: Artificial Intelligence and Statistics, pp. 823–831 (2014)
23. Jia, Y., et al.: Caffe: convolutional architecture for fast feature embedding. In: Proceedings of the 22nd ACM International Conference on Multimedia, pp. 675–678. ACM (2014)
24. Blei, D.M., Ng, A.Y., Jordan, M.I.: Latent Dirichlet allocation. J. Mach. Learn. Res. 3(Jan), 993–1022 (2003)
25. Rehman, S.U., Tu, S., Huang, Y., Liu, G.: CSFL: a novel unsupervised convolution neural network approach for visual pattern classification. AI Commun. 30(5), 311–324 (2017)
26. Rehman, S.U., Tu, S., Huang, Y., Yang, Z.: Face recognition: a novel un-supervised convolutional neural network method. In: IEEE International Conference of Online Analysis and Computing Science (ICOACS), pp. 139–144. IEEE (2016)
27. Rehman, S., et al.: Optimization of CNN through novel training strategy for visual classification problems. Entropy 20(4), 290 (2018)
28. Damer, N., Opel, A., Nouak, A.: CMC curve properties and biometric source weighting in multi-biometric score-level fusion. In: 2014 17th International Conference on Information Fusion (FUSION), pp. 1–6. IEEE (2014)
29. Seha, S., Hatzinakos, D.: Human recognition using transient auditory evoked potentials: a preliminary study. IET Biometrics, IET 7, 242–250 (2018)

Adversarial Active Learning in the Presence of Weak and Malicious Oracles

Yan Zhou[1(✉)], Murat Kantarcioglu[1], and Bowei Xi[2]

[1] School of Engineering and Computer Science, University of Texas at Dallas,
Richardson, TX 75080, USA
{yan.zhou2,muratk}@utdallas.edu
[2] Department of Statistics, Purdue University, West Lafayette, IN 47907, USA
xbw@stat.purdue.edu

Abstract. We present a robust active learning technique for situations where there are weak and adversarial oracles. Our work falls under the general umbrella of active learning in which training data is insufficient and oracles are queried to supply labels for the most informative samples to expand the training set. On top of that, we consider problems where a large percentage of oracles may be strategically lying, as in adversarial settings. We present an adversarial active learning technique that explores the duality between oracle modeling and data modeling. We demonstrate on real datasets that our adversarial active learning technique is superior to not only the heuristic majority-voting technique but one of the state-of-the-art adversarial crowdsourcing technique— *Generative model of Labels, Abilities, and Difficulties* (GLAD), when genuine oracles are outnumbered by weak oracles and malicious oracles, and even in the extreme cases where all the oracles are either weak or malicious. To put our technique under more rigorous tests, we compare our adversarial active learner to the *ideal active learner* that always receives correct labels. We demonstrate that our technique is as effective as the ideal active learner when only one third of the oracles are genuine.

1 Introduction

Active learning techniques use strategical sampling techniques to select the most influential instances for training. These techniques are of great importance when labeled data is expensive or difficult to obtain, such as malware samples and intrusion instances that require comprehensive security domain knowledge and are often laborious and time-consuming to label. The working philosophy of active learning is based on the observation that some samples are more likely to help reduce overall uncertainty than others.

Standard active learning techniques assume that correct labels are provided by trustworthy oracles. This assumption is often falsified in real applications, especially in the security domain where some oracles may maliciously return incorrect labels to influence the learning outcome [10]. In this paper, we address

© Springer Nature Switzerland AG 2019
L. H. U and H. W. Lauw (Eds.): PAKDD 2019 Workshops, LNAI 11607, pp. 77–89, 2019.
https://doi.org/10.1007/978-3-030-26142-9_8

this type of *adversarial active learning* problem. While learning with insufficient training data leads to poor predictive performance, learning with abundant labeled samples with just a few incorrectly labeled may exacerbate the situation. Some active learning models are extended to take into account noisy oracles [12,16]. In these techniques, a probability of incorrect labels returned by an oracle is assumed. These techniques can address the adversarial active learning problem only when the adversary's malicious behavior can be interpreted as a random noise defined by a probability density function. However, in reality malicious oracles often lie selectively. The reasons are twofold. In some cases, malicious oracles typically lie when examples are from a specific class. For example, in spam filtering malicious oracles may lie only when a spam message is promoting counterfeited drugs. In other cases, malicious oracles choose to lie when the sample is in the vicinity of the separating boundary of different classes. This lying tactic can prevent detection of these malicious oracles since they only lie for a small number of difficult samples for which there is a great uncertainty in deciding their class memberships.

In this paper, we consider adversarial active learning with a group of oracles, among which some are noisy and some are malicious. Our work has two key differences from the existing body of research on crowdsourcing: (1.) in crowdsourcing each correctly labeled input is perceived to be of equally high quality, without taking into account its informativeness to learning a hypothesis; and (2.) malicious annotators in crowdsourcing are not assumed to have the oracular power to know the true decision boundary. Unlike the existing work in crowdsourcing either under the random worker paradigm [11,14] or assuming large enough labeling activities for reputation assessment [5], our problem considers cases where labeling activity is limited and selectively malicious. In addition, in our work high quality samples are not simply those correctly labeled, but samples that improve the hypothesis most. We present an adversarial active learning model that differentiates genuine oracles from noisy and malicious ones. We compute the behavior profile of the oracles from a bipartite graph between the oracles and the data. In doing so, we establish a duality between two problems—data modeling and oracle modeling. We build a mixture model for a given set of N i.i.d. data points $X_{n\in[1,N]}$, with a latent variable $Z_{n\in[1,N]}$ that indicates which of the K components of the mixture X_n is from. We group data points corresponding to each component of the mixture into a subset $c_{j\in[1,K]}$. To establish a behavior profile for each oracle $O_{i\in[1,M]}$, we let O_i label data points in a subset c_j and estimate its accuracy e_j. After looping through all subsets, we have a behavior profile $E_i = [e_{i1}, \ldots, e_{iK}]$ for O_i. Next we build an oracle mixture model on the behavior matrix $[E_1, E_2, \ldots, E_M]^T$ of all the oracles. Component oracles that have the

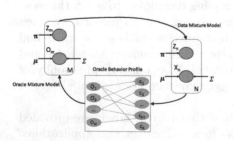

Fig. 1. The duality of the oracle mixture model and the data mixture model.

highest accuracy are chosen as the *genuine* oracles, and are used for labeling the selected sample in the current iteration of active learning. Consequently, the newly labeled sample updates the data mixture model. The process continues as shown in Fig. 1.

The main contribution of our work is a robust active learning model that explicitly filters out noisy and malicious oracles, even when the majority of the oracles are either weak or malicious. Our technique is robust against not only random malicious noise, but more realistic lying behavior of malicious oracles that mitigate their risk of being detected in the wild.

The rest of the paper is organized as follows. Section 2 presents the related work on active learning with noisy oracles. Section 3 formally defines the adversarial active learning problem and presents our adversarial active learning technique. Section 4 presents experimental results on five real datasets. Section 5 concludes our work and discusses future directions.

2   Related Work

Miller et al. [10] discuss potential challenges for adversarial active learning in the security domain. They explore potential vulnerabilities of active learning in terms of query strategy, oracle maliciousness, and inherent vulnerabilities in existing machine learning algorithms. However, no specific defense strategies are presented in their discussions.

There are several existing active learning models that consider noisy oracles in their design. The Agnostic Active learning algorithm [1,2] works in the presence of arbitrary forms of noise. Agnostic active learning does not specifically penalize the noisy oracle for lying. It only guarantees that a hypothesis with the minimum error rate, with respect to the sample distribution labeled by the noisy oracle, never gets removed from the version space.

There are also extensive studies on learning from imperfect labels. Sheng et al. [12] present a repeated labeling strategy for imperfect labeling. They show that repeated-labeling can sometimes, but not always, improve data quality and data mining results. French [4] presents a Bayesian perspective on combining multiple opinions. There is also extensive research on latent variable models for inferring hidden truth in the field of medical diagnosis [13]. Whitehill et. al. [16] present a probabilistic model—Generative model of Labels, Abilities, and Difficulties (GLAD)—to optimally combine labels from different labelers. Instead of assuming a uniform probability on the error rate of the labelers, their model can simultaneously infer the label and the difficulty of data given that the accuracy of each labeler is different. GLAD is used for comparison in our experiments.

The problem of noisy labels has also been studied in crowdsourcing [8,9, 15] where worker quality and the truth of labels are updated iteratively until convergence. This body of research perceives correctly labeled input to be of equally high quality without taking into account its informativeness to learning a hypothesis, and the workers are not assumed to have the oracular power to know the decision boundary of the hypothesis. They also ignore the fact that in adversarial settings malicious oracles do not have to be disobedient all the time. They only lie when mislabeling the selected sample allows them to craft targeted classification errors. More importantly, none of the existing work has been shown to be effective in situations where the majority of the oracles are either weak or malicious. In contrast to the existing research, we consider all these aspects in our design of the adversarial active learning technique.

3   Adversarial Active Learning

We consider the active learning problem in adversarial settings where a subset of the oracles is malicious. For omniscient malicious oracles (knowing the true decision boundary), an effective lying tactic is to return incorrect labels only for data points that are close to the true decision boundary. Besides genuine oracles and malicious oracles, in practice we also come across incompetent oracles that return random noisy labels. A robust active learning algorithm should be designed to be resilient to both malicious oracles and noisy oracles.

3.1   Problem Definition

Let $\mathcal{D}$ be a distribution over $\mathcal{X} \times \{\pm 1\}$ where $\mathcal{X}$ is the input space of a binary classification problem. Let (X, Y) be a pair of random variables with a joint distribution $\mathcal{D}$ where $X \in \mathcal{X}$ and $Y \in \{\pm 1\}$. Let $O = O_g \cup O_w \cup O_m = \{O_1, \ldots, O_M\}$ be a set of M oracles, where O_g is a set of *genuine* oracles that returns correct labels most of the time, O_w is a set of *weak* oracles that returns random noisy labels with a certain probability, and O_m is a set of *malicious* oracles that selectively returns incorrect labels. More specifically, let $S = \{X_1, X_2, \ldots\}$ be a set of i.i.d. samples of X. For an instance $X_i \in S$ with a true label Y_i, an oracle $O_j \in O$ returns a label Y_i'. There are three cases:

(a.) If O_j is a *genuine* oracle, i.e. $O_j \in O_g$, $\Pr(Y_i = Y_i' | X_i) \approx 1$;
(b.) If O_j is a *weak* oracle, i.e. $O_j \in O_w$, $0.5 < \Pr(Y_i = Y_i' | X_i) \ll 1$;
(c.) If O_j is a *malicious* oracle, i.e. $O_j \in O_m$, $\Pr(Y_i = Y_i' | X_i \in S^+) \ll \Pr(Y_i = Y_i' | X_i \in S \setminus S^+)$, where $S^+ \subset S$ is a special subset of instances that the adversary is particularly interested in mislabeling.

Let A be a generic active learning algorithm. If A always receives correct labels during active learning, we refer to it as A^*. Let H^* be the set of hypotheses learned with A^*. Our objective is to find a subset of oracles $O' \subset O$ such that for any instance X_i, the set of hypothesis H' learned with A, that receives labels provided by O', satisfies the constraint $\Pr(H^*(X_i) \neq H'(X_i) \mid |S'| > t) < \delta$ where S' is the set of samples labeled by the oracles O', $0 < t < |S|$, and $\delta \in (0, 1)$ is an arbitrarily small constant. In other words, predictions for X_i by H^* and H' are arbitrarily close after t samples have been labeled.

3.2   Mixture Distributions of Oracles

Consider each oracle $O_i \in O$ as a data point in some input space. With the premise that oracles are sampled from three different distributions of *genuine*, *weak*, and *malicious* oracles, we can define a mixture model of three components over all observed oracles, and a latent variable Z that can be interpreted as assigning an oracle to a specific component of the mixture. Therefore, Z has three possible states, corresponding to the assignment to *genuine*, *weak* and *malicious*, respectively. Our objective of searching for a subset $O' \subset O$ that minimizes the gap between H^* and H' is equivalent to learning and identifying the component that generates the *genuine* oracles in the mixture model. Let θ be the set of all model parameters, the log likelihood function is given by

$$\ln \Pr(X|\theta) = \ln\left\{ \sum_Z \Pr(X, Z|\theta) \right\},$$

where $\Pr(X, Z|\theta)$ is the complete-data likelihood function. Since Z is unknown, model parameters can be found using EM by maximizing the expectation of the complete-data log-likelihood function given by

$$\sum_Z \Pr(Z|X, \theta) \ln \Pr(X, Z|\theta).$$

Unfortunately, there is no existing observed data about oracles that allows us to infer the oracle mixture model. In the next section, we consider a duality of the oracle modeling problem, from which an oracle behavior profile will be computed for each oracle. The behavior profiles are later treated as observations of the oracles, and are used to learn the oracle mixture model.

3.3   Mixture Distributions of Data

Assume data points are generated from a mixture model for which class labels are generated randomly, conditional on the mixture component. Given a set of instances $X = \{X_i\}_{i=1}^N$, under the assumption of sampling from a uniform distribution of the K Gaussians and identity covariance matrices, the expectation of the complete-data log-likelihood function is:

$$\mathbf{E}_{Z|X,\theta}[\log \Pr(X, Z|\theta)] = \sum_{k=1}^K \sum_{i=1}^N \Pr(z_k|X_i, \theta) \cdot \log\left(\pi_k \cdot \frac{1}{(2\pi)^{1/2}} e^{-||X_i - \mu_k||^2} \right)$$

$$= -\sum_{k=1}^K \sum_{i=1}^N \Pr(z_k|X_i, \theta) \cdot ||X_i - \mu_k||^2 + const$$

where θ is the cluster means $(\mu_1, \ldots, \mu_K)$ and π_k is the mixture coefficient. In our adversarial active learning, we set $K \leq |L_0|$—the size of the instances in the initial labeled dataset $|L_0|$, and for the set of instances c_j generated by the jth

Gaussian component, its class label Y_j is determined by its nearest neighbor in the labeled dataset L given by

$$X_L^* = \arg\min_{X' \in L} \sum_{i=1}^{|c_j|} D(X_i - X'),$$

where $D(\cdot, \cdot)$ is a distance measure. Thus $Y_j = Y_L^*$.

3.4   Oracle Behavior Profile

Ideally, if we have the solution to the mixture model for classifying given data, a closed form solution becomes available to the oracle modeling problem because we would know for each oracle whether it is genuine by verifying its labeling on each category of the data. In other words, the latent variables in the oracle mixture model become known. Vice versa, if we have the oracle mixture model,

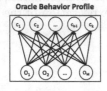

Oracle Behavior Profile

the data categorization problem simply reduces to label querying. This suggests a duality between oracle modeling and data modeling. Although there are hardly perfect mixture models for given data, an informed clustering would be sufficient for creating behavior profiles for the oracles.

Fig. 2. The oracle behavior profile illustrated in a bipartite graph.

Initially, given a small set of labeled data L_0 and a large set of unlabeled data U, we build a Gaussian mixture model with $K \leq |L_0|$ components as described in Sect. 3.3. Suppose we have a bipartite graph with one set of nodes $\{O_{i \in [1,M]}\}$ representing oracles and the other set $\{c_{j \in [1,K]}\}$ representing data generated by each of the K (in our case

$K = 3$) Gaussian components, as shown in Fig. 2. There is an edge e_{ij} between every pair of (O_i, c_j), weighted by the accuracy estimate of O_i on instances in c_j given by $w_{ij} = \Pr(Y_{ij} = Y_j)$, where Y_{ij} is the set of labels returned by oracle O_i for instances in c_j, and Y_j is the label of c_j. The accuracy of oracle $O_{i \in [1,M]}$ on instances in dataset $c_{j \in [1,K]}$ is estimated by comparing the labels assigned by O_i and the labels assigned to instances in c_j. For each $O_{i \in [1,M]}$ of the M oracles, its behavior profile is defined as $w_i = [w_{i1}, \ldots, w_{iK}]$. Therefore, the behavior profile of the entire set of oracles is an $M \times K$ matrix, in which each row is a behavior profile of an oracle. This behavior profile matrix is used to derive the oracle mixture model as discussed in Sect. 3.2. To limit the instances for the oracles to label to a small quantity, we randomly select a set of samples $\overline{U}$ from the unlabeled dataset $U = \bigcup_{j=1}^{K} c_j \setminus L_0$, where $|\overline{U}| \leq |L_0|$. Instances in both L_0 and $\overline{U}$ are included in the query and are labeled only once by each oracle at the initialization stage.

4   Experimental Results

We experiment on five real
datasets—*Hate Speech* [3], *Spam,*
Banknote Authentication, Occu-
pancy Detection [7], and *Web-*
spam [6], all binary classification
problems. Each dataset is randomly divided into two disjoint
sets, with one for active learning and the other for independent
testing. The detailed statistics of

Table 1. Statistics of the datasets.

Dataset	Training		Test
	L_0	U	T
HateSpeech	25	4960	19842
Spam	12	228	4561
Webspam	18	175098	174884
Banknote Authentication	34	652	686
Occupancy Detection	40	774	7329

the datasets is shown in Table 1. Except for *Hate Speech* and *Webspam*, each
dataset uses 5% of the training data as the initially randomly selected labeled
data L_0 for active learning. The rest of the training data is treated as the set
of unlabeled data U, from which samples are selected for labeling by the oracles
during active learning. For *Webspam* and *Hate Speech*, only 18 instances (0.01%)
and 25 instances (0.5%) are needed to get the active learning to start at approximately 50% accuracy. Instances in the test set are unseen and are used only for
evaluation.

We use 30 oracles in our experiment, including a different percentage of
genuine, *weak*, and *malicious* oracles. The *genuine* oracles provide correct labels
at least 90% of the time; the *weak* oracles return labels with accuracy between
50–79%; and the *malicious* oracles return incorrect labels for positive instances
that are less than the average distance away from the true decision boundary,
and return correct labels with an accuracy of at least 90% for everything else.
Note that the setup of oracles is strictly used for testing purpose. This
information is not modeled in any of the algorithms we are testing.

We compare our adversarial active learning technique to (1.) the majority-
voting heuristic; (2.) one of the state-of-the-art crowdsourcing techniques—
GLAD[1]; and (3.) the ideal active learner that always receives correct labels
for training. We use SVM with a Gaussian kernel with $C = \infty$ and $\gamma = 1$ as the
underlying learning algorithm in the active learning process. We set the number
of *genuine* oracles to 5, 10, and 15 out of the total 30 oracles, respectively. The
rest of the oracles are evenly split between *weak* oracles and *malicious* oracles.
Each experiment is repeated 10 times and the results are averaged over the 10
runs.

4.1   Results on the Hate Speech Dataset

Only 5.8% of tweets are labeled as *hate speech*, 77.4% are labeled as *offensive*
language, and the rest 16.8% is *neither*. To keep the dataset relatively balanced,
we keep all the tweets in *hate speech* and *neither*, and select from *offensive*
language an equal number of instances as in *hate speech*, and the learning task

---

[1] https://s3.amazonaws.com/mplabsites/Sites/OptimalLabelingRelease1.0.3.tar.gz.

is to differentiate *hate/offensive speech* from *neither*. We use 20% of the data for training and 80% for independent testing. We sample 0.5% of the training data as the initially labeled data. Figure 3 shows the accuracy of the four algorithms on the independent test set when the number of *genuine* oracles is 5, 10, and 15, respectively. The difference of all mean standard deviations is less than 0.022.

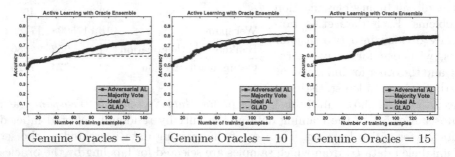

Fig. 3. Adversarial active learning on the *Hate Speech* dataset when the number of genuine oracles is 5, 10, and 15 out of the 30 oracles.

As can be observed, when there are only five *genuine* oracles, the *ideal AL* has the best performance as expected since it always receives the true labels. Meanwhile, our *Adversarial AL* significantly outperforms the *Majority Vote* and *GLAD* algorithms. When there are 10 *genuine* oracles, the gap among the four algorithms becomes much smaller. When the number of *genuine* oracles is greater than or equal to 15, the difference among all four algorithms disappears.

Next, we demonstrate the impact of malicious oracles. Given the number of *genuine* oracles, we compare two cases when the rest of the oracles are: (1.) only weak oracles and (2.) only malicious oracles. We show that *Majority Vote* and *GLAD* perform equally well as *Ideal AL* and *Adversarial AL* when there are only *weak* oracles but no *malicious* oracles, as shown in Fig. 4. On the other hand, when the rest of the oracles are all malicious, both *Majority Vote* and *GLAD* fail miserably, as shown in Fig. 5.

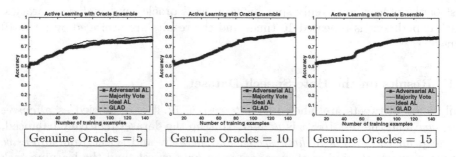

Fig. 4. Adversarial active learning on the *Hate Speech* dataset when there are no malicious oracles.

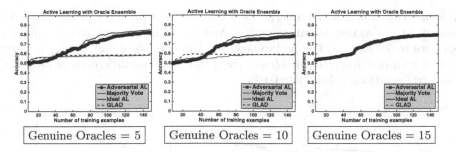

Fig. 5. Adversarial active learning on the *Hate Speech* dataset when the rest are all malicious oracles

4.2 Results on the Spam Dataset

We use 5% of the data for training and the rest for independent testing. 5% of the training data is used as the initially labeled data L_0 for active learning. Figures 6 shows the adversarial active learning results on the *Spam* dataset when there are 5, 10, and 15 genuine oracles.

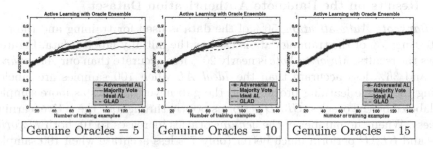

Fig. 6. Adversarial active learning on the *Spam* dataset.

When the number of *genuine* oracles is as small as five, both *Majority Vote* and *GLAD* perform poorly compared to our *Adversarial AL* and the *Ideal AL*. When the number of selected samples reaches 100, the accuracy of *Majority Vote* is approximately 9% lower than our *Adversarial AL* and 16% lower than the *Ideal AL*. Our *Adversarial AL* consistently outperforms *Majority Vote* and *GLAD* throughout the active learning process. The performance of *Majority Vote* and *GLAD* improves significantly when the number of *genuine* oracles increases to 10 and 15. The difference of all mean standard deviations is less than 0.044.

4.3 Results on the Webspam Dataset

We use 50% of the data for independent testing, and sample 18 instances (approximately 0.01% of the training set) as the initially labeled data for active learning. We observe similar results on the *Webspam* dataset as shown in Fig. 7. Both

Majority Vote and *GLAD* underperform our *Adversarial AL* and the *Ideal AL* throughout the active learning process. When the number of *genuine* oracles increases to 10 and 15, the performance of *Majority Vote* and *GLAD* is more in sync with our *Adversarial AL* and the *Ideal AL*. The difference of all mean standard deviations is less than 0.033.

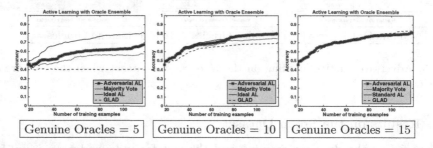

Fig. 7. Adversarial active learning on the *Webspam* dataset.

4.4   Results on the Banknote Authentication Dataset

For *Banknote Authentication*, 50% of the data is used for training and the rest for testing; 5% of the training data is used as the initially labeled data. Figure 8 shows the results. *Majority Vote* is nearly 20% less accurate than our *Adversarial AL*, and 36% less accurate than the *Ideal AL* when 100 samples are labeled during the active learning process, and the gap never narrows as more samples are labeled by the oracles. *GLAD* performs poorly throughout the active learning process. When the number of the *genuine* oracles increases to 10, both *Majority Vote* and *GLAD* perform much better (only 7% less accurate when 100 samples have been labeled). Our *Adversarial AL* is as accurate as the *Ideal AL* during the entire active learning process. When there are 15 *genuine* oracles (50% of the total), the gap among all active learners is completely bridged. The difference of all mean standard deviations is less than 0.024.

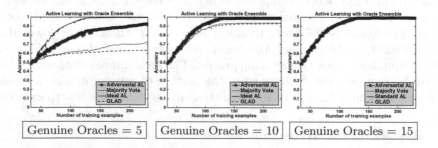

Fig. 8. Adversarial active learning on the *Banknote Authentication* dataset.

4.5 Results on the Occupancy Detection Dataset

The *Occupancy Detection* dataset is largely imbalanced and nearly 80% of the instances fall in the class that we assume the adversary aims to attack. The impact of maliciously mislabeled samples is so severe that labeling more data only make the majority-voting active learner weaker when there are only a few genuine oracles, as shown in Fig. 9. We use 10% of the data for training and the rest for testing; 5% of the training data is used as the initially labeled data. The difference of all mean standard deviations is less than 0.015.

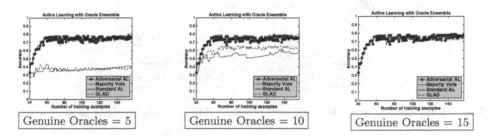

Fig. 9. Adversarial active learning on the *Occupancy Detection* dataset when the number of genuine oracles is 5 and 10 out of the 30 oracles.

When the number of *genuine* oracles is five, the accuracy of *Majority Vote* is nearly 40% lower than our *Adversarial AL* and the *Ideal AL* after 50 samples have been labeled. *GLAD* is as weak as *Majority Vote*. Even when the number of *genuine* oracle increases to 10, the performance of *Majority Vote* is disappointing as more data is labeled by the oracles. *GLAD* is 7% better than *Majority Vote*.

4.6 Impact of Genuine Oracles

We also tested the extreme cases where the accuracy of the *genuine* oracles is also mediocre, between 60% and 89%. In other words, there are no real experts in the group of given oracles. The best an oracle can label for the active learners is less than 90% accurate. Figure 10 shows the active learning results on the five real datasets when 60% of the oracles are *weak* and the other 40% are *malicious*. As can be observed in Fig. 10, our *Adversarial AL* outperforms *Majority Vote* and *GLAD* on all five datasets when there are no *genuine* oracles in the group.

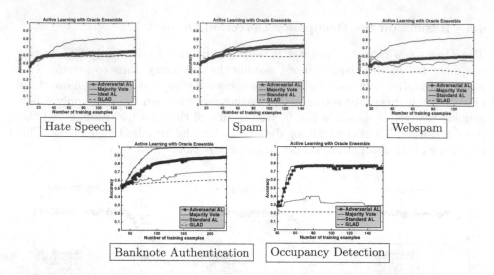

Fig. 10. Adversarial active learning results when all the oracles are weak or malicious.

5 Conclusions

We present an adversarial active learning model that is resistant to noisy oracles and robust against adversarial labeling. Our model takes advantage of the duality between learning the mixture model of the data and the mixture model of the oracles. Our technique is clearly superior to *GLAD* and the *Majority Vote* active learners when the majority of the oracles are noisy or malicious. In addition, our *Adversarial* active learning model demonstrates impressive performance even when there are no *genuine* experts in the given oracles, while *Majority Vote* and *GLAD* are devastated on some datasets.

Acknowledgement. The research reported herein was supported in part by NIH award 1R01HG006844, NSF awards CICI- 1547324, IIS-1633331, CNS-1837627, OAC-1828467 and ARO award W911NF-17-1-0356.

References

1. Balcan, M., Beygelzimer, A., Langford, J.: Agnostic active learning. In: ICML, pp. 65–72 (2006)
2. Beygelzimer, A., Langford, J., Tong, Z., Hsu, D.J.: Agnostic active learning without constraints. In: Advances in Neural Information Processing Systems, vol. 23, pp. 199–207. Curran Associates, Inc. (2010)
3. Davidson, T., Warmsley, D., Macy, M., Weber, I.: Automated hate speech detection and the problem of offensive language. In: Proceedings of the 11th International AAAI Conference on Web and Social Media, ICWSM 2017 (2017)
4. French, S.: Group consensus probability distributions: a critical survey. Bayesian Stat. **2**, 183–202 (1985)

5. Jagabathula, S., Subramanian, L., Venkataraman, A.: Reputation-based worker filtering in crowdsourcing. In: NIPS, pp. 2492–2500 (2014)
6. LIBSVM: LIBSVM Data: Classification, Regression, and Multi-label (2014). http://www.csie.ntu.edu.tw/~cjlin/libsvmtools/datasets/
7. Lichman, M.: UCI machine learning repository (2013). http://archive.ics.uci.edu/ml
8. Liu, Q., Peng, J., Ihler, A.T.: Variational inference for crowdsourcing. In: Advances In Neural Information Processing Systems, pp. 692–700 (2012)
9. Ma, F., et al.: Faitcrowd: fine grained truth discovery for crowdsourced data aggregation. In: Proceedings of the 21th ACM SIGKDD, pp. 745–754 (2015)
10. Miller, B., et al.: Adversarial active learning. In: Proceedings of the 2014 AISec Workshop, pp. 3–14 (2014)
11. Raykar, V.C., Yu, S.: Eliminating spammers and ranking annotators for crowdsourced labeling tasks. J. Mach. Learn. Res. 13, 491–518 (2012)
12. Sheng, V.S., Provost, F., Ipeirotis, P.G.: Get another label? Improving data quality and data mining using multiple, noisy labelers. In: Proceedings of the 14th ACM SIGKDD, pp. 614–622 (2008)
13. Uebersax, J.S.: Statistical modeling of expert ratings on medical treatment appropriateness. J. Am. Stat. Assoc. 88, 421–427 (1993)
14. Vuurens, J.B., de Vries, A.P.: Obtaining high-quality relevance judgments using crowdsourcing. IEEE Internet Comput. 16, 20–27 (2012)
15. Welinder, P., Branson, S., Perona, P., Belongie, S.J.: The multidimensional wisdom of crowds. In: NIPS, pp. 2424–2432 (2010)
16. Whitehill, J., Ruvolo, P., Wu, T., Bergsma, J., Movellan, J.R.: Whose vote should count more: optimal integration of labels from labelers of unknown expertise. In: NIPS, pp. 2035–2043 (2009)

The Most Related Knowledge First: A Progressive Domain Adaptation Method

Yunyun Wang[1,2]([✉]), Dan Zhao[1,2], Yun Li[1,2], Kejia Chen[1,2], and Hui Xue[3]

[1] Department of Computer Science and Engineering,
Nanjing University of Posts and Telecommunications, Nanjing 210046,
People's Republic of China
wangyunyun@njupt.edu.cn
[2] Jiangsu Key Laboratory of Big Data Security and Intelligent Processing,
Nanjing 210023, People's Republic of China
[3] School of Computer Science and Engineering, Southeast University,
Nanjing 210096, People's Republic of China

Abstract. In domain adaptation, how to select and transfer related knowledge is critical for learning. Inspired by the fact that human usually transfer from the more related experience to the less related one, in this paper, we propose a novel progressive domain adaptation (PDA) model, which attempts to transfer source knowledge by considering the transfer order based on relevance. Specifically, PDA transfers source instances iteratively from the most related ones to the least related ones, until all related source instances have been adopted. It is an iterative learning process, source instances adopted in each iteration are determined by a gradually annealed weight such that the later iteration will introduce more source instances. Further, a reverse classification performance is used to set the termination of iteration. Experiments on real datasets demonstrate the competiveness of PDA compared with the state-of-arts.

Keywords: Domain adaptation · Progressive transfer · Iteration ·
Reverse classification

1 Introduction

Domain adaptation, which aims to use knowledge in some related domain (source domain) to help the learning of current domain (target domain), has attracted much attention recently. The source and target domains usually have different data distributions, as a result, domain adaptation makes it possible that knowledge can be transferred across different distributions.

There are lots of domain adaptation methods developed in literature. Those methods can mainly be divided into four categories [1]: instance-based method [2], feature-based method [3], parameter-based method [4] and relational-based method. Although the source instances have different distribution from the target ones, there are certain parts that can still be used to help training a target classifier. Considering that, instance-based methods attempt to select or weight source instances for transfer, such

L. H. U and H. W. Lauw (Eds.): PAKDD 2019 Workshops, LNAI 11607, pp. 90–102, 2019.
https://doi.org/10.1007/978-3-030-26142-9_9

as TradaBoost [5] and Kernel Mean Matching (KMM). Feature-based methods can be further categorized into two kinds: asymmetric and symmetric methods. The asymmetric methods transform features of the source domain so that it can be more closely match the target domain, typical including SCL [6] and ARTL [7]. The symmetric methods aim at discovering a common latent feature space to minimize the distribution difference between domains and maintain the intrinsic structure of original domains simultaneously, such as TCA and BDA [8]. The parameter-based methods transfer knowledge by sharing certain characters for parameters in both domains. Finally, the relational-based methods transfer knowledge by utilizing the relationship of logistical network in the source domain.

In fact, only part knowledge from the source domain can be helpful for learning in the target domain. As a result, how to find and utilize the related source knowledge is critical for transfer. From human cognition, one transfers the previous experience to the current one, and in most cases, transfers from the more related experience to the less related one. For example, when one learns the violin, one can transfer skills from the related string instruments first, such as viola or cello, then the others if necessary. When learning "dog" with the help of "cat" shown in Fig. 1, one usually transfers the more related Siamese first, and then the others in a meaningful order. As a result, we attempt to borrow such learning paradigm, in order to transfer from the most related source knowledge to the least related one, until all related knowledge has been utilized. Specifically, we develop a progressive domain adaptation (PDA) model, in which the transfer order of source knowledge is considered in terms of relevance. It learns in an iterative process, and the source instances adopted in each iteration are determined by a weight, which is gradually annealed such that later iterations will introduce more source instances. Further, we adopt the reverse classification performance to set the iteration termination, in order to judge if all related knowledge has been adopted.

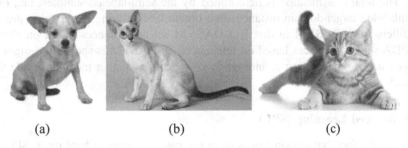

(a)   (b)   (c)

Fig. 1. Learning (a) "dog" with the help of "cat", one usually transfers the more related (b) Siamese first, then (c) the others in a meaningful order

The main contributions of the paper are summarized as follows,

- A novel PDA model is proposed by iteratively transfer source instances from the most related ones to the least related ones, until all related source knowledge has been adopted.

- The reverse classification performance is used to determine the termination of iteration, such that only the related source knowledge is transferred.
- Empirical results on real datasets demonstrate that by progressively transferring source instances, PDA can achieve encouraging performance compared with the state-of-the-arts.

2 Related Works

2.1 Iterative Domain Adaptation Method

There are already iterative domain adaptation methods in literature. Extended from AdaBoost, TrAdaBoost uses a small amount of labeled target instances along with the source instances to learn a target model. TrAdaBoost follows an iterative learning type. In each round, if a target instance is misclassified, then a larger weight will be assigned such that it will be given more importance in the next round. At the same time, if a source instance is misclassified, then a smaller weight will be assigned such that it will have less importance in the next iteration. PDA differs from TrAdaboost in that: (1) TrAdaBoost selects instances based on classification error, while PDA selects instances based on relevance. (2) The target instances should be labeled in TrAdaBoost, while in PDA, there can be no labeled instances.

In [9], a domain adaptation method DASVM in the framework of SVMs has been developed. It also learns in an iterative style. Specifically, labeled source instances are used for determining an initial solution for the target domain. Then in each round, those unlabeled target instances, which lie between margin with the maximum and minimum decision function values, are exploited for properly adjusting the decision function. While labeled source instances, which are far from the decision function, are gradually erased. The final classification is determined by the semi-labeled samples, i.e., originally unlabeled target-domain instances that obtain labels during the learning process. PDA differs from DASVM in that: (1) DASVM selects instances based on margin, while PDA selects instances based on relevance. (2) DASVM is specially designed in the large margin scheme, thus cannot easily be extended to other methods, while there is no such restriction for PDA.

2.2 Self-paced Learning (SPL)

Inspired by the fact that human learns from the easy concepts to hard ones, SPL [10] selects instances for learning in terms of easiness, i.e., learns by iteratively adding training instances from easy ones to hard ones. It iteratively trains the classifier and selects the easy instances for training. Further, SPLD [11] considers the diversity to avoid the possibility that all "easy" samples may come from the same class. PDA differs from SPL(D) in that SPL(D) is designed for supervised problems by iteratively selecting training instances, while PDA is for domain adaptation by iteratively transferring labeled source instances. Furthermore, PDA has an early termination in order to transfer only related source instances.

3 Progressive Domain Adaptation

3.1 Problem Formulation

Given source domain D_s with instances $\{x_i^s, y_i^s\}_{i=1}^{n_s}$ where $x_i^s \in R^d$ and $y_i^s \in R^C$, C is the number of classes, target domain D_t with instances $\{x_j^t\}_{j=1}^{n_t}$. The optimization problem of PDA is formulated as

$$
\min_{f,v_i} \sum_{i=1}^{n_s} v_i \ell(f(x_i), y_i) + \sigma \|f\|_K^2 + \lambda \bar{D}_{f,K}(J_s, J_t) + \gamma \bar{M}_{f,K}(P_s, P_t) + \varphi(v, \delta, \beta) \tag{1}
$$
$$
s.t. \; v_i \in \{0, 1\}, i = 1 \ldots n_s
$$

Where $\ell(\cdot, \cdot)$ is the loss function, v_i is the weight for source instance x_i, $i = 1 \ldots n_s$, $\|f\|_K^2$ controls the complexity of $f(x)$. $\bar{D}_{f,K}(J_s, J_t) = \bar{D}_{f,K}(P_s, P_t) + \sum_{c=1}^{C} \bar{D}_{f,K}^{(c)}(Q_s, Q_t)$ describes the joint distribution adaptation, where the marginal and conditional distribution adaptations are formulated as

$$
\bar{D}_{f,K}(P_s, P_t) = \left\| \frac{1}{n_s} \sum_{i=1}^{n_s} v_i f(x_i) - \frac{1}{n_t} \sum_{j=1}^{n_t} f(x_j) \right\|_H^2 \tag{2}
$$

And

$$
\bar{D}_{f,K}^{(c)}(Q_s, Q_t) = \left\| \frac{1}{n_s^{(c)}} \sum_{x_i \in D_s^{(c)}} v_i f(x_i) - \frac{1}{n_t^{(c)}} \sum_{x_j \in D_t^{(c)}} f(x_j) \right\|_H^2 \tag{3}
$$

As a result,

$$
\bar{D}_{f,K}(J_s, J_t) = \mathrm{tr}(\alpha^\mathsf{T} K \bar{M}_0 K \alpha) + \sum_{c=1}^{C} \mathrm{tr}(\alpha^\mathsf{T} K \bar{M}_c K \alpha) = \mathrm{tr}(\alpha^\mathsf{T} K \bar{M} K \alpha) \tag{4}
$$

Where $K \in R^{(n_s+n_t) \times (n_s+n_t)}$ is kernel matrix, and $\bar{M} = \sum_{c=0}^{C} \bar{M}_c$, $\bar{M}_c$ are revised MMD matrices [7, 12] formulated as

$$
(\bar{M}_c)_{ij} =
\begin{cases}
\frac{v_i v_j}{n_s^{(c)} \times n_s^{(c)}}, & x_i, x_j \in D_s^{(c)} \\[2mm]
\frac{1}{n_t^{(c)} \times n_t^{(c)}}, & x_i, x_j \in D_t^{(c)} \\[2mm]
\frac{-v_i(orv_j)}{n_s^{(c)} \times n_t^{(c)}}, & x_i \in D_s^{(c)}, x_j \in D_t^{(c)} \, or \, x_i \in D_t^{(c)}, x_j \in D_s^{(c)} \\[2mm]
0, & otherwise
\end{cases} \tag{5}
$$

The fourth item is the revised manifold regularizer formulated as

$$
\bar{M}_{f,k}(P_s, P_t) = \sum_{i,j=1}^{n_s} \left(v_i f(x_i) - v_j f(x_j) \right)^2 W_{ij} + 2 \sum_{i,j=1}^{n_s+n_t} \left(v_i f(x_i) - f(x_j) \right)^2 W_{ij}
$$
$$
\sum_{i,j=1}^{n_t} \left(f(x_i) - f(x_j) \right)^2 W_{ij} = \mathrm{tr}(\alpha^\mathsf{T} K \bar{L} K \alpha) \tag{6}
$$

Where W_{ij} denotes the similarity of instances over the manifold [7, 13]. $\bar{\mathbf{L}} = \bar{\mathbf{D}}^{-\frac{1}{2}}(\bar{\mathbf{D}} - \bar{W})\bar{\mathbf{D}}^{-\frac{1}{2}}$, $\bar{\mathbf{D}}$ is a diagonal matrix with the diagonal component given by $\bar{D}_{ii} = \sum_{j=1}^{n_s + n_t} \bar{W}_{ij}$. $\bar{\mathbf{W}}$ is defined as

$$\bar{W}_{ij} = \begin{cases} v_i v_j W_{ij}, & x_i, x_j \in D_s \\ W_{ij}, & x_i, x_j \in D_t \\ v_i W_{ij} \ or \ v_j W_{ij}, & x_i \in D_s, x_j \in D_t \ or \ x_i \in D_t, x_j \in D_s \end{cases} \tag{7}$$

$\varphi(v, \delta, \beta)$ is a regularizer for the weights of source instance,

$$\varphi(v, \delta, \beta) = -\delta\|v\|_1 - \beta\|v\|_{2,1} \tag{8}$$

Where the l_1-norm controls the relevance degree of source instances in each round with δ as the parameter. When gradually increasing δ as the learning proceeds, the weights will generally become increasingly higher. This can gradually involve less related source instances for transferring. The $l_{2,1}$-norm controls the group sparsity such that the selected source instances can be from all classes if possible. When all $v_i = 1$, $i = 1 \dots n_s$, PDA degenerates to ARTL.

3.2   Problem Optimization

By using the squared loss function, a progressive transfer regularized least square classifier (PTRLS) is obtained, the optimization problem is written as

$$\min_{f, v_i} \sum_{i=1}^{n_s} v_i (f(x_i) - y_i)^2 + \sigma\|f\|_K^2 + \lambda \bar{D}_{f,K}(J_s, J_t) + \gamma \bar{M}_{f,K}(P_s, P_t) + \varphi(v, \delta, \beta) \tag{9}$$

$$s.t. \ v_i \in \{0, 1\}, \ i = 1 \dots n_s$$

The decision function $f(x)$ and the weights v_i are optimized at the same time, and alternative convex search is adopted here. Specifically, with fixed weights, the optimization problem for decision function can be formulated as

$$\min_{f} \sum_{i=1}^{n_s} v_i (f(x_i) - y_i)^2 + \sigma\|f\|_K^2 + \lambda \bar{D}_{f,K}(J_s, J_t) + \gamma \bar{M}_{f,K}(P_s, P_t) \tag{10}$$

Based on the representation theory [12], we have

$$\min_{\alpha} J(\alpha) = \text{tr}((\mathbf{K}_s \alpha - \mathbf{Y}_s)^{\text{T}} \mathbf{V}(\mathbf{K}_s \alpha - \mathbf{Y}_s)) + \text{tr}(\alpha^{\text{T}} \mathbf{K} \alpha) + \lambda \text{tr}(\alpha^{\text{T}} \mathbf{K} \bar{\mathbf{M}} \mathbf{K} \alpha) \\ + \gamma \text{tr}(\alpha^{\text{T}} \mathbf{K} \bar{\mathbf{L}} \mathbf{K} \alpha) \tag{11}$$

Where $\mathbf{K}_s$ is the kernel matrix of source domain and $\mathbf{Y}_s$ is a vector including labels of $\mathbf{X}_s$, $\mathbf{V}$ is a diagonal matrix with $V_{ii} = v_i$. By setting the derivation of J with respect α to 0, we have

$$2\mathbf{K}_s^T\mathbf{V}(\mathbf{K}_s\alpha - \mathbf{Y}_s) + 2\mathbf{K}\alpha + 2\lambda\mathbf{K}\bar{M}K\alpha + 2\gamma\mathbf{K}\bar{L}K\alpha = 0 \tag{12}$$

Thus,

$$\alpha = \left(\mathbf{K}_s^T\mathbf{V}\mathbf{K}_s + \mathbf{K} + \lambda\mathbf{K}\bar{M}K + \gamma\mathbf{K}\bar{L}K\right)^{-1}\mathbf{K}_s^T\mathbf{V}\mathbf{Y}_s \tag{13}$$

At the same time, with fixed $f(x)$, the source instance weights are obtained by

$$\min_{v_i} \sum_{i=1}^{n_s} v_i(f(x_i) - y_i)^2 + \lambda\bar{D}_{f,K}(J_s, J_t) + \gamma\bar{M}_{f,K}(P_s, P_t) + \varphi(v, \delta, \beta) \tag{14}$$
$$s.t.\ v_i \in \{0,1\}, i = 1\ldots n_s$$

Following SPLD [14], the update criteria for v_i in each iteration are obtained as follows:

For source instances in each class,

- Instances with $R\left(y_i^{(p)}, f\left(x_i^{(p)}\right)\right) < \delta + \beta\frac{1}{\sqrt{i}+\sqrt{i-1}}$ will be selected in training, $v_i^{(p)} = 1$.

- Instances with $R\left(y_i^{(p)}, f\left(x_i^{(p)}\right)\right) \geq \delta + \beta\frac{1}{\sqrt{i}+\sqrt{i-1}}$ will not be selected in training, $v_i^{(p)} = 0$.

Where $R\left(y_i^{(p)}, f\left(x_i^{(p)}\right)\right) = L + D + M$ describes the "relevance" of source instance

x_i, $L = \|f(x_i) - y_i\|^2$ is the empirical loss, $D = \left\|f(x_i) - \frac{\sum_{x_j \in D_t}f(x_j)}{n_t}\right\|^2 + \left\|f\left(x_i^{(c)}\right) - \frac{\sum_{x_j \in D_t \wedge \hat{y}(x_j)=c}f(x_j)}{n_t^{(c)}}\right\|^2$ is the joint distribution adaptation, and $M = \left\|\frac{\sum_{x_j \in D_t}(f(x_i)-f(x_j))^2W_{ij}}{n_t}\right\|^2$

is the manifold consistency. A source instance is "related" if it is similar to the target instances in distribution, at the same time, it is consistent in manifold structure, and easy to be classified.

3.3   Iteration Termination Based on Reverse Classification Performance

In order to use only the related source instances, we design an early termination strategy for PTRLS. Since there are limited or even no labeled instances in the target domain, we refer to a reverse classification performance [15] to indirectly validate the learning of target domain based on the source set. Specifically, a circular validation strategy is adopted to validate the solution in each iteration. Firstly, the current classifier is used to get the pseudo-label $\mathbf{Y}_p^t$ for the target data. Then target instances along with pseudo-labels $\{\{\mathbf{X}_T, \mathbf{Y}_p^t\}\}$ are used to train a classifier for the source instances. Finally, classification accuracy $\frac{|x:x\in D_s \bigwedge \mathbf{Y}_s=\mathbf{Y}_p^s|}{|x:x\in D_s|}$ over the source instances is used for performance assessment over the target domain indirectly, where $\mathbf{Y}_s$ are the given labels of $\mathbf{X}_s$ while $\mathbf{Y}_p^s$ are the labels predicted.

Through above steps, a group of classifiers can be obtained with respect to different iterations, along with the corresponding reverse accuracies. Finally, the classifier with the best accuracy is chosen to predict the target instances.

3.4 Algorithm Description

The following algorithm describes the learning of PTRLS. It alternatively seeks the target decision function and transfers the related source instances in each class from the most related ones to the least related ones. Finally, classifier with the best reverse classification is adopted for predicting target instances. The alternative search converges as the objective function is monotonically decreasing and bounded from below.

Algorithm 1: PTRLS

input: Source instances $\mathbf{X}_s = \{x_i^s, y_i^s\}_{i=1}^{n_s}$ with b classes, target instances $\mathbf{X}_t = \{x_j^t\}_{j=1}^{n_t}$, parameters $\lambda, \gamma, \delta, \beta, \mu_1$ and μ_2

output: target classifier w

train initial classifier w_0 with all source instances;

repeat

for $p = 1$ to b **do** //for each class

    sort source instances in the pth class in ascending order by relevance R;

    for $i = 1$ to n_p **do** //related samples first

        if $R(y_i^{(p)}, f(x_i^{(p)}, w)) < \delta + \beta \frac{1}{\sqrt{i} + \sqrt{i-1}}$ then $v_i^{(p)} = 1$; //transfer this instance

        else $v_i^{(p)} = 0$; //do not transfer this instance

        end

    end

end

update w by (13) and compute the corresponding reverse performance;

$\delta = \mu_1 \cdot \delta, \beta = \beta \cdot \mu_2$;

until $v_i = 1, \forall i$ //all source instances are adopted

return w with the best reverse classification performance;

---

4 Experiments

4.1 Datasets

Several image datasets are used here, which have been broadly adopted in compute vision and pattern recognition, including USPS, MNIST, MSRC, VOC2007, Office [16, 17] and Caltech-256 [18]. The description can be seen in Fig. 2.

USPS dataset is composed of 7291 training images and 2007 testing images of size 16×16. MNIST dataset is composed of 60000 training examples and 10000 testing examples of size 28×28. USPS and MNIST [7] follow different distributions and share 10 semantic classes. We construct USPS_vs_MNIST by randomly selecting 1800 images in USPS as the source domain and 2000 images in MNIST as the target domain. Meanwhile, we switch the source/target domain to get another dataset MNIST_vs_USPS.

MSRC dataset contains 4323 images labeled by 18 classes. VOC2007 dataset contains 5011 images annotated with 20 concepts. MSRC and VOC2007 [19] follow different distributions and share 6 semantic classes. We construct one dataset MSRC_vs_VOC by randomly selecting 1269 images in MSRC as the source domain and 1530 images in VOC2007 as the target domain. Then, we switch the source/target domain to get another dataset VOC_vs_MSRC.

Office is a benchmark database for visual domain adaptation. The database contains three object domains, i.e., Webcam (low-resolution images by a web camera), and DSLR (high-resolution images by a digital SLR camera). Caltech-256 is a standard database for object recognition, it consists of 30607 images and 256 categories.

We adopt the smaller Office + Caltech10 datasets. They share 10 common classes. More specifically, there are three domains, W (Webcam), D (DSLR), and C (Caltech10). By selecting some of these three different domains, we construct 4 cross-domain object datasets, e.g., W_vs_D, D_vs_C, C_vs_D, D_vs_W.

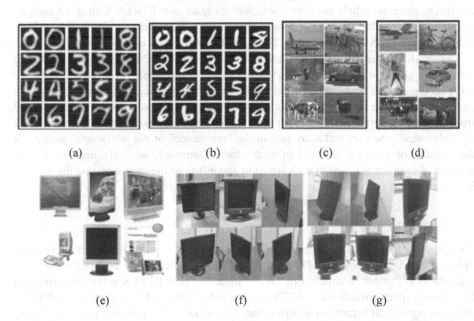

(a)   (b)   (c)   (d)

(e)   (f)   (g)

Fig. 2. Image datasets (a) USPS, (b) MNIST, (c) MSRC, (d) VOC2007, (e) Caltech-256, (f) DSLR, (g) Webcam, where DSLR and Webcam images are from Office.

Finally, the datasets used are list in the following Table 1.

Table 1. Datasets in our experiments

Datasets	Source domain	Target domain
USPS vs MNIST	USPS	MNIST
	MNIST	USPS
MSRC vs VOC	MSRC	VOC
	VOC	MSRC
Office and Caltech10	Caltech10	DSLR
	DSLR	Caltech10
	Webcam	DSLR
	DSLR	Webcam

4.2 Experimental Setup

In experiments, both source and target datasets are divided into two parts, one for training while the other for testing. We use reverse classification method to validate classifiers since no labels are contained. We compare our PTRLS with three domain adaptation methods, including Joint Distribution Adaptation (JDA) [3], ARTL (ARRLS) and DASVM, along with SVM on the target data as the baseline.

As in ARTL, the shrinkage regularization parameter σ, MMD regularization λ, and manifold regularization γ are fixed to 0.1, 10 and 1, respectively. Following SPLD, the self-paced parameters and are updated by absolute values of μ_1, μ_2 ($\mu_1, \mu_2 \geq 1$) at the end of every iteration. In practice, we actually specify the number of samples selected from all classes during each iteration [14]. Specifically, we sort samples in each class by "relevance", then set and according to the "relevance" of the rth sample, where r is the number of samples selected in each class. Therefore, we only need to set the number of samples selected in each iteration according to the data size, and the number increases with a ratio of 5% in the next iteration.

4.3 Comparison Results

The experimental results are given in Table 2. Each row gives the accuracy of individual method over each dataset, and the bold value indicates the best performance in each dataset. Further, a superscript "W/L" indicates that PTRLS achieves significant better/worse performance than ARTL, and a subscript "W/L" indicates that PTRLS achieves significant better/worse performance than DASVM. The next to last row gives cases PTRLS performs better than DASVM/ARRLS, and the last row gives the average performances of methods on all datasets.

Table 2. Performance comparison on real datasets

Dataset	SVM	JDA	DASVM	ARRLS	PTRLS
USPS_vs_MNIST	26.80	50.50	47.32	54.25	**55.63**$_W$
MNIST_vs_USPS	54.22	62.28	63.84	66.72	**71.89**$_W^W$
MSRC_vs_VOC	32.66	27.39	**37.13**	33.92	32.27L
VOC_vs_MSRC	39.01	47.04	54.90	51.61	**62.78**$_W^W$
D_vs_C	51.48	51.85	**56.83**	52.70	52.05L
C_vs_D	52.58	50.58	**55.24**	50.86	53.05L
W_vs_D	57.77	72.78	70.32	73.33	**78.22**$_W^W$
D_vs_W	68.88	**78.87**	67.21	76.00	**78.22**$_W^W$
(DASVM/ARRLS)	-	-	-	-	5/4
Average	47.93	55.16	56.59	57.42	60.51

From Table 2, we can get several observations as follows

- When PTRLS is adopted, the performance can be further improved over most cases. Therefore, it is reasonable to transfer source instances iteratively from the most related ones to the least related ones.
- Compared with iterative DASVM, PTRLS performs better over 5 datasets. As a result, iteratively selecting source instances based on relevance, PTRLS can achieve encouraging performance.
- Compared with ARRLS, PTRLS performs better over 4 datasets, and comparable over the other 4 datasets, validating again the progressive learning of PTRLS. The reason for insignificant improvements over some datasets can be that the termination by reverse classification cannot detect the best performance.

4.4 Performance of PTRLS with Different Iterations

We show the iterative progress of PTRLS in Fig. 3, with the performances of reverse classification over 6 datasets, along with ARRLS as the baseline.

From Fig. 3, we can find that

- As iteration number increases, the performance of PTRLS usually increases first while decreases later. It is consistent with our motivation that transferring the related source instances can help learning for the target domain, while using unrelated source knowledge can do no help or even harm for learning. As a result, it is reasonable to transfer just the related source knowledge.
- PTRLS terminated by reverse classification can usually achieve a better performance than ARRLS. However, the improvement is not so insignificant over some datasets. As a result, some better termination strategy is still worth studying in the future.

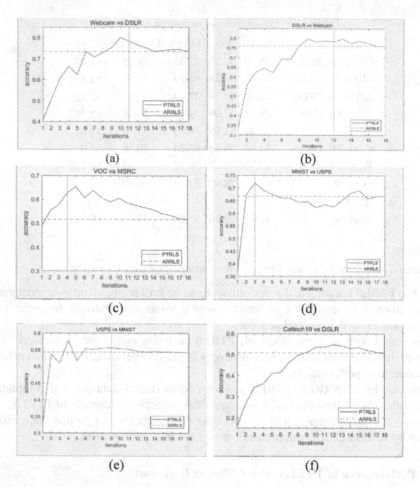

Fig. 3. The iteration of PTRLS over (a) Webcam_vs_DSLR (b) DSLR_vs_Webcam (c) VOC_vs_MSRC (d) MNIST_vs_USPS (e) USPS_vs_MNIST (f) Caltech10_vs_DSLR with ARRLS as the baseline. The green line represents the iteration selected by reverse classification (Color figure online)

5   Conclusion

In this paper, we develop a progressive domain adaptation (PDA) model, in which the source instances are transferred iteratively from the most related ones to the least related ones, until all related source instances have been adopted. It is an iterative learning process, and the reverse classification performance is used to set the terminating criterion, in order to judge if all related source knowledge has been adopted. Experiments on several real datasets validate PTRLS compared with several state-of-arts. It is noted that PDA is a data selecting process for the source domain, thus it can be combined with other feature-based domain adaptation method to further boost the performance. Furthermore, it is still worth to improve the iteration termination strategy, such that PDA can detect the "best" performance in the iteration process.

Acknowledgement. This work was supported by the National Natural Science Foundation of China under Grant Nos. 61876091 and 61772284.

References

1. Pan, S.J., Yang, Q.: A survey on transfer learning. IEEE Trans. Knowl. Data Eng. **22**(10), 1345–1359 (2010)
2. Taylor, M.E., Stone, P.: Transfer learning for reinforcement learning domains: a survey. J Mach Learn Res **10**(10), 1633–1685 (2009)
3. Long, M., Wang, J., Ding, G., Sun, J., Yu, P.S.: Transfer feature learning with joint distribution adaptation. In: Proceedings of the IEEE International Conference on Computer Vision, pp. 2200–2207. IEEE, Sydney (2013)
4. Gao, J., Fan, W., Jiang, J., Han, J.: Knowledge transfer via multiple model local structure mapping. In: Proceedings of the 14th ACM SIGKDD International Conference on Knowledge Discovery and Data Mining, pp. 283–291. ACM, Las Vegas (2008)
5. Dai, W., Yang, Q., Xue, G.R., Yu, Y.: Boosting for transfer learning. In: Proceedings of the 24th International Conference on Machine Learning, pp. 193–200. ACM, Corvalis (2007)
6. Blitzer, J., McDonald, R., Pereira, F.: Domain adaptation with structural correspondence learning. In: Proceedings of the 2006 Conference on Empirical Methods in Natural Language Processing, pp. 120–128. Association for Computational Linguistics, Sydney (2006)
7. Long, M., Wang, J., Ding, G., Pan, S.J., Yu, P.S.: Adaptation regularization: a general framework for transfer learning. IEEE Trans. Knowl. Data Eng. **26**(5), 1076–1089 (2014)
8. Wang, J., Chen, Y., Hao, S., Feng, W., Shen, Z.: Balanced distribution adaptation for transfer learning. In: Proceedings of the IEEE International Conference on Data Mining (ICDM), pp. 1129–1134. IEEE, New Orleans (2017)
9. Bruzzone, L., Marconcini, M.: Domain adaptation problems: a DASVM classification technique and a circular validation strategy. IEEE Trans. Pattern Anal. Mach. Intell. **32**(5), 770–787 (2010)
10. Kumar, M.P., Packer, B., Koller, D.: Self-paced learning for latent variable models. In: Proceedings of the 24th Annual Conference on Neural Information Processing Systems. Curran Associates Inc, Vancouver (2010)
11. Lu, J., Meng, D., Yu, S., Lan, Z., Shan, S., Hauptmann, A.: Self-paced learning with diversity. In: Proceedings of the 28th Annual Conference on Neural Information Processing Systems, Montreal (2014)
12. Schölkopf, B., Herbrich, R., Smola, A.J.: A generalized representer theorem. In: Helmbold, D., Williamson, B. (eds.) Computational Learning Theory 2001. LNCS, vol. 2111, pp. 416–426. Springer, Heidelberg (2001). https://doi.org/10.1007/3-540-44581-1_27
13. Belkin, M., Niyogi, P., Sindhwani, V.: Manifold regularization: a geometric framework for learning from labeled and unlabeled examples. J. Mach. Learn. Res. **7**(1), 2399–2434 (2006)
14. Lu, J., Meng, D., Yu, S., Lan, Z., Shan, S., Hauptmann, A.: Supplementary materials: self-paced learning with diversity. In: Proceedings of the 28th Annual Conference on Neural Information Processing Systems, Montreal (2014)
15. Valindria, V.V., Lavdas, I., Bai, W., et al.: Reverse classification accuracy: predicting segmentation performance in the absence of ground truth. IEEE Trans. Med. Imaging **36**(8), 1597–1606 (2017)

16. Saenko, K., Kulis, B., Fritz, M., Darrell, T.: Adapting visual category models to new domains. In: Daniilidis, K., Maragos, P., Paragios, N. (eds.) European Conference on Computer Vision 2010. LNCS, vol. 6314, pp. 213–226. Springer, Heidelberg (2010). https://doi.org/10.1007/978-3-642-15561-1_16

17. Li, W., Duan, L., Xu, D., Tsang, I.W.: Learning with augmented features for supervised and semi-supervised heterogeneous domain adaptation. IEEE Trans. Pattern Anal. Mach. Intell. **36**(6), 1134–1148 (2013)

18. Long, M., Wang, J., Sun, J., Yu, P.S.: Domain invariant transfer kernel learning. IEEE Trans. Knowl. Data Eng. **27**(6), 1519–1532 (2015)

19. Long, M., Ding, G., Wang, J., Sun, J., Guo, Y., Yu, P.S.: Transfer sparse coding for robust image representation. In: Proceedings of the IEEE Conference on Computer Vision and Pattern Recognition, pp. 407–414. IEEE, Portland (2013)

Learning Data Representation for Clustering (LDRC 2019)

Deep Architectures for Joint Clustering and Visualization with Self-organizing Maps

Florent Forest[1,2](✉) ⓘ, Mustapha Lebbah[1] ⓘ, Hanane Azzag[1] ⓘ,
and Jérôme Lacaille[2]

[1] Université Paris 13, Laboratoire d'Informatique de Paris-Nord (LIPN),
93430 Villetaneuse, France
forest@lipn.univ-paris13.fr
[2] Safran Aircraft Engines, 77550 Moissy-Cramayel, France

Abstract. Recent research has demonstrated how deep neural networks are able to learn representations to improve data clustering. By considering representation learning and clustering as a joint task, models learn clustering-friendly spaces and achieve superior performance, compared with standard two-stage approaches where dimensionality reduction and clustering are performed separately. We extend this idea to topology-preserving clustering models, known as self-organizing maps (SOM). First, we present the Deep Embedded Self-Organizing Map (DESOM), a model composed of a fully-connected autoencoder and a custom SOM layer, where the SOM code vectors are learnt jointly with the autoencoder weights. Then, we show that this generic architecture can be extended to image and sequence data by using convolutional and recurrent architectures, and present variants of these models. First results demonstrate advantages of the DESOM architecture in terms of clustering performance, visualization and training time.

Keywords: Clustering · Self-organizing map ·
Representation learning · Deep learning · Autoencoder

1 Introduction and Related Work

1.1 Joint Representation Learning and Clustering

Representations learned by deep neural networks are successful in a wide range of supervised learning tasks such as classification. Recent research has demonstrated how deep neural networks are able to learn representations to improve unsupervised tasks, such as clustering, in particular for high-dimensional data where traditional clustering algorithms tend to be ineffective. Most clustering algorithms, the most well-know example being the k-means algorithm, rely on similarity metrics (e.g. euclidean distance) that become meaningless in very high-dimensional spaces. The standard solution is to first reduce the dimensionality

L. H. U and H. W. Lauw (Eds.): PAKDD 2019 Workshops, LNAI 11607, pp. 105–116, 2019.
https://doi.org/10.1007/978-3-030-26142-9_10

and then cluster the data in a low-dimensional space. This can be achieved by using, for example, linear dimensionality reduction techniques as Principal Component Analysis (PCA), or models with more expressive power such as deep autoencoder neural networks (AE). In other words, this standard approach first optimizes a pure information loss criterion between data points and their low-dimensional embeddings (generally via a reconstruction loss between a data point and its reconstruction), and then optimize a pure clustering criterion (e.g. k-means quantization error). In contrast, recent *deep clustering* approaches treat representation learning and clustering as a joint task and focus on learning representations that are *clustering-friendly*, i.e. that preserve the prior knowledge of cluster structure.

One of the early approaches, Deep Embedded Clustering (DEC) [17], jointly learns representations and soft cluster assignments by optimizing a KL-divergence that minimizes within-cluster distance; IDEC [5] improves on this approach by optimizing the reconstruction loss jointly with the KL-divergence. The Deep Clustering Network (DCN) [18] combines representation learning with k-means clustering using an alternating training procedure to alternately update the autoencoder weights, cluster assignments and centroid vectors. A review of deep clustering is available in [1]. More recently, [3] overcame the non-differentiability of hard cluster assignments by introducing a smoothed version of the k-means loss.

Most recent approaches perform clustering using generative models such as variational autoencoders (VAE) with a gaussian mixture model (GMM) prior [8] or Wasserstein generative adversarial networks (WGAN) with GMM prior [6] and achieve state-of-the-art performance.

While the previously mentioned work do not make specific assumptions on the type of data and its regularities, other methods focus on specific types of data. For image data, it is common to use architectures based on convolutional neural networks (CNN) that leverage the two-dimensional regularity of images to share weights across spatial locations, as in [19] or [1]. Convolutional architectures can also be used for data with a one-dimensional regularity, such as (multivariate) time series. In this case, one-dimensional *causal* convolutions (also called *temporal* convolutions) are adapted. In particular, *dilated* convolutions are particularly successful in learning long-term dependencies, and even outperform recurrent LSTM networks on various tasks [2]. We are not aware of any application of these architectures for clustering, and will expose this idea in the last section. On the other hand, deep clustering of time series using an LSTM-based architecture was tackled in a recent unpublished work, Deep Temporal Clustering [14].

1.2   Joint Representation Learning and Self-organization

We focus on a specific family of clustering algorithms called self-organizing maps, which perform simultaneous clustering and visualization by projecting high-dimensional data onto a low-dimensional map (typically two-dimensional for visualization purpose) having a grid topology. The grid is composed of *units*, also called *neurons* or *cells*. Each map unit is associated with a *prototype vector* from the original data space (also called *code vector*). Self-organizing map algorithms enforce a topological constraint on the map, so that neighboring units on the map correspond to prototype vectors that are close in the original high-dimensional space. The most well-known self-organizing map model is Kohonen's self-organizing map (SOM) [10, 11].

In this work, we propose several architectures for joint representation learning and self-organization with SOM. The main goals are to:

1. Learn the feature space and the SOM code vectors simultaneously, without using a two-stage approach.
2. Find a *SOM-friendly* space (using the term coined by [18]), i.e. a latent space that is more adapted to the SOM algorithm, according to some quality metric.

The SOM prototypes are learned in the latent space. To learn this new representation, we use an autoencoder neural network, composed of an encoder network that maps data points to the latent space, and a decoder network that reconstructs latent points into vectors of the original data space. For visualization and interpretation of the map, we need the prototypes to lie in the original feature space, so we reconstruct them using the decoder part of the autoencoder network. This approach very much resembles joint representation learning and clustering, but with an additional topology constraint. Our experiments show that using autoencoders with sufficiently high capacity yields meaningful low-dimensional representations of high-dimensional data that facilitate SOM learning and improve clustering performance, and that self-organization and representation learning can be achieved in a single joint task, thus cutting down overall training time.

To the best of our knowledge, the only other work performing joint representation learning with a SOM is the SOM-VAE model introduced in a recent unpublished work [4]. Their model is based on the VQ-VAE (Vector Quantization Variational Autoencoder) model which enables to train variational autoencoders (VAEs) with a discrete latent space [15]. [4] have added a topology constraint on the discrete latent space by modifying the loss function of VQ-VAE. However, there are many important differences between our DESOM model and SOM-VAE. First, SOM-VAE utilizes a discrete latent space to represent the SOM prototypes, whereas in DESOM, the SOM is learned in a continuous latent space. Secondly, they use a fixed window neighborhood to update the map prototypes, whereas we use a gaussian neighborhood with exponential radius decay. Finally, the DESOM model presented in this work is based on a deterministic autoencoder and not a VAE. Using a VAE in DESOM is left as future work.

We will first present our model with a generic, fully-connected, feed-forward autoencoder. The last sections will extend it to convolutional and recurrent architectures. Code is available at https://github.com/FlorentF9/DESOM.

2   DESOM: Deep Embedded SOM

We propose an approach where self-organization of the SOM prototypes and representation learning through a deterministic autoencoder are performed jointly by stochastic gradient descent. The architecture of DESOM, in the case of a fully-connected AE dimensioned for the MNIST dataset, a 10-dimensional latent space and an 8×8 map, is illustrated in Fig. 1.

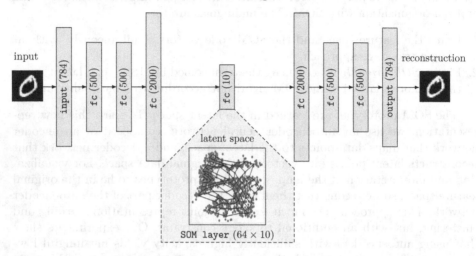

Fig. 1. Architecture of DESOM layers with an 8×8 map and 10-dimensional latent

2.1   Loss Function

We note $\mathbb{X} = \{\mathbf{x}_i\}_{1 \leq i \leq N}$ the data samples. The self-organizing map is composed of K units, associated with the set of prototype vectors $\{\mathbf{m}_k\}_{1 \leq k \leq K}$. $\delta(k, l)$ is the topographic distance between units k and l on the map (Manhattan distance for a 2D grid). We define the neighborhood function of the SOM and a temperature parameter T, controlling the radius of the neighborhood. In this work, we adopt a gaussian neighborhood: $\mathcal{K}^T(d) = e^{-\frac{d^2}{T^2}}$, and exponential temperature decay.

The encoder and decoder parameter weights are respectively noted $\mathbf{W_e}$ and $\mathbf{W_d}$. The encoding function is denoted by $\mathbf{f_{W_e}}$ and the decoding function by $\mathbf{g_{W_d}}$. Thus, $\mathbf{z}_i = \mathbf{f_{W_e}}(\mathbf{x}_i)$ is the embedded version of $\mathbf{x}_i$ in the intermediate latent space, and $\tilde{\mathbf{x}}_i = \mathbf{g_{W_d}}(\mathbf{f_{W_e}}(\mathbf{x}_i))$ is its reconstruction by the decoder. Our goal is to jointly optimize the autoencoder network weights and the SOM prototype vectors. For this task, we define a loss function composed of two terms, that can be written as:

$$\mathcal{L}(\mathbf{W_e}, \mathbf{W_d}, \mathbf{m}_1, \ldots, \mathbf{m}_K, \chi) = \mathcal{L}_r(\mathbf{W_e}, \mathbf{W_d}) + \gamma \mathcal{L}_{som}(\mathbf{W_e}, \mathbf{m}_1, \ldots, \mathbf{m}_K, \chi) \quad (1)$$

The first term $\mathcal{L}_r$ is the autoencoder reconstruction loss, chosen to be a simple least squares loss:

$$\mathcal{L}_r(\mathbf{W_e}, \mathbf{W_d}) = \sum_i ||\tilde{\mathbf{x}}_i - \mathbf{x}_i||^2 \quad (2)$$

The second term is the self-organizing map loss, denoted $\mathcal{L}_{som}$. It depends on the set of parameters $\{\mathbf{m}_k\}_{1 \leq k \leq K}$ and on the assignment function, denoted χ, assigning a data point to its closest prototype according to euclidean distance, i.e.:

$$\chi(\mathbf{z}) = \underset{k}{\mathrm{argmin}} ||\mathbf{z} - \mathbf{m}_k||^2 \tag{3}$$

The expression of the self-organizing map loss is:

$$\mathcal{L}_{som}(\mathbf{W_e}, \mathbf{m}_1, \dots, \mathbf{m}_K, \chi) = \sum_i \sum_{k=1}^{K} \mathcal{K}^T \left(\delta(\chi(\mathbf{z}_i), k) \right) ||\mathbf{z}_i - \mathbf{m}_k||^2 \tag{4}$$

Note that when T converges towards zero, the SOM loss becomes identical to a k-means loss, thus our model converges towards a model equivalent to DCN [18] or DKM [3]. Finally, the hyperparameter γ trades off between minimizing the autoencoder reconstruction loss and the self-organizing map loss.

2.2 Gradients and Training

We use a joint training procedure optimizes both the network parameters and the prototypes using stochastic gradient descent (with the Adam optimization scheme [9]), as the $\mathcal{L}_r$ loss is differentiable; the only non-differentiable parts are the weighting terms $w_{i,k} \equiv \mathcal{K}^T \left(\delta(\chi(\mathbf{z}_i), k) \right)$ of the SOM loss. To alleviate this, we compute the best matching units for the current (encoded) batch and fix the assignment function χ between each optimization step. Thus, these terms $w_{i,k}$ become constant with respect to the network parameters and the prototypes. This requires to compute the pairwise euclidean distances between the map prototypes and the current batch of (encoded) samples between each SGD step. The gradients of the loss function $\mathcal{L}$ w.r.t. autoencoder weights and prototypes are easy to derive if we consider the assignment function to be fixed at each step:

$$\frac{\partial \mathcal{L}}{\partial \mathbf{W_e}} = \frac{\partial \mathcal{L}_r}{\partial \mathbf{W_e}} + \gamma \frac{\partial \mathcal{L}_{som}}{\partial \mathbf{W_e}} \tag{5}$$

$$\frac{\partial \mathcal{L}}{\partial \mathbf{W_d}} = \frac{\partial \mathcal{L}_r}{\partial \mathbf{W_d}} \tag{6}$$

$$\frac{\partial \mathcal{L}}{\partial \mathbf{m}_k} = \gamma \frac{\partial \mathcal{L}_{som}}{\partial \mathbf{m}_k} \tag{7}$$

The gradients for a single data point $\mathbf{x}_i$ are:

$$\frac{\partial \mathcal{L}_r^i}{\partial \mathbf{W_e}} = 2(\mathbf{gw_d}(\mathbf{fw_e}(\mathbf{x}_i)) - \mathbf{x}_i) \frac{\partial \mathbf{gw_d}(\mathbf{fw_e}(\mathbf{x}_i))}{\partial \mathbf{W_e}} \tag{8}$$

$$\frac{\partial \mathcal{L}_r^i}{\partial \mathbf{W_d}} = 2(\mathbf{gw_d}(\mathbf{fw_e}(\mathbf{x}_i)) - \mathbf{x}_i) \frac{\partial \mathbf{gw_d}(\mathbf{fw_e}(\mathbf{x}_i))}{\partial \mathbf{W_d}} \tag{9}$$

$$\frac{\partial \mathcal{L}_{som}^i}{\partial \mathbf{W_e}} = 2 \sum_{k=1}^{K} w_{i,k}(\mathbf{fw_e}(\mathbf{x}_i) - \mathbf{m}_k) \frac{\partial \mathbf{fw_e}(\mathbf{x}_i)}{\partial \mathbf{W_e}} \tag{10}$$

$$\frac{\partial \mathcal{L}_{som}^i}{\partial \mathbf{m}_k} = 2w_{i,k}(\mathbf{m}_k - \mathbf{fw_e}(\mathbf{x}_i)) \tag{11}$$

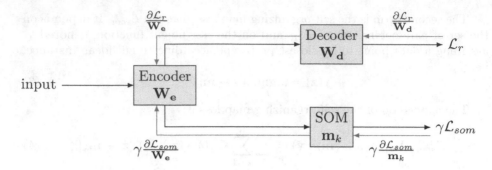

Fig. 2. Path of gradients in the DESOM model

The paths of the gradients of the loss function are illustrated on Fig. 2. We optimize 1 using backpropagation and minibatch stochastic gradient descent (SGD), with a learning rate l_r (in our experiments Adam is used instead, but the equations are derived for vanilla SGD). Given a batch of n_b samples, the encoder's weights are updated by:

$$\mathbf{W_e} \leftarrow \mathbf{W_e} - \frac{l_r}{n_b} \sum_{i=1}^{n_b} \left(\frac{\partial \mathcal{L}_r^i}{\partial \mathbf{W_e}} + \gamma \frac{\partial \mathcal{L}_{som}^i}{\partial \mathbf{W_e}} \right) \tag{12}$$

The decoder's weights are updated by:

$$\mathbf{W_d} \leftarrow \mathbf{W_d} - \frac{l_r}{n_b} \sum_{i=1}^{n_b} \frac{\partial \mathcal{L}_r^i}{\partial \mathbf{W_d}} \tag{13}$$

And finally, the map prototypes are updated by the following update rule:

$$\mathbf{m}_k \leftarrow \mathbf{m}_k - \frac{l_r}{n_b} \sum_{i=1}^{n_b} \gamma \frac{\partial \mathcal{L}_{som}^i}{\partial \mathbf{m}_k} \tag{14}$$

2.3 Training Procedure

The training procedure is detailed in Algorithm 1.

Input: training set $\mathbb{X}$; SOM dimensions (m, n); initial and final temperatures T_{max}, T_{min}; number of iterations *iterations*; batch size *batchSize*
Output: AE weights $\mathbf{W_e}, \mathbf{W_d}$; SOM code vectors $\{\mathbf{m}_k\}$
Initialize AE weights $\mathbf{W_e}, \mathbf{W_d}$ (Glorot uniform scheme) ;
Initialize SOM parameters $\{\mathbf{m}_k\}$ (with random data sample) ;
for $iter = 1, \ldots, iterations$ **do**

$\quad\quad T \leftarrow T_{max} \left(\frac{T_{min}}{T_{max}}\right)^{iter/iterations}$;

$\quad\quad$ Load next training batch ;
$\quad\quad$ Encode current batch and compute weights $w_{i,k}$;
$\quad\quad$ Train DESOM on batch by taking a SGD step (by 12, 13 and 14 ;
end

Algorithm 1: DESOM training procedure

3   Evaluation

We evaluated the clustering and visualization performance of our model on standard classification benchmark datasets. In this section, we will detail our evaluation methodology and the results on the MNIST and REUTERS-10k datasets.

Datasets. The MNIST dataset [12] consists in 70000 grayscale images of handwritten digits, of size 28-by-28 pixels. We divided pixel intensities by 255 to obtain a 0–1 range and flattened the images into 784-dimensional vectors. REUTERS-10k [13] is a text dataset built from the RCV1-v2 corpus. REUTERS-10k is created by restricting the documents to 4 classes, sampling a subset of 10000 examples and computing TF-IDF features on the 2000 most frequently occurring words. We used the same preparation code as in [5].

Qualitative Evaluation. We assessed that, just like a standard SOM, our model produces a topologically organized map for efficient visualization, and that the decoded code vectors are of high quality. An example of DESOM map for MNIST can be seen on Fig. 3. We verified that the capacity of the AE (number of layers and units) was directly linked with the visual quality of the prototypes. In particular, using standard SOM directly on this kind of data produces blurred prototype images, due to averaging in original space.

Quantitative Evaluation. Then, we evaluated the clustering quality of DESOM by measuring two common external clustering indices, purity and NMI (Normalized Mutual Information). We compared DESOM with 5 other SOM-based models:

- **minisom**: a standard SOM (from minisom module[1]).
- **kerasom**: our implementation of a SOM in Keras (equivalent to DESOM with identity encoder and decoder) and trained by SGD.

---
[1] https://github.com/JustGlowing/minisom.

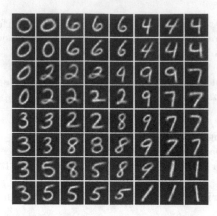

Fig. 3. Decoded prototypes visualized on a DESOM map for MNIST

- **AE+minisom**: a two-stage approach where minisom is fit on the encoded dataset using an autoencoder with the same architecture as the one used in DESOM.
- **AE+kerasom**: the same two-stage approach but with our kerasom model, resulting in DESOM without joint optimization of AE and SOM.
- **SOM-VAE**: results from the author's paper [4] (only for MNIST).
- **DESOM**: our proposed DESOM model with joint representation learning and self-organization.

In all models, the AE has a $[500, 500, 2000, 10]$ architecture and the map has 8×8 units. The γ parameter was fixed empirically to 0.001, number of iterations is 10000 with a batch size of 256. Results are summarized in Table 1. Results on MNIST show that reducing dimensionality with an autoencoder improves clustering quality. DESOM and AE+kerasom perform best and have similar results, but DESOM requires no pre-training. However, the AE struggles to find good representations on REUTERS-10k, and DESOM outperforms all other models by a large margin, suggesting that joint training with a self-organizing map prior has enabled to learn SOM-friendly representations.

Training Time. An advantage of joint training is reduced training time of DESOM compared with AE+kerasom (other models cannot be compared due to difference in implementation). Indeed, to obtain the results listed in Table 1, kerasom and DESOM were trained for the same number of iterations and required the same training time (about 2 min on MNIST on a laptop GPU). If we add the AE pretraining time, the overall training time of AE+kerasom nearly doubles (we pretrained for 200 epochs). As a conclusion, the joint representation learning adds almost no training time overhead.

Table 1. Clustering performance of SOM-based models according to purity and NMI (mean and standard deviation on 10 runs). Best performance in bold.

Method	MNIST		REUTERS-10k	
	Purity	NMI	Purity	NMI
minisom (8 × 8)	0.637 ± 0.023	0.430 ± 0.016	0.690 ± 0.028	0.230 ± 0.024
kerasom (8 × 8)	0.826 ± 0.005	0.565 ± 0.003	0.697 ± 0.067	0.324 ± 0.051
AE+minisom (8 × 8)	0.872 ± 0.017	0.616 ± 0.010	0.690 ± 0.021	0.235 ± 0.015
AE+kerasom (8 × 8)	**0.939** ± 0.003	**0.661** ± 0.002	0.777 ± 0.012	0.306 ± 0.010
SOM-VAE (8 × 8)	0.868 ± 0.003	0.595 ± 0.002	-	-
DESOM (8 × 8)	**0.939** ± 0.004	0.657 ± 0.004	**0.849** ± 0.011	**0.381** ± 0.009

4   Architecture Variants

In this section, we present extensions of the generic DESOM architecture for data with structural regularities such as images and sequential data (e.g. multivariate timeseries). The advantage of deep architectures is that we only need to change the representation learning part of the model (autoencoder), that maps the input to its latent embedding; the SOM layer, training procedure and loss function remain the same.

4.1   ConvDESOM: Convolutions for Images and Sequences

For image data, we can replace the fully-connected autoencoder with a convolutional autoencoder and a typical architecture for image recognition, composed of alternating 2D convolutions and pooling operations. An example of such an architecture, that we call ConvDESOM, is represented on Fig. 4. The output of the encoder is now a 2D feature map that is flattened before serving as input to the SOM layer. The decoder is composed of convolutional and up-sampling layers.

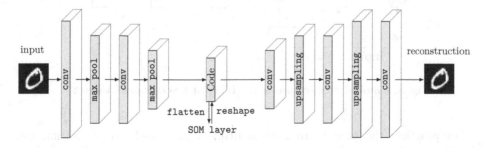

Fig. 4. Architecture of the ConvDESOM model variant (example for 2D images).

For sequence data and time series in particular, the same type of architecture can be used, but with 1D convolutions instead. The exact architecture depends on the use case:

– Pooling layers will mitigate location dependance.
– Convolutions may be causal and/or dilated (dilation allows to increase the receptive field exponentially with the network depth while keeping the number of parameters low, thus allowing for long effective memory [2]).
– Convolution kernel size, dilation, padding policy, pooling and the number of layers have a direct influence on the dimensionality of the latent code used by the SOM layer.

We are still conducting experiments on these architectures.

4.2 LSTM-DESOM: Recurrent Architecture for Sequences

In this last section, we propose a recurrent variant of DESOM, called LSTM-DESOM, based on Long Short-Term Memory (LSTM) cells [7]. This architecture is targeting sequential data, and like causal convolutions, recurrent networks can handle sequences of arbitrary length. It is based on an LSTM autoencoder [16], which is a particular case of the LSTM encoder-decoder architecture that learns a code representation from an input sequence and then reconstructs this sequence. Similarly to the standard DESOM presented in the second section, the latent representation is used to learn the SOM code vectors. An unrolled illustration of a basic LSTM-DESOM architecture is represented in Fig. 5.

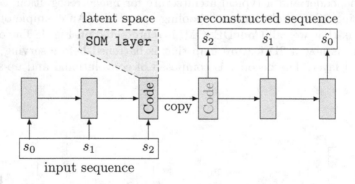

Fig. 5. Unrolled architecture of the LSTM-DESOM model variant.

In practice, the model can have multiple layers, and may condition the decoder on the reversed input sequence for reconstruction (see [16]). A slightly more complex architecture is used in [14] for joint representation learning and clustering of time series. Again, no experiments with this architecture have been conducted yet.

5 Conclusion and Future Work

The Deep Embedded Self-Organizing Map extends the ideas of joint representation learning and clustering to topology-preserving clustering with self-organizing maps. It can be used to explore and visualize large, high-dimensional datasets, and the architecture can be adapted for various types of data, including images and sequences. Compared with other SOM-based algorithms, it shows similar or superior performance. By combining representation learning and self-organization in a joint task, it reduces overall training time compared with traditional two-stage approaches. The specific properties of the *SOM-friendly* latent space learned by DESOM need to be studied more thoroughly in future work. Future work will also include the study and evaluation of the convolutional and recurrent variants of DESOM.

References

1. Aljalbout, E., Golkov, V., Siddiqui, Y., Cremers, D.: Clustering with Deep Learning: Taxonomy and New Methods (2018). arXiv:1801.07648
2. Bai, S., Kolter, J.Z., Koltun, V.: An Empirical Evaluation of Generic Convolutional and Recurrent Networks for Sequence Modeling (2018). https://doi.org/10.1016/S0925-5273(03)00047-1, arXiv:1803.01271
3. Fard, M.M., Thonet, T., Gaussier, E.: Deep k-Means: Jointly Clustering with k-Means and Learning Representations (2018). arXiv:1806.10069
4. Fortuin, V., Hüser, M., Locatello, F., Strathmann, H., Rätsch, G.: Deep Self-Organization: Interpretable Discrete Representation Learning on Time Series (2018). arXiv:1806.02199
5. Guo, X., Gao, L., Liu, X., Yin, J.: Improved deep embedded clustering with local structure preservation. In: IJCAI, pp. 1753–1759 (2017)
6. Harchaoui, W., Mattei, P., Alamansa, A., Bouveyron, C.: Wasserstein Adversarial Mixture Clustering (2018). https://hal.archives-ouvertes.fr/hal-01827775/
7. Hochreiter, S., Schmidhuber, J.: Long short-term memory. Neural Comput. **9**(8), 1735–1780 (1997)
8. Jiang, Z., Zheng, Y., Tan, H., Tang, B., Zhou, H.: Variational deep embedding: an unsupervised and generative approach to clustering. In: IJCAI, pp. 1965–1972 (2017). https://doi.org/10.24963/ijcai.2017/273
9. Kingma, D.P., Ba, J.L.: Adam: a method for stochastic optimization. In: ICLR (2015). arXiv:1412.6980
10. Kohonen, T.: Self-organized formation of topologically correct feature maps. Biol. Cybern. **43**(1), 59–69 (1982). https://doi.org/10.1007/BF00337288
11. Kohonen, T.: The self-organizing map. Proc. IEEE **78**, 1464–1480 (1990). https://doi.org/10.1109/5.58325
12. LeCun, Y., Bottou, L., Bengio, Y., Haffner, P.: Gradient-based learning applied to document recognition. In: Proceedings of the IEEE (1998). https://doi.org/10.1109/5.726791
13. Lewis, D.D., Yang, Y., Rose, T.G., Li, F.: RCV1: a new benchmark collection for text categorization research. J. Mach. Learn. Res. **5**, 361–397 (2004). http://dl.acm.org/citation.cfm?id=1005332.1005345

14. Madiraju, N.S., Sadat, S.M., Fisher, D., Karimabadi, H.: Deep Temporal Clustering: Fully Unsupervised Learning of Time-Domain Features, pp. 1–11 (2018). arXiv:1802.01059

15. van den Oord, A., Vinyals, O., Kavukcuoglu, K.: Neural discrete representation learning. In: NIPS (2017). arXiv:1711.00937

16. Srivastava, N., Mansimov, E., Salakhutdinov, R.: Unsupervised Learning of Video Representations using LSTMs (2015). arXiv:1502.04681

17. Xie, J., Girshick, R., Farhadi, A.: Unsupervised deep embedding for clustering analysis. In: ICML, vol. 48 (2015). https://doi.org/10.1007/JHEP01(2016)157, arXiv:1511.06335

18. Yang, B., Fu, X., Sidiropoulos, N.D., Hong, M.: Towards K-means-friendly spaces: simultaneous deep learning and clustering. In: ICML (2016). arXiv:1610.04794

19. Yang, J., Parikh, D., Batra, D.: Joint Unsupervised Learning of Deep Representations and Image Clusters (2016). https://doi.org/10.1109/CVPR.2016.556, arXiv:1604.03628

Deep Cascade of Extra Trees

Abdelkader Berrouachedi[✉], Rakia Jaziri, and Gilles Bernard

LIASD research Lab, University of Paris VIII, Paris, France
{aberrouachedi,rjaziri,gb}@ai.univ-paris8.fr

Abstract. Deep neural networks have recently become popular because of their success in such domains as image and speech recognition, which has lead many to wonder whether other learners could benefit from deep, layered architectures. In this paper, we propose the Deep Cascade of Extra Trees (DCET) model. Representation learning in deep neural networks mostly relies on the layer-by-layer processing of raw features. Inspired by this, DCET uses a deep cascade of decision forests structure, where the cascade in each level receives the best feature information processed by the cascade of forests of its preceding level. Experiments show that its performance is quite robust regarding hyper-parameter settings; in most cases, even across different datasets from different domains, it is able to get excellent performance by using the same default setting.

Keywords: Machine learning · Extra-Trees · Deep Forest · Deep learning

1 Introduction

Deep learning currently provides the best solution [20] to many problems in image [21] and speech recognition [13], and natural language processing [8]. It applies to such problems as anomaly detection, classifying or clustering.

Current deep models are usually neural networks, with layers of parametrized differentiable nonlinear modules trained by backpropagation. Though powerful, they have drawbacks, as model complexity or the powerful computing facilities needed in training. On performance itself, there are two drawbacks: it is dependant on careful tuning of many hyper-parameters; most models present a risk of overfitting the training data.

Their success has lead researchers to wonder whether other learners could benefit from deep, layered architectures. As others, we believe that in order to tackle complicated learning tasks, it is likely that learning models have to go deep. It is interesting to consider whether deep learning can be realized with other models, because they have their own advantages and may exhibit great potential if being able to go deep.

In a recent research [1], we proposed the Deep Extra Tree model, inspired from extremely randomized trees, that was able to perform as well as deep neural networks in a broad range of classification tasks, while being easier to tune and

© Springer Nature Switzerland AG 2019
L. H. U and H. W. Lauw (Eds.): PAKDD 2019 Workshops, LNAI 11607, pp. 117–129, 2019.
https://doi.org/10.1007/978-3-030-26142-9_11

much faster. Here we present a more complex deep model, Deep Cascade Extra Tree (DCET), whose performances are also on par with deep neural networks in classification tasks. Though slightly slower than the DET model, DCET is still way faster than deep neural networks; in half of the datasets, results are enhanced over the DET ones.

The extremely randomized trees (Extra-Trees) model operates by strongly randomizing both attribute and cut-point choices while splitting a tree node. Our approach unifies decision trees with the representation learning functionality inspired from deep neural networks. It also addresses the risk of overfitting.

In the next section we will present related works, before introducing DCET, the experiments, and end with our conclusion.

2 Related Work

In 2005 [12] proposed the Deep Neural Decision Trees, a hybrid architecture where the output of a deep neural network is fed to a random forest [3]. The performance of their model is higher than the deep neural network alone, but the drawback is that it is highly time consuming.

A very interesting new method devised to be an alternative to deep neural networks is gcForest, a deep forest proposed in 2017 [24]. It is a multi-layered structure where each layer is a set of random forests, without back-propagation. However, this approach is slower than conventional approaches because of the high number of random forests on each layer.

[22] proposes the Siamese Deep Forest, based on gcForest but modified in order to develop a structure solving the task of metric learning, inspired by the Siamese Neural Network [11].

[23] proposes a neural network operating similarly to random forests, Neural Network with Random Forest (NNRF). Like random forests, for each input vector, NNRF only activates one path and thus efficiently performs forward and backward propagation. In addition, the one path property also makes the NNRF able to deal with small datasets, as the number of parameters for each path is relatively small. Unlike random forests, NNRF learns complex multivariate functions in each node in order to choose between relevant paths, and is thus able to learn more complex datasets. Extensive experiments on real-world datasets from different domains demonstrate the effectiveness of this model. Further experiments are conducted to analyze the sensitivity of the hyper-parameters.

[18] introduces the Forward Thinking Deep Random Forest (FTDRF) close to gcForest; one of the fundamental differences is that while gcForest passes the output of whole random forests, concatenated with the original data, to subsequents layers, FTDRF passes only the outputs of the individual trees. This diminishes the number of trees needed while yielding a similar performance. They provide a general mathematical formulation of Forward Thinking that allows for other types of deep learning models to be considered. The proof of concept is demonstrated by applying FTDRF on the MNIST dataset.

3   The Proposed Approach

In this section, we will first present the extra trees which constitute the elements of our DCET structure, then the global architecture and operation of our model, ending with its hyper-parameters.

3.1   Extra Tree

The Extra Tree model was proposed in [6], with the main objective of further randomizing tree building in the context of numerical input features. The choice of optimal cut-points is responsible for a large proportion of the variance of the induced tree. With respect to random forests, the method drops the idea of using bootstrap copies of the learning sample, and instead of trying to find an optimal cut-point for each one of the randomly chosen features at each node, it selects a cut-point at random, which appears to be times faster.

Forests of Extra-Trees, usually called Extra-Trees (it would be clearer to call it Extra-Forest) can be used as random forests. They estimate class distribution by counting the percentage of each class at the leaf node containing the concerned instances, and then averaging across all trees (Fig. 1).

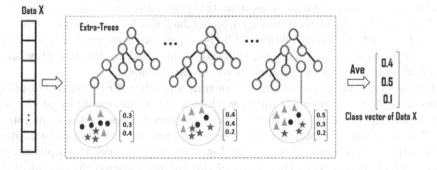

Fig. 1. Computation of class vectors; symbols label the classes

3.2   Deep Cascade Extra Trees Structure

The main difference between the structure presented here and the DET structure [1] is that while in the last each level in the deep structure is a layer of extra-trees forests, here each level is a cascade of layers of extra-trees forests (Fig. 2).

For the sake of clarity, we differentiate levels and layers, calling "level" the outside layers and reserving the term "layer" for the inside layers. Each level is composed of a sequence of layers and each layer is a set of extra-trees forests.

The training set is divided in as many batches of vectors as they are forests in each layer. To reduce the risk of overfitting, the batches produced for each forest are generated by k-fold cross validation repeated at each level. With cross

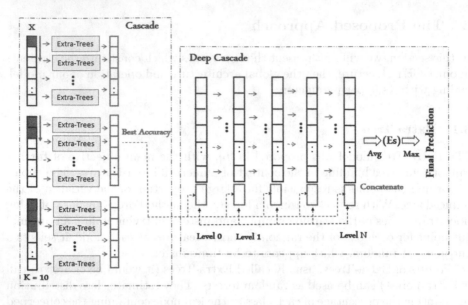

Fig. 2. Deep Cascade Extra-Trees structure

validation, each input vector is used as training data $k - 1$ times, resulting in $k - 1$ vectors, which are then averaged to produce the final vector of the layer.

Each extra-trees forest contains 100 trees, generated by randomly selecting a feature at each node of the tree, splitting it, and making the tree grow until each leaf node contains only instances of the same class.

Given an instance, each forest will output an estimate of class distribution, by counting the percentage of different classes of training examples at the leaf node where the instance falls, then averaging across all trees in the same forest.

Thus the output of each forest is a vector representing the class distribution of its batch vectors. For each first layer forest of the first level, the input simply is its batch of vectors. The other layers of the first level are fed the output of the first layer concatenated with the input vector. For every other level, the input for each layer is the input vector concatenated with the best output vector of the preceeding level (the one with the best accuracy).

The principle of operation at each level has the following stages:

1. generate k-1 batches of data across the forests;
2. if not first level, concatenate the output of previous level;
3. else if first layer do not concatenate;
4. else concatenate first layer output;
5. compute the output of each forest;
6. average all the outputs;
7. compute accuracy;
8. if accuracy is worsening by a threshold, terminate;
9. then transmit the output vector with best accuracy to next level
10. else transmit the input vector to next layer.

Number of levels and of layers in each level is automatically determined by the data, as a new level or a new layer is produced only if accuracy is not at its maximum value. In contrast to most deep neural networks whose depth is previously fixed, DCET, like DET, adaptively decides its depth by terminating training when performance cannot be bettered, as is also the case for instance with FTDRF [18]. Thus it can be applied to training data of different scales.

The main difference of operation between DET and DCET is that while in the first model output vectors are concatenated with the input at each layer, in DCET they are concatenated with the input only once at each level, enabling a more thorough exploration of the hypotheses space.

The parameters of DCET are the following, with their default values (as one can see, less parameters are needed than for deep neural models):

- k for k-fold: as is usual in the literature, we take $k = 10$;
- number of forests: $k - 1$;
- number of extra-trees in forest: 500;
- accuracy threshold: 0.3%.

If accuracy gets higher or does not lower significantly, we keep going on, else we stop. Significance is measured with the accuracy threshold; if the lowering of accuracy is less than the threshold, it is deemed unsignificant.

4   Experiments

4.1   Configuration

In this section, we compare DCET with deep multi-layer perceptrons (coined simply MLP hereafter) and several other state-of the art algorithms. We have used 9 different gold standard datasets for classification problems, that cover a wide range of conditions: number of classes (between 2 and 10), learning sample size (between 280 and 60,000), number of attributes. These datasets are publicly available. We have chosen datasets which have no missing values and only numerical attributes. Except for IMDB, we selected 70% of samples for training (and validation), and 30% for testing.

We compare, for all datasets, DCET with MLP and DET. DCET has the default value for its parameters (same for DET), as given above (those values have been determined in our first experiments), but for small data sets, where we choose 100 trees instead of 500 in the forest. MLP has been implemented with Keras and Tensorflow, implementation available at[1]. Otherwise explicitly indicated, it has two fully connected layers with 1024 and 512 units, respectively, and a sigmoid layer appended. ReLU is used as activation function, categorical cross-entropy as loss function, Adam for optimization.

For the other models, *scikit-learn* library was used (https://scikit-learn.org/). For Decision Tree, the CART model was chosen.

---

[1] https://github.com/KaderBerrouachedi/Deep-Models/tree/master/DeepCascadeExtraTrees.

4.2 Results

Our experiments use a PC Intel Xeon E5-2686 v4 with 64 vCPUs and 256 GiB of RAM. We present three types of results: tables of performances, figures for the trace of growing depth with accuracy values for the biggest depth and the smallest depth, and the time comparison between MLP and DCET.

ALIO dataset is a set of images by Geusebroek [7] for outlier detection. It has 27 numeric attributes (HSB histograms) and 50,000 instances, 1508 outliers (3.04%) and 48492 inliers (96.98%). Table 1 shows DCET has best accuracy, and Fig. 3 shows that it stopped at depth 5 (it is the optimal depth), adapting to the small-scale data.

Table 1. Performances on ALIO

	Accuracy	Precision	Recall	F1-score
Deep Cascade Extra-Trees (DCET)	**98.00%**	98%	98%	98%
Deep Extra-Trees (DET)	97.77%	98%	98%	98%
Extra-Trees	97.32%	97%	97%	97%
MLP	97.19%	/	/	/
Random Forest	97.14%	97%	97%	97%
AdaBoost	97.00%	96%	97%	96%
Decision Tree	95.60%	95%	96%	96%
Gaussian Naive Bayes	76.72%	73%	67%	70%

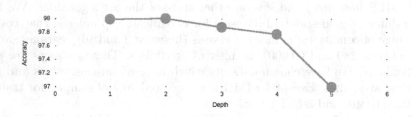

Fig. 3. Accuracy and depth for Alio dataset

PageBlocks is a collection of blocks of page layout of 54 documents detected by a segmentation process [17], from the UCI Machine Learning Repository [5], for separating text and picture. It has 5473 instances. All attributes are numeric. Textual content is labelled as inlier, the rest as outlier. Table 2 shows that DCET has best accuracy and F-score equal to the others (same configuration as above).

Table 2. Performances on PageBlocks

	Accuracy	Precision	Recall	F1-score
Deep Cascade Extra-Trees (DCET)	**99.26%**	99%	99%	99%
Deep Extra-Trees (DET)	99.09%	99%	99%	99%
gcForest	98.79%	99%	99%	99%
AdaBoost	98.79%	99%	99%	99%
Decision Tree	98.32%	98%	98%	98%
MLP	98.53%	/	/	/
Extra-Trees	97.06%	99%	99%	99%
Gaussian Naive Bayes	94.31%	98%	94%	96%

SpamBase is a dataset build from emails coming from postmasters and individuals who had filed spam [5], created by Blake and Merz [2]. It has 58 attributes and 4601 instances. 1813 are spam (39.4%), labelled as outliers, the remaining as inliers. Table 3 shows performances nearly identical for all models except Naive Bayes, even if DCET has a (very) slight advantage in accuracy.

Table 3. Performances on SpamBase

	Accuracy	Precision	Recall	F1-score
Deep Cascade Extra-Trees (DCET)	**95.33%**	95%	95%	95%
Deep Extra-Trees (DET)	95.25%	95%	95%	95%
Extra-Trees	95.01%	95%	95%	95%
gcForest	94.61%	95%	95%	95%
Random Forest	94.45%	94%	94%	94%
AdaBoost	94.37%	94%	94%	94%
MLP	93.91%	/	/	/
Decision Tree	90.41%	90%	90%	90%
Gaussian Naive Bayes	81.63%	86%	82%	82%

Waveform dataset contains three classes of waves, with 33% of each; class 0 was defined here as an outlier class and down-sampled to 100 objects. It has 21 numeric attributes and 3443 instances, with 100 outliers (2.9%) and 3343 inliers (97.1%) [25]. Table 4 contains the results.

Wilt is a collection of images segments generated by segmenting pansharpened images of land cover, with spectral information from the QuickBird multispectral image bands and texture information from the panchromatic image band. It comes from a remote sensing study by Johnson et al. [9], that involved detecting

Table 4. Performances on Waveform

	Accuracy	Precision	Recall	F1-score
Deep Cascade Extra-Trees (DCET)	**97.87%**	98%	98%	97%
Deep Extra-Trees (DET)	**98.06%**	98%	98%	98%
gcForest	97.96%	98%	98%	97%
Random Forest	97.57%	97%	97%	96%
Extra-Trees	97.28%	97%	97%	96%
MLP	97.05%	/	/	/
AdaBoost	96.70%	96%	97%	96%
Decision Tree	96.32%	96%	96%	96%
Gaussian Naive Bayes	93.90%	96%	94%	95%

diseased trees in QuickBird imagery. Segments containing such trees are outliers. It has 4,819 instances. Table 5 shows all models have equivalent performances, except Naive Bayes, with a slight advantage to DET over DCET as second.

Table 5. Performances on Wilt

	Accuracy	Precision	Recall	F-score
Deep Cascade Extra-Trees (DCET)	**98.34%**	98%	98%	98%
Deep Extra-Trees (DET)	**98.74%**	99%	99%	99%
gcForest	98.20%	98%	98%	98%
Extra-Trees	97.57%	98%	98%	97%
AdaBoost	97.51%	97%	98%	97%
Random Forest	97.23%	97%	97%	97%
Decision Tree	96.68%	96%	97%	97%
MLP	93.91%	/	/	/
Gaussian Naive Bayes	90.52%	93%	91%	92%

HTRU2 describes a sample of pulsar candidates collected during the High Time Resolution Universe Survey [10]. It contains 17,898 spurious samples caused by RFI/noise, manually checked, with 8 continuous attributes computed from the pulse folded profile, describing a longitude-resolved version of the signal averaged in time and frequency (see [15] for details). It is labelled positive or negative. Data is stored in CSV and ARFF formats. Table 6 shows the performances of the usual algorithms (MLP, gcForest, Extra Trees, Random Forest, Decision Tree, Adaboost, Naive Bayes), and the performances of GH-VFDT as reported in [14]. They keep close to each other, Naive Bayes being behind. The complexity of the dataset is shown by the depth 17 (Fig. 4).

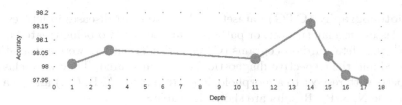

Fig. 4. Accuracy and depth for HTRU2

Table 6. Performances on HTRU2

	Accuracy	Precision	Recall	F1-score
Deep Cascade Extra Trees (DCET)	**98.16**%	98%	98%	98%
Deep Extra-Trees (DET)	**98.19**%	98%	98%	98%
Extra-Trees	98.06%	98%	98%	98%
gcForest	98.06%	98%	98%	98%
Random Forest	98.10%	98%	98%	98%
GH-VFDT	97.80% [14]	/	/	/
AdaBoost	97.80%	98%	98%	98%
MLP	97.54%	/	/	/
Decision Tree	96.75%	97%	97%	97%
Gaussian Naive Bayes	94.89%	96%	95%	95%

Table 7. Performances on ORL

	Accuracy	Precision	Recall	F1-score
Deep Cascade Extra-Trees (DCET)	**95.00**%	97%	95%	95%
Deep Extra-Trees (DET)	**97.50**%	98%	97%	98%
gcForest	92.50%	95%	93%	93%
Extra-Trees	86.66%	91%	87%	87%
Random Forest	85.00%	91%	85%	85%
Gaussian Naive Bayes	74.16%	90%	74%	76%
Decision Tree	47.50%	56%	47%	47%
MLP	25.00%	/	/	/
AdaBoost	14.16%	19%	14%	15%

ORL dataset [19] contains 400 gray-scale facial images taken from 40 persons (10 each). Pictures were taken at different times, varying lighting, facial expressions and facial details (glasses/no glasses), against a dark homogeneous background with the subjects in an upright, frontal position (with tolerance for some side movement). Table 7 shows DET has the best performances, DCET is second.

Cardiotocography (CTG) dataset is related to heart diseases in fetuses, with three classes: normal, suspect, or pathological, the last two being treated as outliers [4]. 1688 fetal cardiotocograms (CTGs) with 34 outliers were automatically processed and the respective diagnostic features measured. They were classified by experts with respect to a morphologic pattern (code A, B, C...) and to a fetal state (code N, S, P). Results are shown in Table 8.

Table 8. Performances on CTG

	Accuracy	Precision	Recall	F1-score
Deep Cascade Extra-Trees (DCET)	**98.03**%	98%	98%	97%
Deep Extra-Trees (DET)	**99.1**%	99%	99%	99%
gcForest	97.83%	97%	98%	97%
Extra-Trees	97.63%	98%	98%	98%
Random Forest	97.43%	96%	97%	97%
AdaBoost	97.23%	96%	97%	97%
Decision Tree	96.63%	97%	98%	97%
Gaussian Naive Bayes	93.29%	96%	93%	95%

Table 9. Comparison of test accuracy on IMDB

	Accuracy	Precision	Recall	F1-score
Deep Cascade Extra-Trees (DCET)	**53.23**%	53%	53%	53%
Deep Extra-Trees (DET)	53.17%	53%	53%	53%
Extra-Trees	51.82%	52%	52%	52%
Random Forest	53.06%	53%	53%	53%
MLP	50.02%	/	/	/
Decision Tree	50.72%	51%	51%	51%
Gaussian Naive Bayes	50.41%	51%	50%	46%

IMDB. We used a subset of 50,000 movie reviews from IMDB, compiled by Andrew Maas [16] labeled by sentiment (positive/negative). The data is split evenly with 25k reviews intended for training and 25k for testing your classifier. Moreover, each set has 12.5k positive and 12.5k negative reviews. IMDB lets users rate movies on a scale from 1 to 10. To label these reviews the curator of the data labeled anything with less than 4 stars as negative and anything with more than 7 stars as positive. Reviews with 5 or 6 stars were left out. This dataset has been first converted to vectors using Word2Vec Skipgram before being fed to the models (Table 9); this probably explains the globally poor results.

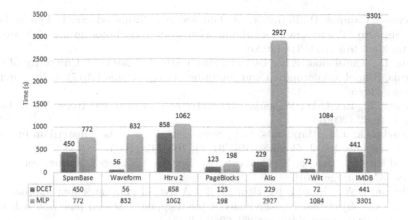

Fig. 5. Running time.

5   Conclusion

One first observation is that DCET, for all datasets, is at least on par with the best algorithms, often slightly better, specially on accuracy. Except for the ORL dataset, where DCET has a clear advantage on all other models except DET, the advantage in performance is not such that it cannot be attributed to bias.

These models have been trained on a representative list of datasets, balanced and unbalanced, small-scale and large-scale, with different contexts of use, in order to limit the effect of data bias on the results (Fig. 5).

Our experiments show that DCET is a good alternative to deep neural networks in classification tasks, all the more so as it is very easy to configure. With the DET experiments, this confirms that deep learning can benefit to non-neural models. Comparing DCET, which has been devised in order to cope with bigger volumes of data, and DET, more simple model, one can observe that the biggest datasets (e.g. ALIO or IMDB) give best results with DCET, and the smaller datasets (e.g. ORL) or those with the lowest number of attributes (e.g. Hilt or HTRU2) give best results with DET.

We hope these results will encourage other researches on bringing deep-layered architectures to other learning models.

References

1. Abdelkader, B., Rakia, J., Gilles, B.: Deep extremely randomized trees. In: International Joint Conference on Neural Networks (2019, to submitted)
2. Blake, C., Merz, C.: UCI repository of machine learning databases. Department of information and computer science. University of California, Irvine (1998). http://www.ics.uci.edu/~mlearn/mlrepository.html
3. Breiman, L.: Random forests. Mach. Learn. **45**(1), 5–32 (2001)

4. Ayres-de Campos, D., Bernardes, J., Garrido, A., Marques-de Sa, J., Pereira-Leite, L.: Sisporto 2.0: a program for automated analysis of cardiotocograms. J. Maternal-Fetal Med. **9**(5), 311–318 (2000)
5. Dua, D., Taniskidou, E.K.: UCI machine learning repository. University of California, School of information and computer science, Irvine (2017). http://archive.ics.uci.edu/ml
6. Geurts, P., Ernst, D., Wehenkel, L.: Extremely randomized trees. Mach. Learn. **63**(1), 3–42 (2006)
7. Geusebroek, J.M., Burghouts, G.J., Smeulders, A.W.: The Amsterdam library of object images. Int. J. Comput. Vis. **61**(1), 103–112 (2005)
8. Goodfellow, I., Bengio, Y., Courville, A., Bengio, Y.: Deep Learning, vol. 1. MIT Press, Cambridge (2016)
9. Johnson, B.A., Tateishi, R., Hoan, N.T.: A hybrid pansharpening approach and multiscale object-based image analysis for mapping diseased pine and oak trees. Int. J. Remote Sens. **34**(20), 6969–6982 (2013)
10. Keith, M., et al.: The high time resolution universe pulsar survey-i. System configuration and initial discoveries. Mon. Not. Roy. Astron. Soc. **409**(2), 619–627 (2010)
11. Koch, G., Zemel, R., Salakhutdinov, R.: Siamese neural networks for one-shot image recognition. In: ICML Deep Learning Workshop (2015)
12. Kontschieder, P., Fiterau, M., Criminisi, A., Rota Bulo, S.: Deep neural decision forests. In: Proceedings of the IEEE International Conference on Computer Vision, pp. 1467–1475 (2015)
13. Krizhevsky, A., Sutskever, I., Hinton, G.E.: Imagenet classification with deep convolutional neural networks. In: Advances in Neural Information Processing Systems, p. 2012 (2012)
14. Lyon, R.J., Stappers, B., Cooper, S., Brooke, J., Knowles, J.: Fifty years of pulsar candidate selection: from simple filters to a new principled real-time classification approach. Mon. Not. Roy. Astron. Soc. **459**(1), 1104–1123 (2016)
15. Lyon, R.J.: Why are pulsars hard to find? Ph.D. thesis, The University of Manchester, Manchester (2016)
16. Maas, A.L., Daly, R.E., Pham, P.T., Huang, D., Ng, A.Y., Potts, C.: Learning word vectors for sentiment analysis. In: Proceedings of the 49th Annual Meeting of the Association for Computational Linguistics: Human Language Technologies, pp. 142–150. Association for Computational Linguistics, Portland, June 2011. http://www.aclweb.org/anthology/P11-1015
17. Malerba, D., Esposito, F., Semeraro, G.: A further comparison of simplification methods for decision-tree induction. In: Fisher, D., Lenz, HJ. (eds.) Learning from Data. LNS, vol. 112, pp. 365–374. Springer, Heidelberg (1996). https://doi.org/10.1007/978-1-4612-2404-4_35
18. Miller, K., Hettinger, C., Humpherys, J., Jarvis, T., Kartchner, D.: Forward thinking: Building deep random forests. arXiv preprint arXiv:1705.07366 (2017)
19. Samaria, F.S., Harter, A.C.: Parameterisation of a stochastic model for human face identification. In: Proceedings of the Second IEEE Workshop on Applications of Computer Vision, pp. 138–142. IEEE (1994)
20. Schmidhuber, J.: Deep learning in neural networks: an overview. Neural Netw. **61**, 85–117 (2015)
21. Simonyan, K., Zisserman, A.: Very deep convolutional networks for large-scale image recognition. CoRR abs/1409.1556 (2014). http://arxiv.org/abs/1409.1556
22. Utkin, L.V., Ryabinin, M.A.: A siamese deep forest. arXiv preprint arXiv:1704.08715 (2017)

23. Wang, S., Aggarwal, C., Liu, H.: Using a random forest to inspire a neural network and improving on it. In: Proceedings of the 2017 SIAM International Conference on Data Mining, pp. 1–9. SIAM (2017)
24. Zhou, Z.H., Feng, J.: Deep forest: Towards an alternative to deep neural networks. arXiv preprint arXiv:1702.08835 (2017)
25. Zimek, A., Gaudet, M., Campello, R.J., Sander, J.: Subsampling for efficient and effective unsupervised outlier detection ensembles. In: Proceedings of the 19th ACM SIGKDD International Conference on Knowledge Discovery and Data Mining, pp. 428–436. ACM (2013)

Algorithms for an Efficient Tensor Biclustering

Dina Faneva Andriantsiory[1,2]([✉]) [iD], Mustapha Lebbah[2]([✉]) [iD],
Hanane Azzag[2]([✉]) [iD], and Gael Beck[2]([✉]) [iD]

[1] African Institute for Mathematical Sciences (AIMS),
Km2 Route de Joal, Centre IRD, BP 1418, Mbour, Senegal
dina.f.andriantsiory@aims-senegal.org
[2] Computer Science Laboratory of Paris North (LIPN, CNRS UMR 7030),
University of Paris 13, 93430 Villetaneuse, France
{lebbah,azzag,beck}@lipn.univ-paris13.fr

Abstract. Consider a data set collected by *(individuals-features)* pairs
in different times. It can be represented as a tensor of three dimensions *(Individuals, features and times)*. The tensor biclustering problem
computes a subset of individuals and a subset of features whose signal trajectories over time lie in a low-dimensional subspace, modeling
similarity among the signal trajectories while allowing different scalings
across different individuals or different features. This approach are based
on spectral decomposition in order to build the desired biclusters. We
evaluate the quality of the results from each algorithms with both synthetic and real data set.

Keywords: Multilinear algebra · Tensor decomposition ·
Principal component analysis

1 Introduction

Clustering analysis has become a fundamental tool in statistics and machine
learning. Many clustering algorithms have been developed with the general idea
of seeking groups among different individuals in all space of features. Biclustering consists of simultaneous partitioning of a set of observations and a set
of their features into subsets often called bicluster. Consequently, a subset of
rows exhibiting significant coherence within a subset of columns in the matrix
can be extracted, which corresponds to a specific coherent pattern [2,8]. Nowadays, there is a new type of data collection, in which we may collect data by
individual-feature pair at multiple times. The variation of a couple *(individual-feature)* at different instants is called trajectory. This data can be represented
as a three dimensional object called tensor $\mathcal{T} \in \mathbb{R}^{n_1 \times n_2 \times m}$, where n_1 and n_2 are
respectively the size of observations and features at m different times. Tensor
biclustering selects a subset of individual indices and a subset of features indices
whose trajectories are highly correlated. Grouping those trajectories according

© Springer Nature Switzerland AG 2019
L. H. U and H. W. Lauw (Eds.): PAKDD 2019 Workshops, LNAI 11607, pp. 130–138, 2019.
https://doi.org/10.1007/978-3-030-26142-9_12

to the correlation or similarity behaviour between them is useful in different area such as decision making, but it is still a very challenging topic in research.

In [7], the authors proposed different methods based on the spectral decomposition of matrix and the length of trajectory, although they provide a unique bicluster. The goal of this article is to provide two algorithms for extracting r biclusters in the tensor datasets ($r \geq 2$). Many tools on tensor manipulation already exist in literature to solve this tensor biclustering problem [1,3–6]. Our algorithms are based on a spectral decomposition as proposed in [7]. This article is structured as follows. In Sect. 2, we start by a brief summary of problem formulation. Section 3 introduces our algorithm extensions. Section 4 is related to the experiments. We make some concluding remarks in Sect. 5.

2   Problem Formulation

We use the common notation where $\mathcal{T}$, $\mathcal{X}$ and $\mathcal{Z}$ are used respectively to denote input, signal and noise tensors. For any set J, $|\bar{J}|$ denotes its cardinality. $[n]$ denotes the set $\{1, 2, \ldots, n\}$. $|\bar{J}| = [n] - J$. $\|x\|_2 = (x^t x)^{1/2}$ is the second norm of the vector x. $x \otimes y$ is the Kronecker product of two vectors x and y. We also use Matlab notation to denote the elements in tensor. Specifically, $\mathcal{T}(:, :, i)$, $\mathcal{T}(:, i, :)$ and $\mathcal{T}(i, :, :)$ are respectively the $i - th$ frontal, lateral and horizontal slice. $\mathcal{T}(:, i, j)$, $\mathcal{T}(i, :, j)$ and $\mathcal{T}(i, j, :)$ denote respectively the mode $- 1$, mode $- 2$ and mode $- 3$ fiber. Let $\mathcal{T} \in \mathbb{R}^{n_1 \times n_2 \times m}$ a third-order tensor, $\mathcal{T} = \mathcal{X} + \mathcal{Z}$ where $\mathcal{X}$ is the signal tensor and $\mathcal{Z}$ is the noise tensor. Consider

$$\mathcal{T} = \mathcal{X} + \mathcal{Z} = \sum_{r=1}^{q} \sigma_r (u_r^{J_1^{(r)}} \otimes w_r^{J_2^{(r)}} \otimes v_r) + \mathcal{Z}, \tag{1}$$

where $J_1^{(i)}$ and $J_2^{(i)}$ are respectively the sets of observations indices and features indices in the $i - th$ bicluster and $u_r \in \mathbb{R}^{n_1}$, $w_r \in \mathbb{R}^{n_2}$ and $v_r \in \mathbb{R}^m$ are unit vectors. We assume that $u_i^{J_1^{(i)}}$ and $w_i^{J_2^{(i)}}$ have zero entries outside of $J_1^{(i)}$ and $J_2^{(i)}$ respectively for $i \in \{1, 2 \cdots, q\}$ and $\sigma_1 \geq \sigma_2 \geq \cdots \geq \sigma_q > 0$. We define $J_1 = \bigcup_i J_1^{(i)}$ and $J_2 = \bigcup_i J_2^{(i)}$. Under this model, trajectories $\mathcal{X}(J_1, J_2, :)$ form at most q dimensional subspace.

Concerning the noise model, if $(j_1, j_2) \notin J_1 \times J_2$, we assume that entries of the noise trajectory $\mathcal{Z}(j_1, j_2, :)$ are independent and identically distributed (i.i.d) and each entry has a standard normal distribution. If $(j_1, j_2) \in J_1 \times J_2$, we assume that entries of $\mathcal{Z}(j_1, j_2, :)$ are i.i.d and each entry has a Gaussian distribution with zero means and σ_z^2 variance. We analyse tensor biclustering problem under two variances models of the noise trajectory:

– **Noise Model I:** in this model, we assume $\sigma_z^2 = 1$, i.e., the variance of the noise within and outside of the clustering is assumed to be the same. Although this model simplifies the analysis, it has the following drawback: under this noise model, for every value of σ_1, the average trajectory lengths in the bicluster is larger than the average trajectory lengths outside the bicluster.

Indeed, let $T_1 \in \mathbb{R}^{m \times k^2}$ be a matrix whose columns include trajectories $\mathcal{T}(j_1, j_2, :\ )$ for $(j_1, j_2) \in J_1 \times J_2$ (i.e T_1 is the unfolded $\mathcal{T}(j_1, j_2, :)$). We can write $T_1 = X_1 + Z_1$ where X_1 and Z_1 are unfolded $\mathcal{X}(j_1, j_2, :)$ and $\mathcal{Z}(j_1, j_2, :)$, respectively. The squared Frobinius norm of X_1 is equal to $\|X_1\|_F^2 = \sigma_1^2$. Morever, the squared Frobenius norm of Z_1 has a Chi-squared distribution with mk^2 degrees of freedom i.e $\chi^2(mk^2)$. Thus, the average squared Frobenius norm of T_1 is equal to $\sigma_1^2 + \sigma_z^2 mk^2$. Let $T_2 \in \mathbb{R}^{m \times k^2}$ be a matrix whose columns include only noise trajectories. Using a similar argument, we have $\mathbb{E}[\|T_2\|_F^2] = mk^2$, which is smaller than $\sigma_1^2 + \sigma_z^2 mk^2$.

- **Noise Model II:** in this model, we assume $\sigma_z^2 = \max\left(0, 1-(\sigma_1^2/mk^2)\right)$, i.e., σ_z^2 is modeled to minimize the difference between average trajectory lengths within and outside the bicluster.

  Indeed, if $\sigma_1^2 < mk^2$, without noise, the average trajectory in the bicluster is smaller than the one outside the bicluster. In this regime, having $\sigma_z^2 = 1 - (\sigma_1^2/mk^2)$ makes the average trajectory lengths within and outside the bicluster comparable. This regime is called the low-SNR (signal noise ratio) regime. If $\sigma_1^2 > mk^2$, the average trajectory lengths in the bicluster is larger than the one outside the bicluster. This regime is called high-SNR regime. In this regime, adding noise to signal trajectories increases their lengths and makes solving the tensor biclustering problem easier. Therefore, in this regime we assume $\sigma_z^2 = 0$ to minimize the difference between average trajectory lengths within and outside of the bicluster.

2.1   Tensor Folding and Spectral (FS)

The algorithm and the asymptotic behaviour of Tensor FS method are available in [7]. Under the assumption $q = 1$ and $n = |n_1| = |n_2|$, we drop the subscript (1) from $J_1^{(1)}$ and $J_2^{(1)}$. We assume also that $|J_1| = |J_2| = k$. The author propose to provide only one bicluster. This method separates the selection of the two sets J_1 and J_2 using lateral slice and horizontal slice of the tensor respectively.

$$T_{(j_1, 1)} = \mathcal{T}(j_1, :, :) \quad \text{and} \quad T_{(j_2, 2)} = \mathcal{T}(:, j_2, :) \tag{2}$$

$$C_1 = \sum_{j_2=1}^{n} T_{(j_2,2)}^t T_{(j_2,2)} \quad \text{and} \quad C_2 = \sum_{j_1=1}^{n} T_{(j_1,1)}^t T_{(j_1,1)} \tag{3}$$

The aim is to select the row and column indices whose trajectories are highly correlated. The elements of J_1 and J_2 are the indices of the top k elements of the top eigenvector of the matrix C_1 and C_2 respectively (Algorithm 1). We denoted by $\hat{J}_1$ and $\hat{J}_2$ the subset of individuals and features respectively in the bicluster given from the algorithm.

Tensor FS method have the best performance in both noise models compared to the three another methods (tensor unfolding+spectral, thresholding sum of squared and individual trajectory lengths) proposed by Soheil Feizi, Hamid Javadi, David Tse [7].

Algorithm 1. Tensor folding and spectral

Input: tensor $\mathcal{T}$, and the cardinality of output k
Output: The set of indices $\hat{J}_1$ and $\hat{J}_2$

1 Input: $\mathcal{T}$, k
2 Initialize: C_1, C_2, T_1 and T_2
3 **for** i *in* $[n]$ **do**
4   Compute T_1 according to equation (2)
5   Update C_1 according to equation (3)
6   Compute T_2 according to equation (2)
7   Update C_2 according to equation (3)

8 Compute $\hat{u}_1$, the top eigenvector of C_1
9 Compute $\hat{w}_1$, the top eigenvector of C_2
10 Compute $\hat{J}_1$, set of indices of the k largest values of $|\hat{u}_1|$
11 Compute $\hat{J}_2$, set of indices of the k largest values of $|\hat{w}_1|$
12 **return** $\hat{J}_1$ *and* $\hat{J}_2$

---

3   Extension of Tensor Folding and Spectral

In this section, we aim to extract many biclusters in the tensor data and improve the quality of the result. We propose several methods in order to do this task. However instead of seeking only one bicluster, we assume that in Eq. (1) $q = r \geq 2$ where r is defined by the number of gap in the eigenvalues of the covariance matrix C_1 and C_2 (Eq. (3)).

3.1   Recursive Extension

The classical method extracts one rank of the low dimensional subspace which is not very interesting because it neglect the majority of the data sets. So, Direct improvement of this method is to compute recursively according to the number of gap shown in the eigenvalues (Algorithm 2). In this method, there is no intersection in two different blocks of tensor biclustering.

Algorithm 2. Recursive Extension

Input: tensor $\mathcal{T}$, array cardinality of each bicluster k
Output: The set of all couple set of bicluster

1 $i \longleftarrow 1$
2 **while** $i < k.length + 1$ **do**
3   compute the first bicluster $(\hat{J}_1^{(i)}, \hat{J}_2^{(i)})$ (by using the algorithm 1)
4   keep $(\hat{J}_1^{(i)}, \hat{J}_2^{(i)})$ on the dataset and change the entries to zero. We use it as a new dataset
5     $i \longleftarrow i + 1$

6 **return** $(\hat{J}_1^{(1)}, \hat{J}_2^{(1)})$, $(\hat{J}_1^{(2)}, \hat{J}_2^{(2)}) \cdots$

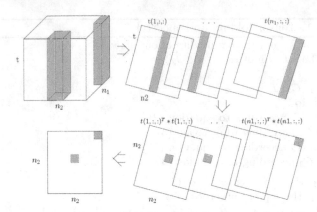

Fig. 1. A visualization of the tensor FS extension algorithm to compute the bicluster index $(J_2^{(i)})_i$. Here we have two biclusters and the sets $J_2^{(1)}$ and $J_2^{(2)}$ do not intersect.

3.2 Multiple Biclusters

This method extract simultaneously the r biclusters in our tensor by using the idea of top r principal component analysis (PCA). The orthogonality of the principal components favor the quality of the result (Algorithm 3).

The illustration step of tensor FS method is showed in the Fig. 1. For each fix individual, we have a horizontal slice of the tensor represented by $m \times n_2$ matrix (Eq. (2)). Then, we compute the covariance matrix for each horizontal slice and their sum give us only one squared matrix of order n_2 (C_2 in Eq. (3)). We apply singular value decomposition (SVD) in C_2, the top r eigenvectors in the matrix C_2 ensure the selection of the elements of the features index set $(J_2^{(i)})_{i \in [r]}$ (Algorithm 3). A similar step is applied to each lateral slice of the tensor to find all the element of the index set $(J_1^{(i)})_{i \in [r]}$.

Since k is a fix parameter, multiple biclusters method allow some trajectory belong to many blocks of tensor biclustering. We call them a boundary of biclus-

---

Algorithm 3. Multiple Biclusters

Input: tensor $\mathcal{T}$, and the list of cardinality of the tensor biclustering k
Output: The set of all couple set of bicluster
1 $r \leftarrow$ length of k
2 Compute the matrices C_1 and C_2 according to equation (3)
3 Compute the top r eigenvectors of C_1 and C_2
4 **for** $i \leftarrow 1$ **to** r **do**
5 $\quad$ Compute $\hat{J_1}^{(i)}$ from eigenvector $|u_i|$
6 $\quad$ Compute $\hat{J_2}^{(i)}$ from eigenvector $|w_i|$
7 Compute $I_1 \longleftarrow \bigcap_i \hat{J_1}^{(i)}$ and $I_2 \longleftarrow \bigcap_i \hat{J_2}^{(i)}$
8 **return** $((\hat{J_1}^{(i)}, \hat{J_2}^{(i)}))_{i \in [r]}$ and (I_1, I_2)

---

ter. Those boundaries are very important as they belong to the intersection of all the biclusters. Thus they have all their properties.

4   Experimentation

4.1   Synthetic Data

We generate synthetic data to evaluate the performance of our methods. In this dataset, we create two biclusters with signal strength σ_1 and σ_2 such that $\sigma_1 > \sigma_2$. We assume that v_1 and v_2 are fixed unit vectors in $\mathbb{R}^m$ and $v_1 = v_2$. We assume also that $J_1^{(1)} \cap J_1^{(2)} = \emptyset$ and $J_2^{(1)} \cap J_2^{(2)} = \emptyset$. We have $n - n_1 - n_2 - 150$, $m = 40$ and $k = |J_1^{(1)}| = |J_1^{(2)}| = |J_2^{(1)}| = |J_2^{(2)}| - 30$. According to Sect. 2, we assume:

$$u_1(j_1) = \begin{cases} 1/\sqrt{k} & \text{for } j_1 \in J_1^{(1)} \\ 0 & \text{if not} \end{cases}, \quad w_1(j_2) = \begin{cases} 1/\sqrt{k} & \text{for } j_2 \in J_2^{(1)} \\ 0 & \text{if not} \end{cases},$$

$$u_2(j_1) = \begin{cases} 1/\sqrt{k} & \text{for } j_1 \in J_1^{(2)} \\ 0 & \text{if not} \end{cases}, \quad w_2(j_2) = \begin{cases} 1/\sqrt{k} & \text{for } j_2 \in J_2^{(2)} \\ 0 & \text{if not} \end{cases}$$

We apply the assumption above to generate the input tensor $\mathcal{T}$ with the noise model II define in Sect. 2. Let $\hat{J}_1^{(1)} \times \hat{J}_2^{(1)}$ and $\hat{J}_1^{(2)} \times \hat{J}_2^{(2)}$ be the two estimated biclusters indices of $J_1^{(1)} \times J_2^{(1)}$ and $J_1^{(2)} \times J_2^{(2)}$ respectively where $|J_1^{(1)}| = |J_2^{(1)}| = |J_1^{(2)}| = |J_2^{(2)}| = k$. We fix the signal strength $\sigma_2 - 2\sigma_1/3$, if the value of $\sigma_1 > 90$, the bar plot of the top five eigenvalues of both covariance matrices tell us that there is two block of tensor biclustering in the data (see Fig. 2).

In this case, we know the value of parameter $k = 30$. To evaluate the inference quality of the result given from the algorithm, we compute the recovery rate:

$$0 \le \frac{|\hat{J}_1^{(1)} \cap J_1^{(1)}|}{4k} + \frac{|\hat{J}_1^{(2)} \cap J_1^{(2)}|}{4k} + \frac{|\hat{J}_2^{(1)} \cap J_2^{(1)}|}{4k} + \frac{|\hat{J}_2^{(2)} \cap J_2^{(2)}|}{4k} \le 1.$$

Recovery rate return zero if the algorithm do not find any of the element of the two biclusters and return one if the algorithm find all the elements of the two biclusters.

We did the experiment with different value of signal strength (σ_1) and for each value of σ_1 we repeat 10 times. Then we compute the average of the recovery rate (Fig. 2(c)).

4.2   Real Data

We apply the both contribution algorithms to an electricity load diagrams data[c] set during four years (2011–2014). This data set contains electricity consumption

---

[c] http://archive.ics.uci.edu/ml/datasets/ElectricityLoadDiagrams20112014.

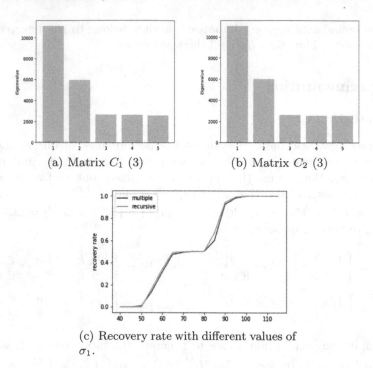

(a) Matrix C_1 (3)   (b) Matrix C_2 (3)

(c) Recovery rate with different values of σ_1.

Fig. 2. Synthetic data sets, $n = 150$, $m = 40$

of 370 clients for each 15 min during four years. After the data prepossessing, we have a tensor $\mathcal{T} \in \mathbb{R}^{n_1 \times n_2 \times m}$ where $n_1 = 365$ is the number of day in one year, $n_2 = 161$ is the number of clients and $m = 4$ is the number of years.

As illustrated in Fig. (3(a), (d)), the gap on the eigenvalues shows the existence of two tensor biclustering (Sect. 3) in the data set. The parameter k cardinality of each index sets are defined from the multiple bicluster method, we choose k with few intersection of two blocks of bicluster $|J_1^{(1)}| = |J_1^{(2)}| = 50$ and $|J_2^{(1)}| = |J_2^{(2)}| = 25$. After compilation, we note that the two individuals sets $J_1^{(1)}$ and $J_1^{(2)}$ are disjoint in both methods. Besides, the two features sets have 22 intersection elements for multiple biclusters method and 19 intersection elements for the recursive method. So, we have two distinct blocks with two distinct subsets of individuals and one subset of feature.

To evaluate the quality of the bicluster given for each algorithm, we compute the total absolute pairwise correlations of the trajectories among each bicluster. With the recursive method, the trajectories in first bicluster is highly correlated but the quality of the second bicluster is a little bit low as seen in Fig. (3(b), (c)). Besides, with multiple bicluster method, the trajectories on both biclusters are highly correlated as seen in Fig. (3(e), (f)).

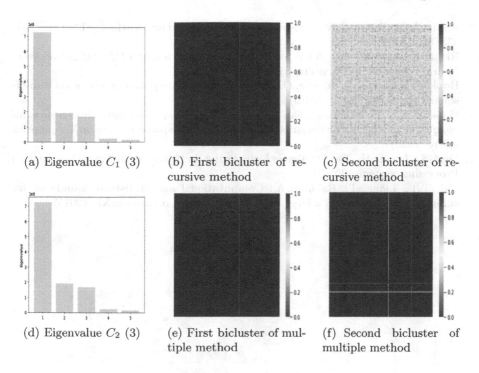

(a) Eigenvalue C_1 (3)

(b) First bicluster of recursive method

(c) Second bicluster of recursive method

(d) Eigenvalue C_2 (3)

(e) First bicluster of multiple method

(f) Second bicluster of multiple method

Fig. 3. Trajectories correlations of each bicluster for each method

5   Conclusion

In this article, we introduced two methods to increase the number of bicluster selected in the tensor data set based on [7], which depends on the number of rank of the low dimensional subspace. The goal is to extract r subsets of tensor ($r \geq 2$) rows and columns such that each block of the trajectories form a low dimensional subspace. We proposed two algorithms to solve this problem, tensor recursive and multiple bicluster. The performance of both algorithms depends on the parameter k, one way to choose this parameter is in the multiple bicluster method. If the parameter chosen gives a lot of index intersections, decreasing the value of k is a good idea to improve the quality of the results.

References

1. Montanari, A., Reichman, D., Zeitouni, O.: On the limitation of spectral methods: from the Gaussian hidden clique problem to rank-one perturbations of Gaussian tensors. In: Advances in Neural Information Processing Systems (2015)
2. Chen, Y., Xu, J.: Statistical-computational tradeoffs in planted problems and submatrix localization with a growing number of clusters and submatrices. arXiv preprint arXiv 1402 (2014)

3. Kolda, T.G., Bader, B.W.: Tensor decompositions and applications. SIAM Rev. **51**, 455–500 (2009)
4. Richard, E., Montanari, A.: A statistical model for tensor PCA. In: Advances in Neural Information Processing Systems (2014)
5. Hopkins, S.B., Shi, J., Steurer, D.: Tensor principal component analysis via sum-of-square proofs. In: COLT (2015)
6. Hopkins, S.B., Schramm, T., Shi, J., Steurer, D.: Fast spectral algorithms from sum-of-squares proofs: tensor decomposition and planted sparse vectors. arXiv preprint arXiv (2015)
7. Feizi, S., Javadi, H., Tse, D.: Tensor biclustering. In: Advances in Neural Information Processing Systems, 30 (2017)
8. Cai, T.T., Liang, T., Rakhlin, A.: Computational and statistical boundaries for submatrix localization in a large noisy matrix. arXiv preprint arXiv (2015)

Change Point Detection in Periodic Panel Data Using a Mixture-Model-Based Approach

Allou Samé[✉] and Milad Leyli-Abadi

IFSTTAR, COSYS, GRETTIA, Université Paris-Est,
77447 Marne la Vallée, France
{allou.same,leyli-abadi.milad}@ifsttar.fr

Abstract. This paper describes a novel method for common change detection in panel data emanating from smart electricity and water networks. The proposed method relies on a representation of the data by classes whose probabilities of occurrence evolve over time. This dynamics is assumed to be piecewise periodic due to the cyclic nature of the studied data, which allows the detection of change points. Our strategy is based on a hierarchical mixture of t-distributions which entails some robustness properties. The parameter estimation is performed using an incremental strategy, which has the advantage to allow the processing of large datasets. The experiments carried out on realistic data showed the full relevance of the proposed method.

Keywords: Change detection · Segmentation-clustering · Robust mixture model · Incremental learning · Smart grid

1 Introduction

Nowadays, with the advent of smart cities, energy and water are monitored in a very fine way, conjointly at different locations of the urban environment. The analysis of the vast amount of data collected, conducted with respect for the privacy, aims to achieve a better balance between supply and demand. It must also lead to savings in resources. The data generally collected in this context are consumption measurements from smart meters, that can be assimilated to panel data. Indeed, these are relative to the dynamics of n entities (individual houses or collective buildings) over time. Exploratory and predictive analyses of these rich data require the implementation of powerful algorithms, based on flexible statistical models. Latent variable models are typical examples.

In this article, our focus is on the (offline) change detection in such data, a task which can also be regarded as the segmentation of the data along the time axis. Several researches have been conducted in this context. A model called random break model has been proposed in [11,12] for change detection in panel data. It assumes that each individual series has its own break point and it is

© Springer Nature Switzerland AG 2019
L. H. U and H. W. Lauw (Eds.): PAKDD 2019 Workshops, LNAI 11607, pp. 139–150, 2019.
https://doi.org/10.1007/978-3-030-26142-9_13

shown that the distribution of the break points can be consistently estimated. Under this random break model, the likelihood function is similar to that of mixture distributions and the maximum likelihood estimates are obtained via the Expectation Maximization (EM) algorithm. Within the framework of panel data models with non-stationary errors, a change detection approach has been proposed in [3,10]. In [2], the authors develop a statistical approach for estimating a common break in the mean and variance of panel data. A quasi-maximum likelihood (QML) method is adopted for this purpose. In [4,9], an approach based on the adaptation of the cumulative sum (CUSUM) method to panel data has been proposed to detect a change point in the mean of panel data. Alternatively, in [14], the authors propose a CUSUM-based statistic to test for a common variance change point in panel data. The problems of single and multiple change point detection in panel data have also been considered in [5]. A double CUSUM statistic is proposed, which uses a cumulative sums of ordered CUSUMs at each point, and a bootstrap procedure is used for test criterion selection.

The approach adopted in this paper consists in representing panel data, which are dynamic in nature, by a set of clusters whose probabilities of occurrence evolve over time. In the perspective of data segmentation while taking into account their cyclic character, the prior probability of classes is assumed to be piecewise periodic, its break times being the desired change points. This approach leads us to propose a dynamic hierarchical mixture of t-distributions which entails also some robustness properties. To estimate the model parameters in a massive data context, an incremental variant of the algorithm, derived from the method of Neal and Hinton [17], is also developed.

The article is organized as follows. Section 1 introduces the proposed hierarchical mixture model. Section 2 shows how this model can be estimated incrementally, and Sect. 3 describes the application of this method to realistic data.

The data considered in this article represent energy, water or gas consumption recorded from n smart meters linked to individual houses or collective buildings in an urban area. Within this framework, it is usually assumed that, for each meter, the consumption is measured at hourly intervals during several days. Since we are interested here in the dynamics of a panel of consumers on a daily time-scale, the data will be denoted as $(\mathbf{x}_1, \ldots, \mathbf{x}_T)$, where $\mathbf{x}_t = (\boldsymbol{x}_{t1}, \ldots, \boldsymbol{x}_{tn})$ is the cross-sectional sample extracted on day $t \leq T$ from the n meters, with $\boldsymbol{x}_{ti} \in \mathbb{R}^d$. Each individual $\boldsymbol{x}_{ti}$ consists of $d = 24$ values (hourly consumption).

2   Robust Mixture Modeling for Change Detection and Clustering

The model proposed for change detection in panel data is a hierarchically structured model involving a finite mixture of multivariate t-distributions whose mixing weights themselves are modeled as a temporal mixture model which is piecewise periodic. This strategy results in a two-way clustering of panel data: a clustering of individual observations $\boldsymbol{x}_{ti}$, and a segmentation of the time instants, which constitutes the detection of change points. The following sections define this model and describe the estimation strategy used.

2.1 Proposed Model

Under the proposed model, each observation x_{ti} is assumed to have originated from the hierarchical mixture distribution

$$p(x_{ti}; \theta) = \sum_{k=1}^{K} \left(\sum_{\ell=1}^{L} \pi_\ell(t; \alpha) \, \pi_{\ell k}(u_t; \beta_\ell) \right) f(x_{ti}; \mu_k, \Sigma_k, \nu_k), \quad (1)$$

where θ is the global parameter vector of the mixture, K and L are the numbers of mixture components at the two levels of the hierarchy, and f is the multivariate t-distribution defined by

$$f(x; \mu, \Sigma, \nu) = \frac{\Gamma(\frac{\nu+d}{2})}{(\pi\nu)^{\frac{1}{2}} |\Sigma|^{1/2} \Gamma(\frac{\nu}{2})} \left(1 + \frac{(x - \mu)' \Sigma^{-1} (x - \mu)}{\nu} \right)^{-\frac{1}{2}(\nu+d)}, \quad (2)$$

with mean $\mu \in \mathbb{R}^d$, scale matrix $\Sigma \in \mathbb{R}^{d \times d}$ and degrees of freedom ν, Γ being the Gamma function. Here, we chose to work with the multivariate t-distribution rather than the basic Gaussian distribution for its more generic character (we obtain the Gaussian density when ν is large) and for its robustness to outliers [16].

For our mixture model, the proportions $p_k(t) = \sum_{\ell=1}^{L} \pi_\ell(t; \alpha) \pi_{\ell k}(u_t; \beta_\ell)$ allow the time axis to be segmented into contiguous parts, while remaining periodic within each segment. Indeed, we define the mixing weights π_ℓ by

$$\pi_\ell(t; \alpha) = \frac{\exp(\beta_{\ell 1} t + \beta_{\ell 0})}{\sum_{h=1}^{L} \exp(\beta_{h1} t + \beta_{h0})}, \quad (3)$$

where $\alpha = (\beta_{\ell 0}, \beta_{\ell 1})_{\ell=1,\dots,L} \in \mathbb{R}^{2L}$ is the parameter vector to be estimated. Using linear functions of time inside the logistic transformation leads to the desired segmentation [18]. For instance, for $L = 2$ segments, the single change point is given by $t^* = -\beta_{\ell 0}/\beta_{\ell 1}$ [18].

For the proportions $\pi_{\ell k}$ (bottom of the hierarchy) to vary periodically in the course of time while taking into account special days (for example public holidays), we define them as

$$\pi_{\ell k}(u_t; \beta_\ell) = \frac{\exp\left(\beta_{\ell k}' u_t\right)}{\sum_{h=1}^{K} \exp\left(\beta_{\ell h}' u_t\right)}, \quad (4)$$

where $\beta_\ell = (\beta_{\ell 1}', \dots, \beta_{\ell K}')'$ is the parameter vector to be estimated, with $\beta_{\ell k} = (\alpha_{\ell k 0}, \dots, \alpha_{\ell, k, 2q+1})' \in \mathbb{R}^{2q+2}$, and $u_t = (1, C_1(t), S_1(t), \cdots, C_q(t), S_q(t), \mathbb{1}_{sd}(t))'$ is the corresponding covariate, with

$$C_j(t) = \cos\left(\frac{2\pi j t}{m}\right) \quad S_j(t) = \sin\left(\frac{2\pi j t}{m}\right) \quad \mathbb{1}_{sd}(t) = \begin{cases} 1 \text{ if } t \text{ is a special day,} \\ 0 \text{ otherwise.} \end{cases}$$

Concerning the analyzed data, we have $m = 7$ (weekly seasonality), and $q(\leq \lceil \frac{m}{2} \rceil)$ is the required number of trigonometric terms, where $\lceil a \rceil$ is the integer part of a. It should be noticed that if there is no special day in the examined period, then we take $\boldsymbol{u}_t = (1, C_1(t), S_1(t), \cdots, C_q(t), S_q(t))'$, the corresponding parameter being given by $\boldsymbol{\beta}_{\ell k} = (\alpha_{\ell k 0}, \ldots, \alpha_{\ell, k, 2q})'$. For identifiability purposes, the parameters (β_{L0}, β_{L1}) and $\boldsymbol{\beta}_{\ell K}$ are set to be the null vector.

2.2 Generative Formulation

The proposed model can be interpreted from a generative point of view as follows:

(i) generate the cluster labels $z_{ti} \in \{1, \ldots, K\}$ of observations $\boldsymbol{x}_{ti}$ as follows:
   (a) randomly draw segment labels $w_{ti} \in \{1, \ldots, L\}$ according to the multi-nomial distribution $\mathcal{M}(1; \pi_1(t; \boldsymbol{\alpha}), \ldots, \pi_L(t; \boldsymbol{\alpha}))$;
   (b) given $w_{ti} = \ell$, randomly draw cluster labels z_{ti} according to the multino-mial distribution $\mathcal{M}(1; \pi_{\ell 1}(\boldsymbol{u}_t; \boldsymbol{\beta}_\ell), \ldots, \pi_{\ell K}(\boldsymbol{u}_t; \boldsymbol{\beta}_\ell))$;
(ii) given $z_{ti} = k$, generate $\boldsymbol{x}_{ti}$ from the t-distribution with parameters $\boldsymbol{\mu}_k$, $\boldsymbol{\Sigma}_k$ and ν_k as follows:
   (a) generate y_{ti} according to the Gamma distribution $\Gamma(\nu_k/2, \nu_k/2)$;
   (b) given y_{ti}, generate $\boldsymbol{x}_{ti}$ according to the distribution $\mathcal{N}(\boldsymbol{\mu}_k, \boldsymbol{\Sigma}_k/y_{ti})$.

Defined in this way, the model involves the discrete latent variables $\mathbf{w} = (\mathbf{w}_1, \ldots, \mathbf{w}_T)$ and $\mathbf{z} = (\mathbf{z}_1, \ldots, \mathbf{z}_T)$, and the continuous latent variables $\mathbf{y} = (\mathbf{y}_1, \ldots, \mathbf{y}_T)$, with $\mathbf{w}_t = (w_{ti})_{i=1,\ldots,n}$, $\mathbf{z}_t = (z_{ti})_{i=1,\ldots,n}$ and $\mathbf{y}_t = (y_{ti})_{i=1,\ldots,n}$. Figure 1 shows the graphical probabilistic representation of the proposed model. Figure 2 displays examples of mixture weights π_ℓ and $\pi_{\ell k}$, and labels z_{ti} generated from these ones, with $(L = 2, K = 3)$.

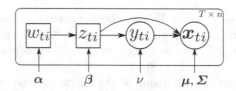

Fig. 1. Graphical probabilistic representation associated to the proposed model

3   Incremental Learning Algorithm

The parameter vector $\boldsymbol{\theta}$ is estimated from the temporal data $(\mathbf{x}_1, \ldots, \mathbf{x}_T)$ and their associated covariates $(\boldsymbol{u}_1, \ldots, \boldsymbol{u}_T)$, by maximizing the log-likelihood

$$\mathcal{L}(\boldsymbol{\theta}) = \sum_{t,i} \log \left(\sum_{k,\ell} \pi_\ell(t; \boldsymbol{\alpha}) \pi_{\ell k}(\boldsymbol{u}_t; \boldsymbol{\beta}_\ell) \, f(\boldsymbol{x}_{ti}; \boldsymbol{\mu}_k, \boldsymbol{\Sigma}_k, \nu_k) \right) \tag{5}$$

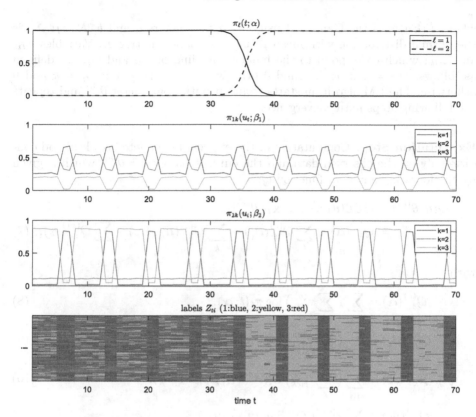

Fig. 2. Examples of mixture proportions $\pi_\ell(t; \alpha)$ and $\pi_{\ell k}(u_t; \beta_\ell)$ for $L = 2$, $K = 3$ and $q = 3$, and labels z_{ti} generated from these probabilities

via the Expectation-Maximization (EM) algorithm [7,15]. For our model, the specificity of this algorithm lies in the estimation of the two categories of mixing weights, in addition to the use of multivariate t-distributions. In the perspective of massive data processing, we propose in this article an incremental version of the EM algorithm. Before introducing this latter version, we start by developing the non-incremental version of the algorithm.

3.1 Parameter Estimation via the EM Algorithm

From the generative formulation of the proposed model, it can easily be shown that the complete log-likelihood is given by

$$
\mathcal{CL}(\theta) = \sum_{t,i,k} w_{ti\ell}\, z_{tik} \log \Big(\pi_\ell(t; \alpha)\, \pi_{\ell k}(u_t; \beta_\ell)
$$

$$
\times \Gamma(y_{ti}; \frac{\nu_k}{2}, \frac{\nu_k}{2})\, \mathcal{N}(x_{ti}; \mu_k, \frac{\Sigma_k}{y_{ti}})\Big), \qquad (6)
$$

where $\Gamma(\cdot; a, b)$ is the Gamma density with parameters a and b, $\mathcal{N}(\cdot; \boldsymbol{\mu}, \boldsymbol{\Sigma})$ is the normal distribution with mean $\boldsymbol{\mu}$ and covariance matrix $\boldsymbol{\Sigma}$. Variables $w_{ti\ell}$ and z_{tik}, which correspond to the binary encoding of w_{ti} and z_{ti}, are defined as follows: $w_{t\ell} = 1$ if $w_t = \ell$ and 0 otherwise, and $z_{tik} = 1$ if $z_{ti} = k$ and 0 otherwise. The EM algorithm starts from an initial parameter $\boldsymbol{\theta}^{(0)}$ and repeats the following steps until convergence:

Expectation Step. Computation of the expected complete log-likelihood conditionally on the observed data and the current parameter $\boldsymbol{\theta}^{(c)}$, which is given by the following auxiliary function Q:

$$Q(\boldsymbol{\theta}, \boldsymbol{\theta}^{(c)}) = \mathbb{E}[\mathcal{CL}(\boldsymbol{\theta}) | \mathbf{x}_1, \ldots, \mathbf{x}_T; \boldsymbol{\theta}^{(c)}]$$
$$= Q_1^{(c)}(\boldsymbol{\alpha}) + \sum_\ell Q_{2\ell}^{(c)}(\boldsymbol{\beta}_\ell) + \sum_k Q_{3k}^{(c)}(\boldsymbol{\mu}_k, \boldsymbol{\Sigma}_k) + \sum_k Q_{4k}^{(c)}(\nu_k), \quad (7)$$

with

$$Q_1^{(c)}(\boldsymbol{\alpha}) = \sum_{t,\ell} \Big(\sum_{i,k} \tau_{ti\ell k}^{(c)}\Big) \log \pi_\ell(t; \boldsymbol{\alpha}), \quad (8)$$

$$Q_{2\ell}^{(c)}(\boldsymbol{\beta}_\ell) = \sum_{t,k} \Big(\sum_i \tau_{ti\ell k}^{(c)}\Big) \log \pi_{\ell k}(\boldsymbol{u}_t; \boldsymbol{\beta}_\ell), \quad (9)$$

$$Q_{3k}^{(c)}(\boldsymbol{\mu}_k, \boldsymbol{\Sigma}_k) = \sum_{t,i} \Big(\sum_\ell \tau_{ti\ell k}^{(c)}\Big) \log \mathcal{N}(\boldsymbol{x}_{ti}; \boldsymbol{\mu}_k, \boldsymbol{\Sigma}_k/\nu_k), \quad (10)$$

$$Q_{4k}^{(c)}(\nu_k) = \sum_{t,i,\ell} \tau_{ti\ell k}^{(c)} \, \Gamma(\lambda_{ti k}^{(c)}; \nu_k/2, \nu_k/2)$$
$$+ \Big(\sum_{t,i,\ell} \tau_{ti\ell k}^{(c)}\Big) \Big(\frac{\nu_k}{2} - 1\Big) \left(\psi\Big(\frac{\nu_k^{(c)} + d}{2}\Big) + \log\Big(\frac{\nu_k^{(c)} + d}{2}\Big)\right), \quad (11)$$

where

$$\tau_{ti\ell k}^{(c)} = \mathbb{E}[w_{ti\ell} \, z_{tik} | \boldsymbol{x}_{ti}; \boldsymbol{\theta}^{(c)}]$$
$$= \frac{\pi_\ell(t; \boldsymbol{\alpha}^{(c)}) \, \pi_{\ell k}(\boldsymbol{u}_t; \boldsymbol{\beta}_\ell^{(c)}) \, f(\boldsymbol{x}_{ti}; \boldsymbol{\mu}_k^{(c)}, \boldsymbol{\Sigma}_k^{(c)}, \nu_k^{(c)})}{\sum_{\ell,k} \pi_\ell(t; \boldsymbol{\alpha}^{(c)}) \, \pi_{\ell k}(\boldsymbol{u}_t; \boldsymbol{\beta}_\ell^{(c)}) \, f(\boldsymbol{x}_{ti}; \boldsymbol{\mu}_k^{(c)}, \boldsymbol{\Sigma}_k^{(c)}, \nu_k^{(c)})} \quad (12)$$

is the posterior probability that $\boldsymbol{x}_{ti}$ originates from the kth cluster and the ℓth segment given the current parameter $\boldsymbol{\theta}^{(c)}$, and

$$\lambda_{tik}^{(c)} = \mathbb{E}[y_{ti} | z_{ti} = k, \boldsymbol{x}_{ti}; \boldsymbol{\theta}^{(c)}]$$
$$= \frac{\nu_k^{(c)} + d}{\nu_k^{(c)} + (\boldsymbol{x}_{ti} - \boldsymbol{\mu}_k^{(c)})' \boldsymbol{\Sigma}_k^{(c)-1} (\boldsymbol{x}_{ti} - \boldsymbol{\mu}_k^{(c)})} \quad (13)$$

is the current conditional expectation of y_{ti}. In Eq. (11), ψ makes reference to the Digamma function, which is defined by $\psi(a) = \partial \log \Gamma(a)/\partial a = \Gamma'(a)/\Gamma(a)$.

Maximization Step. Computation of the parameter $\boldsymbol{\theta}^{(c+1)}$ maximizing the auxiliary function Q w.r.t. $\boldsymbol{\theta}$. The maximizations of $Q_1^{(c)}$ and $Q_{2\ell}^{(c)}$ are logistic regression problems weighted by the probabilities $\sum_{i,k} \tau_{ti\ell k}^{(c)}$ and $\sum_i \tau_{ti\ell k}^{(c)}$. These problems cannot be solved in a closed form. We exploit the well-known Iteratively Reweighted Least Squares (IRLS) algorithm [8,13] to this end.

The maximization of $Q_{3k}^{(c)}$ with respect to $(\boldsymbol{\mu}_k, \boldsymbol{\Sigma}_k)$ is performed in a similar way as for the standard mixture of multivariate Gaussian distributions [15]. We get:

$$\boldsymbol{\mu}_k^{(c+1)} = \frac{\sum_{t,i} (\sum_\ell \tau_{ti\ell k}^{(c)}) \lambda_{tik}^{(c)} \boldsymbol{x}_{ti}}{\sum_{t,i} \sum_\ell (\tau_{ti\ell k}^{(c)}) \lambda_{tik}^{(c)}}, \tag{14}$$

$$\boldsymbol{\Sigma}_k^{(c+1)} = \frac{\sum_{t,i} (\sum_\ell \tau_{ti\ell k}^{(c)}) \lambda_{tik}^{(c)} (\boldsymbol{x}_{ti} - \boldsymbol{\mu}_k^{(c+1)}) (\boldsymbol{x}_{ti} - \boldsymbol{\mu}_k^{(c+1)})'}{\sum_{t,i} (\sum_\ell \tau_{ti\ell k}^{(c)}) \lambda_{tik}^{(c)}}. \tag{15}$$

The maximization of $Q_{4k}^{(c)}$ with respect to ν_k cannot be solved in a closed form, but it can easily be proven that the estimate $\nu_k^{(c+1)}$ is the solution of the non-linear equation

$$\log\left(\frac{\nu_k}{2}\right) - \psi\left(\frac{\nu_k}{2}\right) = \frac{\sum_{t,i,\ell} \tau_{ti\ell k}^{(c)} \left(\lambda_{tik}^{(c)} - \log(\lambda_{tik}^{(c)}) \right)}{\sum_{t,i,\ell} \tau_{ti\ell k}^{(c)}}. \tag{16}$$

3.2   Incremental Version

For massive panel data ($T \times n$ relatively large), as is generally the case with smart meters data, the EM algorithm can become very slow. Indeed, each of its iterations, and more particularly the E-step, requires to process all the data. A solution would consist in distributing the E-step among several process units [6,19], which is made possible by the fact the posterior probabilities τ_{tik} can be computed independently. In this article, we opted for an adaptation to our periodic mixture of the incremental strategy initiated by Neal and Hinton [17]. In order to derive this algorithm, let us emphasize that the EM algorithm previously described can also be developed using exclusively the following sufficient statistics:

$$S_{1\ell k}^{(c)} = \sum_t s_{1t\ell k}^{(c)}, \quad \text{where} \quad s_{1t\ell k}^{(c)} = \sum_i \tau_{ti\ell k}^{(c)}, \tag{17}$$

$$S_{2\ell k}^{(c)} = \sum_t s_{2t\ell k}^{(c)}, \quad \text{where} \quad s_{2t\ell k}^{(c)} = \sum_i \tau_{ti\ell k}^{(c)} \lambda_{tik}^{(c)}, \tag{18}$$

$$S_{3\ell k}^{(c)} = \sum_t s_{2t\ell k}^{(c)}, \quad \text{where} \quad s_{3t\ell k}^{(c)} = \sum_i \tau_{ti\ell k}^{(c)} \log(\lambda_{tik}^{(c)}), \tag{19}$$

$$S_{4\ell k}^{(c)} = \sum_t s_{4t\ell k}^{(c)}, \quad \text{where} \quad s_{4t\ell k}^{(c)} = \sum_i \tau_{ti\ell k}^{(c)} \lambda_{tik}^{(c)} \boldsymbol{x}_{ti}, \tag{20}$$

$$S_{5\ell k}^{(c)} = \sum_t s_{5t\ell k}^{(c)}, \quad \text{where} \quad s_{5t\ell k}^{(c)} = \sum_i \tau_{ti\ell k}^{(c)} \lambda_{tik}^{(c)} \boldsymbol{x}_{ti} \boldsymbol{x}_{ti}'. \tag{21}$$

computed from $\boldsymbol{\theta}^{(c)}$. A mean to accelerate the algorithm without visiting at each E-step the complete data set is to use partial E-steps [17] and to update incrementally both the sufficient statistics and the parameters accordingly. By using the recently computed posterior probabilities to update the parameters generally contributes to accelerate the convergence of the algorithm. In this paper, we opted for a partial E-step, which consists in computing the posterior probabilities $(\tau_{ti\ell k})_{i\ell k}$ only for a day t. Due to their additive nature, the expected sufficient statistics defined by Eqs. (17)–(21) can then be updated incrementally. Algorithm 1 summarizes the main steps of this procedure which is called IEM-SPMIX, where I stands for "incremental", S for "segmental", and P for "periodic". As shown in [17], the resulting incremental algorithm maximizes monotonically the log-likelihood criterion described in Sect. 3.

Algorithm 1. IEM-SPMIX

Input: data $(\mathbf{x}_1, \ldots, \mathbf{x}_T)$ with $\mathbf{x}_t = (\boldsymbol{x}_{ti})_i$, covariates $(\boldsymbol{u}_1, \ldots, \boldsymbol{u}_T)$,
numbers of segments L and local clusters K, initial parameter $\boldsymbol{\theta}^{(0)}$
Set $c = 0$
while *the maximum number of iterations is not reached* **do**

   Choose a time instant t (day)
   E-step
   For the chosen time instant t, compute $\tau_{ti\ell k}^{(c)}$ and $s_{rt\ell k}^{(c)}$ (Eqs. 12, 17–21)

   $\forall t' \neq t$, set $s_{rt'\ell k}^{(c)} = s_{rt'\ell k}^{(c-1)}$

   Update the expected sufficient statistics:

$$S_{r\ell k}^{(c)} = S_{r\ell k}^{(c-1)} - s_{rt\ell k}^{(c-1)} + s_{rt\ell k}^{(c)} \tag{22}$$

   M-step
   From sufficient statistics $S_{1\ell k}^{(c)}$, use the IRLS algorithm to compute

$$\boldsymbol{\alpha}^{(c+1)} = \arg\max_{\boldsymbol{\alpha}} Q_1^{(c)}(\boldsymbol{\alpha}) \quad \text{and} \quad \boldsymbol{\beta}_\ell^{(c+1)} = \arg\max_{\boldsymbol{\beta}_\ell} Q_2^{(c)}(\boldsymbol{\beta}_\ell)$$

   From sufficient statistics $S_{2\ell k}^{(c)}$, $S_{4\ell k}^{(c)}$ and $S_{5\ell k}^{(c)}$, compute

$$\boldsymbol{\mu}_k^{(c+1)} = S_{4\ell k}^{(c)}/S_{2\ell k}^{(c)} \quad \text{and} \quad \boldsymbol{\Sigma}_k^{(c+1)} = S_{5\ell k}^{(c)}/S_{2\ell k}^{(c)} - \boldsymbol{\mu}_k^{(c+1)'}\boldsymbol{\mu}_k^{(c+1)} \tag{23}$$

   From $S_{1\ell k}^{(c)}$, $S_{2\ell k}^{(c)}$ and $S_{3\ell k}^{(c)}$, compute $\nu_k^{(c+1)}$ by solving the equation

$$\log(\nu_k/2) - \psi(\nu_k/2) = \left(\sum_\ell S_{3\ell k}^{(c)} - \sum_\ell S_{2\ell k}^{(c)}\right) \Big/ \sum_\ell S_{1\ell k}^{(c)} \tag{24}$$

   $c \leftarrow c + 1$
end
Output: parameter vector $\widehat{\boldsymbol{\theta}}$

From the parameters $\widehat{\boldsymbol{\theta}}$ estimated by the model, a segmentation of the time instants into contiguous temporal segments E_ℓ $(\ell = 1 \ldots, L)$ can be obtained by setting

$$E_\ell = \left\{ t \in [1; T] \mid \pi_\ell(t; \widehat{\boldsymbol{\alpha}}) = \max_{1 \le h \le L} \pi_h(t; \widehat{\boldsymbol{\alpha}}) \right\}. \tag{25}$$

The estimated change points are therefore the starting time stamp for segments E_ℓ. A partition of the data is deduced by setting

$$\widehat{z}_{ti} = k \iff k = \operatorname{argmax}_{1 \le h \le K} P(z_{ti} = h | \boldsymbol{x}_{ti}; \widehat{\boldsymbol{\theta}}) = \operatorname{argmax}_h \sum_\ell \tau_{ti\ell h}, \tag{26}$$

where $\tau_{ti\ell k}$ is given by Eq. (12).

4 Experiments

In this section, we apply the proposed change detection algorithm to realistic consumption data similar to those collected from real world water networks. This realistic panel data set is obtained using simulations from a specific dynamic model [1]. For this purpose, each observation $\boldsymbol{x}_{ti} \in \mathbb{R}^{24}$ corresponds to the behavior adopted by a user i during a day t (one numeric value per hour). The data set is made of $L = 3$ segments (change at $t = 36$ and $t = 71$) and $K = 8$ clusters. The considered numbers of time instants and individuals are $T = 105$ days (10 weeks) and $n = 100$. The outcome of these data is therefore a tensor of $100 \times 105 \times 24$ numeric values. Figure 3 shows an extract of these data. Each plot represents the set of observations $(\boldsymbol{x}_{ti})_{i=1,\ldots,n}$ of a day t, each observation $\boldsymbol{x}_{ti}$

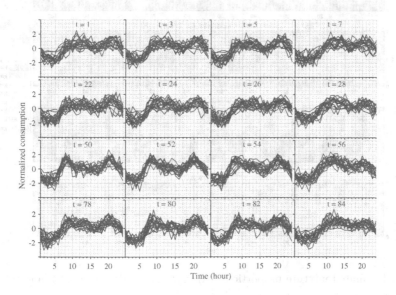

Fig. 3. Extract of a realistic panel data set (20 individuals observed during 16 days)

being represented longitudinally by the series of its 24 numerical values. Since we are mainly concerned in this study with change detection in consumption habits (time series shape) rather than consumption volumes, each series $x_{ti} = (x_{tis})_{s=1,\dots,24}$ has been standardized by setting x_{tis} to $(x_{tis} - \overline{x}_{ti})/\sigma(x_{ti})$, where $\overline{x}_{ti}$ and $\sigma(x_{ti})$ are respectively the mean and standard deviation of x_{ti}.

The algorithm IEM-SPMIX has been launched on these data with $L = 3$ and $K = 8$. Figure 4 shows the estimated mixture proportions. From Fig. 4(a), which gives the proportions $\pi_\ell(t; \widehat{\alpha})$, we get the correct change points $t = 36$ and $t = 71$. In Fig. 4(b–d), the periodic proportions $\pi_{\ell k}(u_t; \widehat{\beta}_\ell)$ are displayed. They reflect the local dynamics of panel data within segments. To have a synthetic view of the global dynamic behaviour of the data, Fig. 4(e) displays, as a function of time, the probability $p_k(t) = P(z_{ti} = k; \widehat{\theta}) = \sum_\ell \pi_\ell(t; \widehat{\alpha}) \pi_{\ell k}(u_t; \widehat{\beta}_\ell)$. This specific plot allows the differences between segments to be clearly distinguished. It can be observed, in particular, that the 4th component has a majority occurrence on the working days of segment 1, while the 2nd component has a majority occurrence on the working days of segment 2.

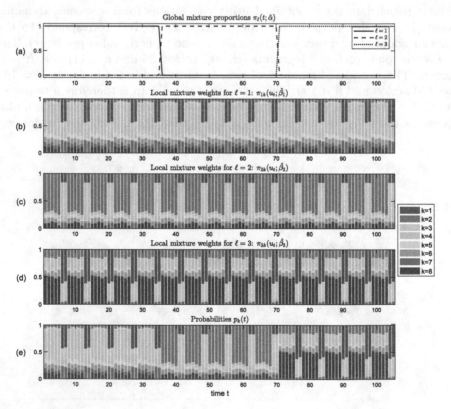

Fig. 4. Estimated mixture proportions: (a) $\pi_\ell(t; \widehat{\alpha})$, (b–d) $\pi_{\ell k}(u_t; \widehat{\beta}_\ell)$, and (e) $p_k(t)$

Figure 5 shows the 8 estimated means $\boldsymbol{\mu}_k (k = 1 \ldots, 8)$ of the t-distributions together with a few observations considered to be noise. An observation $\boldsymbol{x}_{ti}$ is considered as noise if

$$\sum_{\ell,k} \tau_{ti\ell k}(\boldsymbol{x}_{ti} - \widehat{\boldsymbol{\mu}}_k)' \widehat{\boldsymbol{\Sigma}}_k^{-1}(\boldsymbol{x}_{ti} - \widehat{\boldsymbol{\mu}}_k) > \chi^2_{d;0.95}, \tag{27}$$

where $\chi^2_{d;0.95}$ is the 95th percentile of the Chi-square distribution with d degrees of freedom [16]. These means constitute typical consumption behaviour patterns which can help analyzing the data.

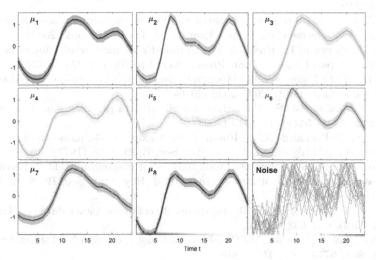

Fig. 5. Estimated means of the t-distributions and a few noise observations obtained from Eq. (27)

5 Conclusion

This article has presented a method for change point detection in periodic panel data such as those collected from smart meters in the electricity and water sectors. The proposed approach is based on a hierarchical mixture of t-distributions whose proportions evolves over time. At the highest level of the hierarchy, the mixture weights, which act as change detectors are logistic regression models involving linear functions of time. At the bottom of the hierarchy, the weights of the mixture are modeled as periodic logistic functions. The parameter estimation is performed incrementally using the expected sufficient statistics associated with the model. This strategy has the advantage of maintaining the monotonic convergence of the log-likelihood. Experiments conducted on realistic simulated data revealed good performances of the proposed method. The perspectives of this work remain the comparison of the proposed method with other strategies, both in terms of change detection and clustering. First of all, we think of a two-step strategy involving clustering into K clusters and then segmentation into L

segments. The relevance of model selection criteria such as BIC, ICL and AIC also deserves to be studied.

References

1. Abadi, M.L., et al.: Predictive classification of water consumption time series using non-homogeneous Markov models. In: IEEE International Conference on Data Science and Advanced Analytics (DSAA), pp. 323–331 (2017)
2. Bai, J.: Common breaks in means and variances for panel data. J. Econ. **157**(1), 78–92 (2010)
3. Bai, J., Carrion-I-Silvestre, J.L.: Structural changes, common stochastic trends, and unit roots in panel data. Rev. Econ. Stud. **76**(2), 471–501 (2009)
4. Chan, J., Horváth, L., Hušková, M.: Darling-Erdős limit results for change-point detection in panel data. J. Stat. Plann. Infer. **143**(5), 955–970 (2013)
5. Cho, H., et al.: Change-point detection in panel data via double CUSUM statistic. Electron. J. Stat. **10**(2), 2000–2038 (2016)
6. Dean, J., Ghemawat, S.: Mapreduce: simplified data processing on large clusters. Commun. ACM **51**(1), 107–113 (2008)
7. Dempster, A.P., Laird, N.M., Rubin, D.B.: Maximum likelihood from incomplete data via the EM algorithm. J. Roy. Stat. Soc. B **39**, 1–38 (1977)
8. Green, P.: Iteratively reweighted least squares for maximum likelihood estimation, and some robust and resistant alternatives. J. Roy. Stat. Soc. B **46**(2), 149–192 (1984)
9. Horváth, L., Hušková, M.: Change-point detection in panel data. J. Time Ser. Anal. **33**(4), 631–648 (2012)
10. Im, K.S., Lee, J., Tieslau, M.: Panel LM unit-root tests with level shifts. Oxf. Bull. Econ. Stat. **67**(3), 393–419 (2005)
11. Joseph, L., Wolfson, D.B.: Estimation in multi-path change-point problems. Commun. Stat. Theory Methods **21**(4), 897–913 (1992)
12. Joseph, L., Wolfson, D.B.: Maximum likelihood estimation in the multi-path change-point problem. Ann. Inst. Stat. Math. **45**(3), 511–530 (1993)
13. Krishnapuram, B., Carin, L., Figueiredo, M.A.T., Hartemink, A.J.: Sparse multinomial logistic regression: fast algorithms and generalization bounds. IEEE Trans. Pattern Anal. Mach. Intell. **27**(6), 957–968 (2005)
14. Li, F., Tian, Z., Xiao, Y., Chen, Z.: Variance change-point detection in panel data models. Econ. Lett. **126**, 140–143 (2015)
15. McLachlan, G.J., Krishnan, T.: The EM Algorithm and Extensions. Wiley, New York (2008)
16. McLachlan, G.J., Ng, S.K., Bean, R.: Robust cluster analysis via mixture models. Austrian J. Stat. **35**(2&3), 157–174 (2016)
17. Neal, R., Hinton, G.: A view of the EM algorithm that justifies incremental, sparse, and other variants. In: Jordan, M.I. (ed.) Learning in Graphical Models, vol. 89. Kluwer, Dordrecht (1998)
18. Samé, A., Chamroukhi, F., Govaert, G., Aknin, P.: Model-based clustering and segmentation of time series with changes in regime. Adv. Data Anal. Classif. **5**(4), 301–321 (2011)
19. Wolfe, J., Haghighi, A., Klein, D.: Fully distributed EM for very large datasets. In: Proceedings of the 25th International Conference on Machine Learning, ICML 2008, pp. 1184–1191. ACM (2008)

The 8th Workshop on Biologically-Inspired Techniques for Knowledge Discovery and Data Mining (BDM 2019)

Neural Network-Based Deep Encoding for Mixed-Attribute Data Classification

Tinglin Huang[1], Yulin He[1,2](✉), Dexin Dai[1], Wenting Wang[1], and Joshua Zhexue Huang[1,2]

[1] College of Computer Science and Software Engineering, Shenzhen University, Shenzhen 518060, China
huangtinglin@email.szu.edu.cn, {yulinhe,wangwt,zx.huang}@szu.edu.cn, 864193052@qq.com
[2] National Engineering Laboratory for Big Data System Computing Technology, Shenzhen University, Shenzhen 518060, China

Abstract. This paper proposes a neural network-based deep encoding (DE) method for the mixed-attribute data classification. DE method first uses the existing one-hot encoding (OE) method to encode the discrete-attribute data. Second, DE method trains an improved neural network to classify the OE-attribute data corresponding to the discrete-attribute data. The loss function of improved neural network not only includes the training error but also considers the uncertainty of hidden-layer output matrix (i.e., DE-attribute data), where the uncertainty is calculated with the re-substitution entropy. Third, the classification task is conducted based on the combination of previous continuous-attribute data and transformed DE-attribute data. Finally, we compare DE method with OE method by training support vector machine (SVM) and deep neural network (DNN) on 4 KEEL mixed-attribute data sets. The experimental results demonstrate the feasibility and effectiveness of DE method and show that DE method can help SVM and DNN obtain the better classification accuracies than the traditional OE method.

Keywords: Mixed-attribute data · Discrete-attribute data · Continuous-attribute data · One-hot encoding · Uncertainty

1 Introduction

The mixed-attribute data classification is a research hotspot in the field of machine learning. The mixed-attribute data are composed of continuous-attribute (or numeric-attribute) data and discrete-attribute (or categorical-attribute) data. Nowadays, the mixed-attribute data become more and more common with the rapid and wide applications of complex and advanced information technologies, e.g., the credit approval data [19], medical examination data [3] and electric consumption data [8]. It is of positive theoretical and practical significance to train an efficient classification model based on the available mixed-attribute data.

© Springer Nature Switzerland AG 2019
L. H. U and H. W. Lauw (Eds.): PAKDD 2019 Workshops, LNAI 11607, pp. 153–163, 2019.
https://doi.org/10.1007/978-3-030-26142-9_14

Currently, there are three main strategies to deal with the classification problems of mixed-attribute data, i.e., constructing the hybrid model, discretizing the continuous-attribute data and encoding the discrete-attribute data. The well-known hybrid models for classifying the mixed-attribute data are hybrid decision tree [21] and extreme learning machine tree [16]. The hybrid model is a tree structure-based classifier in which the neural networks in leaf nodes are used to handle the continuous-attribute data and the decision trees that determine the breach nodes are used to deal with the discrete-attribute data. The main limitation of hybrid model is the high complexity due to the utilization of multiple neural networks. There are many famous discretization methods that can be used to preprocess the continuous-attribute data, e.g., class-dependent discretization [4], Bayes optimal discretization [2] and dynamic discretization [6]. It is unavoidable to loss the data information (e.g., the order relation) in the process of continuous-attribute discretization. The mostly-used method of encoding the discrete-attribute data is the one-hot encoding (OE) [11], which uses a multidimensional 0–1 vector to represent a discrete-attribute. In fact, OE method can not transform the discrete-attribute data into the continuous-attribute data, but uses a numeric trick to represent the discrete-attribute.

This paper proposes a deep encoding (DE) method to tackle the mixed-attribute data classification problem. DE method first uses OE method to encode the discrete-attribute data. Second, DE method trains an improved neural network to classify the OE-attribute data corresponding to the discrete-attribute data. The loss function of improved neural network not only includes the training error but also considers the uncertainty of hidden-layer output matrix (i.e., DE-attribute data), where the uncertainty is calculated with the re-substitution entropy [5]. Third, the classification task is conducted based on the combination of previous continuous-attribute data and transformed DE-attribute data. Finally, we compare DE method with OE method by training support vector machine (SVM) and deep neural network (DNN) on 4 KEEL [14] mixed-attribute data sets. The experimental results demonstrate the feasibility and effectiveness of DE method and show that DE method can help SVM and DNN obtain the better classification accuracies than the traditional OE method.

The remainder of this paper is organized as follows. In Sect. 2, we provide a brief introduction to OE method. In Sect. 3, we present the proposed DE method. In Sect. 4, we report experimental comparisons that demonstrate the feasibility and effectiveness of DE method. Finally, we give our conclusions and future works in Sect. 5.

2   One-Hot Encoding (OE) Method

Assume there is a mixed-attribute data set $\mathbb{D}$ as shown in Table 1, where $\mathcal{M}_1$ and $\mathcal{M}_2$ are the numbers of continuous-attributes and discrete-attributes, respectively; $\mathcal{N}$ is the number of instances; $A_{m_1}^{(C)}$ is the m_1-th continuous-attribute, $x_{n,m_1}^{(C)} \in \Re$, $m_1 = 1, 2, \cdots, \mathcal{M}_1$; $A_{m_2}^{(D)}$ is the m_2-th discrete-attribute of which the attribute values are $\left\{ a_{m_2,1}^{(D)}, a_{m_2,2}^{(D)}, \cdots, a_{m_2,\mathcal{K}_{m_2}}^{(D)} \right\}$, $\mathcal{K}_{m_2}$ is the number of attribute values, $x_{n,m_2}^{(D)} \in \left\{ a_{m_2,1}^{(D)}, a_{m_2,2}^{(D)}, \cdots, a_{m_2,\mathcal{K}_{m_2}}^{(D)} \right\}$, $m_2 = 1, 2, \cdots, \mathcal{M}_2$; y_n is the label of the n-th instance $\bar{x}_n$, $y_n \in \{c_1, c_2, \cdots, c_{\mathcal{L}}\}$, $n = 1, 2, \cdots, \mathcal{N}$; c_l is the l-th class label, $l = 1, 2, \cdots, \mathcal{L}$, $\mathcal{L}$ is the number of labels.

Table 1. The mixed-attribute data set $\mathbb{D}$

$\mathbb{D}$	Continuous-attributes				Discrete-attributes				Label
	$A_1^{(C)}$	$A_2^{(C)}$	$\cdots$	$A_{\mathcal{M}_1}^{(C)}$	$A_1^{(D)}$	$A_2^{(D)}$	$\cdots$	$A_{\mathcal{M}_2}^{(D)}$	$\bar{y}$
$\bar{x}_1$	$x_{11}^{(C)}$	$x_{12}^{(C)}$	$\cdots$	$x_{1\mathcal{M}_1}^{(C)}$	$x_{11}^{(D)}$	$x_{12}^{(D)}$	$\cdots$	$x_{1\mathcal{M}_2}^{(D)}$	y_1
$\bar{x}_2$	$x_{21}^{(C)}$	$x_{22}^{(C)}$	$\cdots$	$x_{2\mathcal{M}_1}^{(C)}$	$x_{21}^{(D)}$	$x_{22}^{(D)}$	$\cdots$	$x_{2\mathcal{M}_2}^{(D)}$	y_2
$\vdots$	$\vdots$	$\vdots$	$\ddots$	$\vdots$	$\vdots$	$\vdots$	$\ddots$	$\vdots$	$\vdots$
$\bar{x}_{\mathcal{N}}$	$x_{\mathcal{N}1}^{(C)}$	$x_{\mathcal{N}2}^{(C)}$	$\cdots$	$x_{\mathcal{N},\mathcal{M}_1}^{(C)}$	$x_{\mathcal{N}1}^{(D)}$	$x_{\mathcal{N}2}^{(D)}$	$\cdots$	$x_{\mathcal{N},\mathcal{M}_2}^{(D)}$	$y_{\mathcal{N}}$

For the discrete-attribute $A_{m_2}^{(D)}$, $m_2 = 1, 2, \cdots, \mathcal{M}_2$ as shown in Table 1, its OE-attribute is represented as

$$\left\{ A_{m_2,1}^{(OE)}, A_{m_2,2}^{(OE)}, \cdots, A_{m_2,\mathcal{K}_{m_2}}^{(OE)} \right\}. \tag{1}$$

The corresponding OE-attribute vector of $x_{n,m_2}^{(D)}$, $n = 1, 2, \cdots, \mathcal{N}$ is

$$\left\{ x_{n,m_2,1}^{(OE)}, x_{n,m_2,2}^{(OE)}, \cdots, x_{n,m_2,\mathcal{K}_{m_2}}^{(OE)} \right\}, \tag{2}$$

where

$$x_{n,m_2,k}^{(OE)} = \begin{cases} 1, \text{ if } x_{n,m_2}^{(D)} = a_{m_2,k}^{(D)} \\ 0, \text{ otherwise} \end{cases}, k = 1, 2, \cdots, \mathcal{K}_{\mathcal{M}_2}. \tag{3}$$

In fact, we find that OE method does not transform the discrete-attribute into continuous-attribute and only extends one discrete-attribute into multiple discrete-attributes, where each discrete-attribute takes the value 0 or 1. For example, the discrete-attribute $A^{(D)}$ having two attribute values a and b is extended into two discrete-attributes $A_1^{(OE)}$ and $A_2^{(OE)}$ which also have two attribute values 0 and 1. In the following example, $\{1, 0, 0, 1\}$ or $\{1, 0, 0, 1\}$ is equivalent to $\{a, b, b, a\}$. This indicates that OE method does not fundamentally transform the discrete-attribute $A^{(D)}$ into the continuous-attribute.

$A^{(D)}$	$\rightarrow$	$A_1^{(OE)}$	$A_2^{(OE)}$
a		1	0
b		0	1
b		0	1
a		1	0

3   Deep Encoding (DE) Method

DE method uses an improved neural network, i.e., encoding neural network (ENN), to classify the OE-attribute data set $\mathbb{D}^{(OE)}$ as shown in the following Table 2.

Table 2. The OE-attribute data set $\mathbb{D}^{(OE)}$

$\mathbb{D}^{(OE)}$	OE-attribute vector of $A_1^{(D)}$			$\cdots$	OE-attribute vector of $A_{\mathcal{M}_2}^{(D)}$			Label
	$A_{11}^{(OE)}$	$A_{12}^{(OE)}$	$\cdots$ $A_{1\mathcal{K}_1}^{(OE)}$	$\cdots$	$A_{\mathcal{M}_2,1}^{(OE)}$	$A_{\mathcal{M}_2,2}^{(OE)}$	$\cdots$ $A_{\mathcal{M}_2,\mathcal{K}_{\mathcal{M}_2}}^{(OE)}$	$\bar{y}$
$\bar{x}_1^{(OE)}$	$x_{111}^{(OE)}$	$x_{112}^{(OE)}$	$\cdots$ $x_{11\mathcal{K}_1}^{(OE)}$	$\cdots$	$x_{1\mathcal{M}_2,1}^{(OE)}$	$x_{1\mathcal{M}_2,2}^{(OE)}$	$\cdots$ $x_{1\mathcal{M}_2,\mathcal{K}_{\mathcal{M}_2}}^{(OE)}$	y_1
$\bar{x}_2^{(OE)}$	$x_{211}^{(OE)}$	$x_{212}^{(OE)}$	$\cdots$ $x_{21\mathcal{K}_1}^{(OE)}$	$\cdots$	$x_{2\mathcal{M}_2,1}^{(OE)}$	$x_{2\mathcal{M}_2,2}^{(OE)}$	$\cdots$ $x_{2\mathcal{M}_2,\mathcal{K}_{\mathcal{M}_2}}^{(OE)}$	y_2
$\vdots$	$\vdots$	$\vdots$	$\ddots$ $\vdots$	$\cdots$	$\vdots$	$\vdots$	$\ddots$ $\vdots$	$\vdots$
$\bar{x}_{\mathcal{N}}^{(OE)}$	$x_{\mathcal{N}11}^{(OE)}$	$x_{\mathcal{N}12}^{(OE)}$	$\cdots$ $x_{\mathcal{N}1\mathcal{K}_1}^{(OE)}$	$\cdots$	$x_{\mathcal{N},\mathcal{M}_2,1}^{(OE)}$	$x_{\mathcal{N},\mathcal{M}_2,2}^{(OE)}$	$\cdots$ $x_{\mathcal{N},\mathcal{M}_2,\mathcal{K}_{\mathcal{M}_2}}^{(OE)}$	$y_{\mathcal{N}}$

There are

$$\mathcal{K} = \sum_{m_2=1}^{\mathcal{M}_2} \mathcal{K}_{m_2} \tag{4}$$

nodes in the input-layer of ENN. The input and output matrices of ENN are

$$\mathbb{X}^{(OE)} = \begin{bmatrix} x_{111}^{(OE)} & x_{112}^{(OE)} & \cdots & x_{11\mathcal{K}_1}^{(OE)} & \cdots & x_{1\mathcal{M}_2,1}^{(OE)} & x_{1\mathcal{M}_2,2}^{(OE)} & \cdots & x_{1\mathcal{M}_2,\mathcal{K}_{\mathcal{M}_2}}^{(OE)} \\ x_{211}^{(OE)} & x_{212}^{(OE)} & \cdots & x_{21\mathcal{K}_1}^{(OE)} & \cdots & x_{2\mathcal{M}_2,1}^{(OE)} & x_{2\mathcal{M}_2,2}^{(OE)} & \cdots & x_{2\mathcal{M}_2,\mathcal{K}_{\mathcal{M}_2}}^{(OE)} \\ \vdots & \vdots & \ddots & \vdots & \cdots & \vdots & \vdots & \ddots & \vdots \\ x_{\mathcal{N}11}^{(OE)} & x_{\mathcal{N}12}^{(OE)} & \cdots & x_{\mathcal{N}1\mathcal{K}_1}^{(OE)} & \cdots & x_{\mathcal{N},\mathcal{M}_2,1}^{(OE)} & x_{\mathcal{N},\mathcal{M}_2,2}^{(OE)} & \cdots & x_{\mathcal{N},\mathcal{M}_2,\mathcal{K}_{\mathcal{M}_2}}^{(OE)} \end{bmatrix} \tag{5}$$

and

$$\mathbb{Y} = \begin{bmatrix} y_{11} & y_{12} & \cdots & y_{1\mathcal{L}} \\ y_{21} & y_{22} & \cdots & y_{2\mathcal{L}} \\ \vdots & \vdots & \ddots & \vdots \\ y_{\mathcal{N}1} & y_{\mathcal{N}2} & \cdots & y_{\mathcal{N}\mathcal{L}} \end{bmatrix}, \tag{6}$$

respectively, where

$$y_{nl} = \begin{cases} 1, & \text{if } y_n = c_l \\ 0, & \text{otherwise} \end{cases}, l = 1, 2, \cdots, \mathcal{L}. \tag{7}$$

ENN has $\mathcal{F}$ hidden layers and the output matrix corresponding to the f-th $(f = 1, 2, \cdots, \mathcal{F})$ hidden-layer is

$$\mathbb{H}^{(f)} = \begin{bmatrix} h_{11}^{(f)} & h_{12}^{(f)} & \cdots & h_{1\mathcal{Q}_f}^{(f)} \\ h_{21}^{(f)} & h_{22}^{(f)} & \cdots & h_{2\mathcal{Q}_f}^{(f)} \\ \vdots & \vdots & \ddots & \vdots \\ h_{\mathcal{N}1}^{(f)} & h_{\mathcal{N}2}^{(f)} & \cdots & h_{\mathcal{N}\mathcal{Q}_f}^{(f)} \end{bmatrix}, \tag{8}$$

where $\mathcal{Q}_f$ is the number of nodes in the f-th hidden-layer. $\mathbb{H}^{(\mathcal{F})}$ is the DE-attribute data corresponding to the discrete-attribute data $\mathbb{X}^{(OE)}$. ENN uses an updated loss function

$$\text{Loss} = \text{Error}\left[\mathbb{D}^{(DE)}\right] - \lambda \times \text{Uncertainty}\left[\mathbb{H}^{(\mathcal{F})}\right] \tag{9}$$

to train the network weights, where $\lambda > 0$ is the enhancement factor. Equation (9) not only includes the training error but also considers the uncertainty of hidden-layer output matrix $\mathbb{H}^{(\mathcal{F})}$. Here, we don't discuss the calculation of Error $\left[\mathbb{D}^{(DE)}\right]$ in detail, because it is the most general training scheme for multi-layer feed-forward neural networks [7,12,13]. For the uncertainty, we use the re-substitution entropy [5] to calculate it as

$$\text{Uncertainty}\left[\mathbb{H}^{(\mathcal{F})}\right] = \frac{1}{\mathcal{Q}_{\mathcal{F}}}\sum_{q=1}^{\mathcal{Q}_{\mathcal{F}}} \text{Entropy}\left[H_q^{(DE)}\right], \tag{10}$$

where

$$\text{Entropy}\left[H_q^{(DE)}\right] = -\frac{1}{\mathcal{N}}\sum_{n=1}^{\mathcal{N}} \ln\left[\hat{p}_{-n}\left(h_{nq}^{(\mathcal{F})}\right)\right] \tag{11}$$

is the re-substitution entropy of one-dimensional DE-attribute data set

$$H_q^{(DE)} = \left[h_{1q}^{(\mathcal{F})}, h_{2q}^{(\mathcal{F})}, \cdots, h_{\mathcal{N}q}^{(\mathcal{F})}\right]^{\mathrm{T}}, q = 1, 2, \cdots, \mathcal{Q}_{\mathcal{F}}.$$

The probability density function in logarithm term of Eq. (11) is estimated as

$$\hat{p}_{-n}\left(h_{nq}^{(\mathcal{F})}\right) = \frac{1}{\mathcal{N}-1}\sum_{m=1,m\neq n}^{\mathcal{N}} \frac{1}{\sqrt{2\pi}b_q}\exp\left[-\frac{1}{2}\left(\frac{h_{nq}^{(\mathcal{F})} - h_{mq}^{(\mathcal{F})}}{b_q}\right)^2\right] \tag{12}$$

with Parzen window method [1,10], where $b_q > 0$ is the bandwidth parameter which is a function with respect to $\mathcal{N}$ and satisfies the following conditions [15,17]:

$$\begin{cases} \lim\limits_{\mathcal{N}\to+\infty} b_q = 0 \\ \lim\limits_{\mathcal{N}\to+\infty} \mathcal{N}b_q = +\infty \end{cases}. \tag{13}$$

When the training of ENN is terminated, DE method transform the discrete-attribute data

$$\left[A_1^{(D)}, A_2^{(D)}, \cdots, A_{\mathcal{M}_2}^{(D)}\right]$$

into the DE-attribute data

$$\left[H_1^{(DE)}, H_2^{(DE)}, \cdots, H_{\mathcal{Q}_{\mathcal{F}}}^{(DE)}\right],$$

where $h_{nq}^{(\mathcal{F})} \in \Re$, $n = 1, 2, \cdots, \mathcal{N}$ is a real number rather than a discrete or categorical value. Based on the data set

$$\left[A_1^{(C)}, A_2^{(C)}, \cdots, A_{\mathcal{M}_1}^{(C)}, H_1^{(DE)}, H_2^{(DE)}, \cdots, H_{\mathcal{Q}_\mathcal{F}}^{(DE)}, \bar{y} \right],$$

we train the final classifier, e.g., support vector machine and neural network, to deal with the mixed-attribute data classification.

4   Experimental Results and Analysis

Three experiments are conducted based on 4 KEEL [14] data sets (1000 instances are randomly selected from the original Adult data set) to demonstrate the feasibility and effectiveness of deep-encoding (DE) method: validating the convergence of encoding neural network (ENN), checking the time consumption of DE method and comparing DE method with one-hot encoding (OE) method.

4.1   Convergency of ENN

We use the ENNs with 1 hidden-layer and 4 hidden-layers to validate the variation tendencies of loss, error and uncertainty with the change of iteration numbers. For the sake of simplicity, the designed ENNs in this experiment are denoted as $\text{ENN}_{\frac{\mathcal{K}}{2}}$ and $\text{ENN}_{3\mathcal{K} \to 2\mathcal{K} \to \mathcal{K} \to \frac{\mathcal{K}}{2}}$. $\text{ENN}_{\frac{\mathcal{K}}{2}}$ has 1 hidden-layer with $\frac{\mathcal{K}}{2}$ nodes and

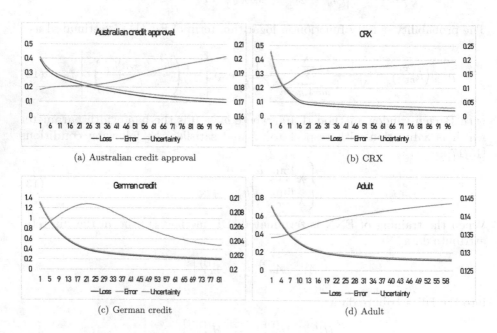

(a) Australian credit approval

(b) CRX

(c) German credit

(d) Adult

Fig. 1. Convergence of $\text{ENN}_{\frac{\mathcal{K}}{2}}$

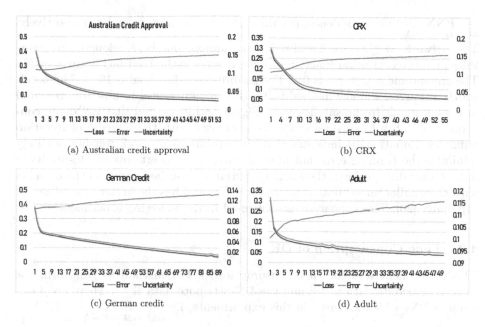

(a) Australian credit approval

(b) CRX

(c) German credit

(d) Adult

Fig. 2. Convergence of $\text{ENN}_{3\mathcal{K}\to2\mathcal{K}\to\mathcal{K}\to\frac{\mathcal{K}}{2}}$

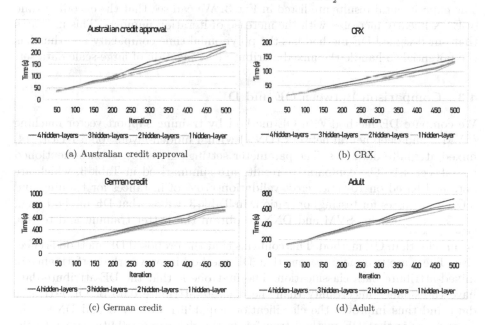

(a) Australian credit approval

(b) CRX

(c) German credit

(d) Adult

Fig. 3. Time consumptions of $\text{ENN}_{3\mathcal{K}\to2\mathcal{K}\to\mathcal{K}\to\frac{\mathcal{K}}{2}}$, $\text{ENN}_{2\mathcal{K}\to\mathcal{K}\to\frac{\mathcal{K}}{2}}$, $\text{ENN}_{\mathcal{K}\to\frac{\mathcal{K}}{2}}$ and $\text{ENN}_{\frac{\mathcal{K}}{2}}$

$\underset{3\mathcal{K}\to2\mathcal{K}\to\mathcal{K}\to\frac{\mathcal{K}}{2}}{\text{ENN}}$ has 4 hidden-layers with $3\mathcal{K}$, $2\mathcal{K}$, $\mathcal{K}$ and $\frac{\mathcal{K}}{2}$ nodes, respectively. The network weights are initialized with random numbers belonging to interval $[0, 1]$. The activation function of hidden-layer is the recitified linear unit [9]. The enhancement factor λ is 0.1. The termination threshold is 10^{-5}. The bandwidth of Parzen window method is 0.5. The experimental results are presented in Figs. 1 and 2. In the sub-figures of Figs. 1 and 2, we find that ENNs are convergent, i.e., the loss values gradually decreases with the increase of iteration numbers. Equation (9) indicates that the optimized network weights not only minimize the training error but also minimize the uncertainty of DE-attribute data. Generally speaking, the learning system established based on the data set with the smaller uncertainty is more stable [18] and has the better generalization capability [20]. This conclusion is confirmed in the following experiment.

4.2   Time Consumption of DE

This experiment test the time consumption of DE method to transform the discrete-attribute data $\mathbb{D}^{(\mathrm{DE})}$ into the DE-attribute data $\mathbb{H}^{(\mathcal{F})}$. There are 4 different ENNs which are used in this experiments, i.e., $\underset{3\mathcal{K}\to2\mathcal{K}\to\mathcal{K}\to\frac{\mathcal{K}}{2}}{\text{ENN}}$, $\underset{2\mathcal{K}\to\mathcal{K}\to\frac{\mathcal{K}}{2}}{\text{ENN}}$, $\underset{\mathcal{K}\to\frac{\mathcal{K}}{2}}{\text{ENN}}$ and $\underset{\frac{\mathcal{K}}{2}}{\text{ENN}}$. The parameter settings are same as the previous experiment. The experimental results are listed in Fig. 3. We can see that the encoding time of ENN linearly increases with the increase of iteration numbers. This indicates that our designed DE method has the polynomial time complexity and thus has the potential to handle the mixed-attribute classification of large-scale data set.

4.3   Comparison Between OE and DE

We compare DE method with OE method by training support vector machine (SVM)[1] and deep neural network (DNN) with 4 hidden-layers[2] on 4 KEEL [14] mixed-attribute data sets. The parameter settings are same as ones mentioned in Subsect. 4.1. The comparative results are summarized in Table 3, which are obtained based on 10-times cross-validation (70% of instances for training and 30% of instances for testing) operation. In Table 3, we see that DE method (e.g., $\underset{3\mathcal{K}\to2\mathcal{K}\to\mathcal{K}\to\frac{\mathcal{K}}{2}}{\text{ENN}}$) helps SVM and DNN to obtain the better training and testing accuracies than OE method. This confirms that the proposed DE-method is effective. There are two main reasons that DE method improves the performance of mixed-attribute data classification. The first one is that the DE-attribute data have the smaller uncertainty than the discrete-attribute data or OE-attribute data and thus improves the classification capabilities of SVM and DNN. The second one is that DE method transforms the discrete-attribute data into the

---

[1] https://www.scipy.org/scipylib/download.html.
[2] https://keras.io/.

Table 3. Comparison between OE and DE based on SVM and DNN classifiers ("Australian credit approval" is simplified as "Australian")

ENN		ENN $\frac{K}{2}$				ENN $K \to \frac{K}{2}$				ENN $2K \to K \to \frac{K}{2}$				ENN $3K \to 2K \to K \to \frac{K}{2}$			
Classifier		SVM		DNN		SVM		DNN		SVM		DNN		SVM		DNN	
Accuracy		Training	Testing	Training	Testing	Training	Testing	Training	Testing	Training	Testing	Training	Testing	Training	Testing	Training	Testing
Australian	OE	0.852	0.870	0.842	0.827	0.863	0.827	0.845	0.844	0.861	0.826	0.849	0.848	0.861	0.850	0.831	0.822
	DE	0.870	0.856	0.810	0.805	0.886	0.805	0.828	0.851	0.899	0.831	0.853	0.847	0.915	0.855	0.900	0.854
CRX	OE	0.862	0.870	0.838	0.826	0.872	0.826	0.843	0.847	0.858	0.826	0.832	0.876	0.856	0.878	0.830	0.835
	DE	0.886	0.866	0.829	0.802	0.890	0.802	0.860	0.849	0.915	0.831	0.879	0.876	0.932	0.865	0.918	0.867
German	OE	0.756	0.699	0.765	0.703	0.758	0.703	0.758	0.720	0.758	0.718	0.747	0.725	0.750	0.719	0.748	0.709
	DE	0.749	0.707	0.739	0.702	0.772	0.702	0.732	0.721	0.867	0.706	0.778	0.757	0.951	0.742	0.922	0.745
Adult	OE	0.840	0.818	0.829	0.802	0.836	0.802	0.820	0.794	0.833	0.784	0.810	0.836	0.827	0.826	0.817	0.802
	DE	0.846	0.818	0.833	0.815	0.853	0.815	0.845	0.803	0.875	0.798	0.834	0.842	0.870	0.852	0.843	0.820

true continuous-attribute data rather than uses the newly-introduced discrete-attributes to represent the previously-existed discrete-attributes. The experimental results in Figs. 1 and 2 confirm that the transformation increases the information amount of data set, i.e., the uncertainty increases with the increase of iteration numbers. The uncertainty of DE-attribute data is represented with re-substitution entropy which is the measurement of information amount.

5   Conclusions and Future Works

In this paper, a new deep encoding (DE) method based on an improved neural network model was proposed to deal with the mixed-attribute data classification. DE method transformed the one-hot encoding (OE)-attribute data corresponding to the discrete-attribute data set into the DE-attribute data set which was the hidden-layer output matrix of improved neural network. When designing the loss function of neural network, the uncertainty of hidden-layer output matrix was considered to ensure the stability and generalization capability of trained learning system. The experimental results demonstrate the feasibility and effectiveness of DE method. DE method will be extended to deal with the big data classification under the distributed computation environment.

Acknowledgments. This paper was supported by National Key R&D Program of China (2017YFC0822604-2), China Postdoctoral Science Foundation (2016T90799), Scientific Research Foundation of Shenzhen University for Newly-introduced Teachers (2018060) and National Training Program of Innovation and Entrepreneurship for Undergraduates (201910590017).

References

1. Babich, G.A., Camps, O.I.: Weighted Parzen windows for pattern classification. IEEE Trans. Pattern Anal. Mach. Intell. **18**(5), 567–570 (1996)
2. Boullé, M.: MODL: a Bayes optimal discretization method for continuous attributes. Mach. Learn. **65**(1), 131–165 (2006)
3. Chen, L., Li, X., Yang, Y., et al.: Personal health indexing based on medical examinations: a data mining approach. Decis. Support Syst. **81**, 54–65 (2016)
4. Ching, J.Y., Wong, A.K.C., Chan, K.C.C.: Class-dependent discretization for inductive learning from continuous and mixed-mode data. IEEE Trans. Pattern Anal. Mach. Intell. **17**(7), 641–651 (1995)
5. He, Y.L., Liu, J.N., Wang, X.Z., et al.: Optimal bandwidth selection for re-substitution entropy estimation. Appl. Math. Comput. **219**(8), 3425–3460 (2012)
6. Hu, H.W., Chen, Y.L., Tang, K.: A dynamic discretization approach for constructing decision trees with a continuous label. IEEE Trans. Knowl. Data Eng. **21**(11), 1505–1514 (2009)
7. Huang, G.B., Zhu, Q.Y., Siew, C.K.: Extreme learning machine: theory and applications. Neurocomputing **70**(1–3), 489–501 (2006)
8. Khan, I., Huang, J.Z., Luo, Z., et al.: CPLP: an algorithm for tracking the changes of power consumption patterns in load profile data over time. Inf. Sci. **429**, 332–348 (2018)

9. Nair, V., Hinton, G.E.: Rectified linear units improve restricted Boltzmann machines. In: Proceedings of the 27th International Conference on Machine Learning, pp. 807–814 (2010)
10. Parzen, E.: On estimation of a probability density function and mode. Ann. Math. Stat. **33**(3), 1065–1076 (1962)
11. Passerini, A., Pontil, M., Frasconi, P.: New results on error correcting output codes of kernel machines. IEEE Trans. Neural Netw. **15**(1), 45–54 (2004)
12. Rowley, H.A., Baluja, S., Kanade, T.: Neural network-based face detection. IEEE Trans. Pattern Anal. Mach. Intell. **20**(1), 23–38 (1998)
13. Svozil, D., Kvasnicka, V., Pospichal, J.: Introduction to multi-layer feed-forward neural networks. Chemometr. Intell. Lab. Syst. **39**(1), 43–62 (1997)
14. Triguero, T., González, S., Moyano, J.M., et al.: KEEL 3.0: an open source software for multi-stage analysis in data mining. Int. J. Comput. Intell. Syst. **10**, 1238–1249 (2017)
15. Wand, M.P., Jones, M.C.: Kernel Smoothing. Chapman and Hall/CRC, Boca Raton (1994)
16. Wang, R., He, Y.L., Chow, C.Y., et al.: Learning ELM-tree from big data based on uncertainty reduction. Fuzzy Sets Syst. **258**, 79–100 (2015)
17. Wang, X.Z., He, Y.L., Wang, D.D.: Non-naive Bayesian classifiers for classification problems with continuous attributes. IEEE Trans. Cybern. **44**(1), 21–39 (2014)
18. Wang, X.Z., He, Y.L.: Learning from uncertainty for big data: future analytical challenges and strategies. IEEE Syst. Man Cybern. Mag. **2**(2), 26–31 (2016)
19. Wang, C.M., Huang, Y.F.: Evolutionary-based feature selection approaches with new criteria for data mining: a case study of credit approval data. Expert Syst. Appl. **36**(3), 5900–5908 (2009)
20. Wang, X.Z., Wang, R., Xu, C.: Discovering the relationship between generalization and uncertainty by incorporating complexity of classification. IEEE Trans. Cybern. **48**(2), 703–715 (2018)
21. Zhou, Z.H., Chen, Z.Q.: Hybrid decision tree. Knowl.-Based Syst. **15**(8), 515–528 (2002)

Protein Complexes Detection Based on Deep Neural Network

Xianchao Zhang[1,2], Peixu Gao[1(✉)], Maohua Sun[3], Linlin Zong[1], and Bo Xu[1,2]

[1] School of Software, Dalian University of Technology, Dalian, China
`peixugao@outlook.com`
[2] Key Laboratory for Ubiquitous Network and Service Software of Liaoning Province, Dalian, China
[3] Dalian Jiaotong University, Dalian, China

Abstract. Protein complexes play an important role for scientists to explore the secrets of cell and life. Most of the existing protein complexes detection methods utilize traditional clustering algorithms on protein-protein interaction (PPI) networks. However, due to the complexity of the network structure, traditional clustering methods cannot capture the network information effectively. Therefore, how to extract information from high-dimensional networks has become a challenge. In this paper, we propose a novel protein complexes detection method called DANE, which uses a deep neural network to maintain the primary information. Furthermore, we use a deep autoencoder framework to implement the embedding process, which preserves the network structure and the additional biological information. Then, we use the clustering method based on the core-attachment principle to get the prediction result. The experiments on six yeast datasets with five other detection methods show that our method gets better performance.

Keywords: Protein complexes detection · Artificial neural network · Deep learning · Network embedding · PPI network

1 Introduction

In the post-genomic era, functional genomics becomes the major task of human life science research. One of the important sub-disciplines is proteomics, which focuses on the study of protein characteristics at a large scale. Protein is of great significance in cellular activities, but this effect is not apparent on a single protein. Existing studies have shown that most proteins accomplish intracellular work by forming complexes with other proteins. Proteins are with close physically interacting in the complexes, these interactions can create, regulate and maintain

This work was supported by National Science Foundation of China (No. 61632019; No. 61876028) and Foundation of Department of Education of Liaoning Province (No. L2015001).

© Springer Nature Switzerland AG 2019
L. H. U and H. W. Lauw (Eds.): PAKDD 2019 Workshops, LNAI 11607, pp. 164–178, 2019.
https://doi.org/10.1007/978-3-030-26142-9_15

specific functions of cells. So, the key to protein complex studies is the interaction between the proteins.

Traditional methods for detecting protein complexes use biological experiments, but suffer from the low efficiency and accuracy. With the widely application of high-throughput techniques in molecular biology, a large amount of physical interactions between the proteins have been found [6,9], which made it possible to build meaningful large PPI networks. Wang et al. [21] proposed a principle that the densely linked regions in the PPI network are more likely to be the real protein complexes. If we analyze the PPI network from a computational perspective, it can be represented as an undirected graph. Vertexes in the graph represent proteins and edges represent the interactions between proteins. Therefore, the problem of detecting protein complexes can be transformed into a computational problem of detecting dense subgraphs from the graph. The large PPI networks facilitate the development of protein complexes detecting by computational methods.

Existing computational methods usually analyze the PPI network in a direct way, which utilize traditional clustering methods in the original network space. However, due to the high non-linearity and sparsity of the network structure, these methods cannot extract the information of the PPI network effectively. The PPI network is in a high-dimensional space, which includes the information of proteins and the relationship between proteins. The relationship are non-linear and difficult to obtain by direct analysis. Therefore, we attempt to map the PPI network into a low-dimensional space for analysis. This is a network embedding process, which maintains the information of the network while reducing the dimension of space. Due to the excellent performance in feature extraction, we utilize a deep neural network to handle the embedding process. Using deep neural networks can map the original data to a non-linear latent space, which better extracts features contained in the data. In the existing network embedding method, SDNE [20] used a deep autoencoder framework to handle the embedding process and performed better than other methods. So we utilize the same architecture based on the deep neural network to embed PPI networks.

Furthermore, we extend the network embedding method by adding the attribute information of the proteins in GO (Gene Ontology). As shown in the traditional computational methods, the use of additional biological information sources can boost the detection performance significantly. GO is currently one of the most comprehensive ontology databases in the bioinformatics community [3]. It provides GO terms to describe three different aspects of gene product features: biological process (Bp), molecular function (Mf), and cellular component (Cc).

In this paper, we propose a protein complexes prediction method called DANE, which combines the deep attributed network embedding process with the clustering process. Firstly, we use a deep neural network to get the vector representation for each protein from GO attributed PPI networks. Secondly, we use a clustering method based on the core-attachment principle to get the final prediction result. We compared with five classic protein complexes detection methods to evaluate the performance of our method, which are COACH [22], CMC [11], MCODE [1], ClusterOne [12] and IPCA [8] on six yeast PPI networks respectively. Experiment results show that our method outperforms the state-of-the-art methods with respect to different evaluate metrics.

In summary, we list the contributions of this paper as follows:

- We make a new attempt that applying a deep network embedding method to the protein complex recognition task, and achieving good results.
- Our embedding method preserves the structural information and vertex attribute information of the network simultaneously. This also makes sense in other network analysis tasks.
- The proposed network representation are able to be integrated with other biological tasks like PPI prediction and disease gene prediction.

The rest of the paper is organized as follows. We reviewed the related work in Sect. 2. In Sect. 3, we introduced the DANE method in detail. In Sect. 4, we compared DANE with five classic protein complex detection methods and showed the experiment results. We summarized this article in Sect. 5.

2   Related Work

2.1   Computational Methods for Protein Complexes Detection

Existing computational methods can be roughly divided into two categories [17]: (1) The methods based on the topological information only. These methods detect dense subgraphs using the original structural information of the graph generated by the PPI network. Most of these methods use traditional graph clustering algorithms to accomplish this task, such as ClusterOne, MCODE, CMC, IPCA and PEWCC [24]. (2) The methods use additional biological information. There are several meaningful biological information like gene expression data and functional data can be used to help the detection process. COACH used the principle that a protein complex conforms to the core-attachment structure to detect complexes. DECAFF [10] generated the predictive complexes by using GO information as the functional information. Ozawa et al. [13] proposed a refinement method to filter predicted complexes based on exclusive and co-operative interactions.

2.2   Network Embedding

Existing methods use a variety of means to achieve the embedding process. Based on the feature vectors, LLE [15] constructed the affinity graph first and then solved the leading eigenvectors as the network representations. DeepWalk [14] used random walk to get the vertex sequence, and adopted the SkipGram model which usually used in the natural language processing task for network embedding. Node2vec [5] introduced two parameters to improve the random walk strategy of DeepWalk, which considered the local and global information simultaneously.

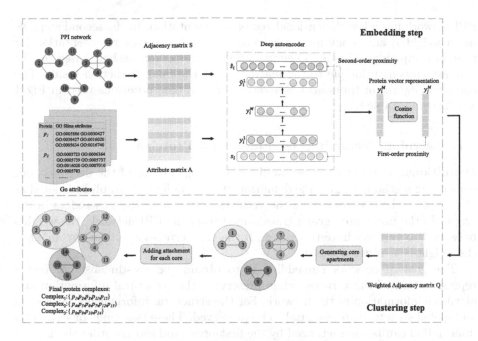

Fig. 1. The basic idea of DANE.

3   Methods

In this section, we explain the details of our algorithm. The DANE algorithm consists of two steps. The first step is to retain the information of the GO attributed PPI network using the deep autoencoder architecture. Then, each protein in the

Table 1. Terms and notations.

Symbol	Definition
N	Number of vertexes
M	Number of layers
W^m	The m-th layer weight matrix of the encoder
$\hat{W}^m$	The m-th layer weight matrix of the decoder
b^m	The m-th layer biases of the encoder
$\hat{b}^m$	The m-th layer biases of the decoder
$S = \{s_1, ..., s_N\}$	The adjacency matrix for the networks
$X = \{x_1, ..., x_N\}$	The input data
$\hat{X} = \{\hat{x}_1, ..., \hat{x}_N\}$	The reconstructed data
$Y^m = \{y_1^m, ..., y_N^m\}$	The m-th layer hidden representations
$\hat{Y}^m = \{\hat{y}_1^m, ..., \hat{y}_N^m\}$	The reconstructed hidden representations

PPI network has a low dimensional vector representation. In the second step, we use a weighted adjacency matrix calculated from the vector representation for protein complex prediction. Then, we apply a clustering method based on the core-attachment principle to the matrix to get the final prediction result. The complete algorithm framework is shown in Fig. 1. The notations used in Fig. 1 are shown in Table 1.

3.1   Learning Vector Representations for Proteins

Definitions. In order to understand the working process of the deep network embedding method, we give the definition of a graph first. A graph is denoted as $G = (V, E)$, where V represents the vertexes and E represents the edges in the graph. In the undirected graph transformed from the PPI network, the weights between two vertexes have two values. If the two vertexes are linked by an edge, the weight is 1, otherwise is 0.

The goal of network embedding is to obtain the low-dimensional vector representation of the vertexes while preserving the structural information and attribute information of the network. For the structural information, both local and global structure are essential to be preserved. These two kinds of structural information can be characterized by the first-order and second-order similarities of vertexes.

The first-order proximity describes the pairwise proximity between vertexes, which represents the local structure. For each pair of vertexes linked by an edge (v_i, v_j), the first-order proximity can be measured by the corresponding element s_{ij} in the adjacency matrix S. If there is no edge between v_i and v_j, their first-order proximity is 0.

The second-order proximity between a pair of vertexes v_i and v_j describes the proximity of the pair's neighborhood structure, which represents the global structure. Let F_i denote the first-order proximity between v_i and other vertexes, the second-order proximity is described as the similarity of F_i and F_j. If there are no coincident vertexes in their neighborhood structure, the second-order proximity between v_i and v_j is 0.

Basis of Deep Autoencoder. DANE uses deep autoencoder as the computational structure, so here we make a brief description. A deep autoencoder consists of two parts, an encoder and a decoder. The two parts are both composed of multiple layers of nonlinear transformation layers. For a given input x, the encoder aims to get a latent representation y in a low-dimensional space, and the decoder aims to reconstruct the input from the latent representation. So, the goal of the deep autoencoder is to get a good reconstruction $\hat{x}$ for input x.

Given the input x_i as y_i^0, the latent representation of the m-th layer can be calculated as follows:

$$y_i^m = sigmoid(W^m y_i^{m-1} + b^m), m = 1, ..., M \tag{1}$$

The reconstruction process is the opposite of the process of getting the latent representation:

$$\hat{y}_i^{m-1} = sigmoid(\hat{W}^m \hat{y}_i^m + \hat{b}^{m-1}), m = M, ..., 1 \qquad (2)$$

where $\hat{y}_i^0$ is the reconstruction result $\hat{x}_i$.

The goal of the autoencoder is to minimize the reconstruction error of the output and input. The loss function is shown as follows:

$$L_e = \sum_{i=1}^{N} ||x_i - \hat{x}_i||_2^2 \qquad (3)$$

Preserving Second-Order Proximity. Based on the loss function of the autoencoder, we first consider the second-order proximity. In the PPI network, the second-order proximity of proteins refers to how similar the neighborhood structure of a pair of proteins are. Here, we use the adjacency matrix S of the PPI network to model the neighborhood of each protein. Each element s_{ij} in S reflects the connection between the protein v_i and v_j. If and only if there is an edge (v_i, v_j) in set E, $s_{ij} = 0$. Therefore, the row vector s_i in S describes the neighborhood structure of the protein v_i and S contains the neighborhood information of all proteins.

Although the loss function of the autoencoder is to make a better reconstruction of the input instead of directly maintaining the similarity between similar inputs. As [16] proved, minimizing the reconstruction loss can capture the data manifolds to maintain structural information about the network. Thus, we use the adjacency matrix S as the input to the autoencoder. Each row vector s_i in the adjacency matrix S represents the neighborhood structure of the protein v_i. During the reconstruction process, the maintenance of the input information will make the low-dimensional representations of proteins similar if they have the similar neighborhood structure. So, we use formula (3) as the loss function of the second-order proximity.

Preserving Attribute Information. Due to the sparsity of the PPI network, the number of links visible is far less than the number of links that may exist. Therefore, the number of non-zero elements in the adjacency matrix S is much smaller than the zero element. If S is reconstructed directly using the autoencoder structure, the entire reconstruction process will pay more attention to the zero elements that account for a larger proportion in S.

To solve this problem, we make a modification to the loss function. A penalty matrix P is added to the loss function to emphasize the non-zero elements in the adjacency matrix S. We use a hyperparameter β to control the size of the penalty and consider the attribute information of the proteins into this penalty matrix. We use GO slims as the attribute information to build the attribute matrix G, in which $g_{ik} = 1$ represents the protein i has a corresponding GO slim attribute k. Here, we removes the Cc part in GO slims because it includes some protein complexes information. We calculate the cosine similarity between each row vector in the matrix G to get the attribute similarity matrix A. Since the adjacency matrix S and the attribute similarity matrix A both characterize the degree of similarity between the proteins, we use the summed result of the matrix as the reference of the penalty matrix P. The calculation of P is as follows:

$$P = (S + A) \cdot (\beta - 1) + 1 \tag{4}$$

Thus, the non-zero elements in S are multiplied by a larger coefficient to increase their proportion in the loss function, which mitigates the effect of zero elements in S on the reconstruction process. The joint loss function of the second-order proximity and attribute information is shown as follows, where $\odot$ denotes the Hadamard product:

$$L_{2+a} = \sum_{i=1}^{N} ||(s_i - \hat{s}_i) \odot p_i||_2^2 = ||(S - \hat{S}) \odot P||_F^2 \tag{5}$$

Preserving First-Order Proximity. For the first-order proximity, we use the latent representations Y that calculated by the encoding process. The first-order proximity considers whether there is a connection between two proteins, which is reflected in the adjacency matrix S of the PPI network. If using S as the input of the autoencoder, the similarity between each latent representation y_i^M can be regarded as the first-order proximity between proteins. The loss function is to minimize a pair of y_i^M of the proteins which are linked in the original network:

$$L_1 = \sum_{i,j=1}^{N} s_{ij} ||(y_i^M - y_j^M)||_2^2 \tag{6}$$

Preventing Over-Fitting. To prevent the over-fitting problem, we add a regularization function to constrain the parameters W and $\hat{W}$ in the layer of autoencoder. We use L2-norm regularizer and the loss function is shown as follows:

$$L_r = \frac{1}{2} \sum_{m=1}^{M} (||W^m||_F^2 + ||\hat{W}^m||_F^2) \tag{7}$$

With combining three parts of loss function together, we preserve the attribute information, the first-order and second-order proximity simultaneously. The joint loss function is as follows, where harmonic factor α balances the weight of the first-order, second-order and attribute proximity between vectors:

$$L = L_{2+a} + L_1 + L_r = ||(S - \hat{S}) \odot P||_F^2$$
$$+ \alpha \cdot \sum_{i,j=1}^{N} s_{ij} ||(y_i^M - y_j^M)||_2^2 + \frac{1}{2} \sum_{m=1}^{M} (||W^m||_F^2 + ||\hat{W}^m||_F^2) \tag{8}$$

3.2   Clustering Based on Core-Attachment Structure

Before clustering begins, we first make a transformation on the vector representation of the protein resulted by the network embedding process. By calculating the cosine similarity between y_i and y_j for the vector representation of each interconnected pair of proteins, we obtain a weighted adjacency matrix Q, which is shown as follows:

$$q_{ij} = \begin{cases} cos\_sim(y_i, y_j) & s_{ij} = 1 \\ 0 & s_{ij} = 0. \end{cases} \tag{9}$$

where cos_sim indicates the cosine similarity of the vector representation between two linked proteins.

Our clustering method is based on the structural principle that proposed by Gavin et al. [4]. They suggest that the structure of protein complexes can be divided into two parts, a core and the attachments. The proteins in the core structure are usually functionally similar and closely related. These proteins are unique, which do not appear in other complexes. The proteins in the attachment structure are not closely related with each other, but they have strong connections with the proteins in the core structure. These proteins are reusable, different protein complexes may share the same attachment structure.

In order to simulate the core-attachment structure, our method divides the formation of protein complexes to two steps. The first step is generating a set of seed cores, the second step is adding the attachments into each core based on their connection strengths.

In the first step, we use the maxima cliques generating algorithm proposed by Tomita et al. [19] to mine the cliques that have at least three proteins in the PPI network. We name these cliques $Core\_c$. Using the following steps to process the $Core\_c$ set, we can get the final $Seed\_c$ set without overlap:

- Sort the cliques in the $Core\_c$ set in descending order of their bio_score, which considers both the inside connective density and biological correlation of each clique:

$$bio\_score(Clique_p) = \sum_{i,j \in Clique_p} q_{ij} \tag{10}$$

- Remove the first item $Clique_1$ from the sorted set $\{Clique_1, Clique_2, ..., Clique_m\}$, add it into $Seed\_c$ set.
- For any other clique $Clique_i \in Core\_c$, if $Clique_i \cap Clique_1 \neq \emptyset$, update $Clique_i$ with $Clique_i - Clique_1$. Then, check whether $|Clique_i| \geq 3$. If not, remove $Clique_i$ from the $Core\_c$ set.

Repeating this process until $Core\_c$ is empty, the cliques in the $Seed\_c$ can be regarded as the real core structure in protein complexes.

In the second step, we use the *con_score* to measure the strength of the connection between core $Clique_j$ and a candidate attachment protein p_i:

$$con\_score(p_i, Clique_j) = \frac{\sum_{k \in Clique_j} q_{ik}}{|Clique_j|} \tag{11}$$

This score considers the topological and biological connectivity simultaneously. We judge the candidate attachment protein by whether the *con_score* is bigger than a threshold value θ. If yes, adding p_i to the attachment structure of core $Clique_j$. After all the attachment structures have been constructed, we combine each core and its attachments as the final protein complex prediction.

4   Experiment Results

4.1   Datasets

For performance comparison, we implemented six real world yeast PPI networks: DIP [23], Krogan-core [7], Krogan14k [7], Biogrid [18], Gavin [4] and Collins [2]. The details of these networks are shown in Table 2. The GO slim information was downloaded from the website https://downloads.yeastgenome.org/curation/literature/go_slim_mapping.tab. To compare the predicted results with reference complexes, we downloaded benchmark complex dataset from a public repository http://wodaklab.org/cyc2008/ and prune out all complexes whose size is less than or equal to 2. The final reference complex dataset contains 231 protein complexes.

Table 2. The PPI datasets used in the experiment.

PPI networks	Number of proteins	Number of interactions
DIP	4928	17201
Krogan-core	2708	7123
Krogan14k	3581	14076
Biogrid	5640	59748
Gavin	1430	6531
Collins	1622	9074

4.2   Evaluation Metrics

For comprehensive comparisons, we adopt several evaluation measures which have mentioned in ClusterOne [12]. First, in order to evaluate the performance of this detection algorithm, we define P as the set of protein complexes detected from detection algorithm and B as the set of reference protein complexes. Here,

Table 3. Performance comparison based on four evaluation metrics on the six yeast datasets.

Datasets	Methods	#predicted complexes	#matched complexes	Precision	Recall	F-score	Acc	F-score + Acc
DIP	COACH	747	253	0.339	0.655	0.447	0.550	0.997
	CMC	543	182	0.335	0.634	0.438	0.519	0.957
	MCODE	59	24	0.407	0.116	0.181	0.301	0.482
	ClusterOne	341	108	0.317	0.418	0.360	0.477	0.837
	IPCA	1236	425	0.344	0.664	0.453	0.542	0.995
	DANE	315	168	0.533	0.606	**0.567**	**0.585**	**1.152**
Krogan-core	COACH	348	190	0.546	0.526	0.536	0.529	1.065
	CMC	177	96	0.542	0.496	0.518	0.534	1.052
	MCODE	71	49	0.690	0.241	0.358	0.395	0.753
	ClusterOne	243	112	0.461	0.478	0.470	0.510	0.980
	IPCA	579	339	0.585	0.534	0.559	0.541	1.100
	DANE	204	145	0.711	0.524	**0.603**	**0.553**	**1.156**
Krogan14K	COACH	570	233	0.409	0.539	0.465	0.527	0.992
	CMC	396	166	0.419	0.552	0.476	0.510	0.986
	MCODE	49	28	0.571	0.129	0.211	0.315	0.526
	ClusterOne	225	90	0.400	0.358	0.378	0.497	0.875
	IPCA	983	444	0.452	0.547	0.495	0.558	1.053
	DANE	247	158	0.640	0.511	**0.568**	**0.560**	**1.128**
Biogrid	COACH	1507	373	0.248	0.772	0.375	0.615	0.990
	CMC	1349	265	0.196	0.815	0.317	0.625	0.942
	MCODE	58	11	0.190	0.052	0.081	0.355	0.436
	ClusterOne	475	160	0.337	0.655	0.445	0.654	1.099
	IPCA	3473	1362	0.392	0.762	0.518	**0.699**	1.217
	DANE	671	285	0.462	0.792	**0.584**	0.672	**1.256**
Gavin	COACH	326	145	0.445	0.453	0.449	0.528	0.977
	CMC	272	119	0.438	0.431	0.434	0.468	0.902
	MCODE	69	46	0.667	0.228	0.340	0.388	0.728
	ClusterOne	243	85	0.350	0.427	0.384	0.534	0.918
	IPCA	557	257	0.461	0.452	0.457	0.525	0.982
	DANE	190	115	0.605	0.424	**0.499**	**0.539**	**1.038**
Collins	COACH	251	154	0.614	0.539	0.574	0.605	1.179
	CMC	146	98	0.671	0.513	0.582	0.615	1.197
	MCODE	111	85	0.766	0.448	0.566	0.558	1.124
	ClusterOne	203	105	0.517	0.539	0.528	0.613	1.141
	IPCA	396	280	0.707	0.530	0.606	0.592	1.198
	DANE	197	141	0.716	0.528	**0.608**	**0.618**	**1.226**

we utilize the neighborhood affinity score $NA(p, b)$ to evaluate whether a predicted protein complex $p \in P$ matches a known protein complex $b \in B$ or not.

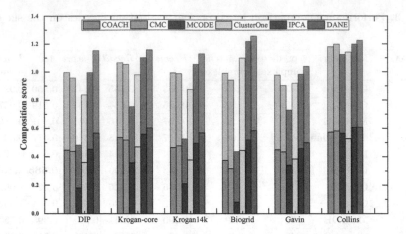

Fig. 2. Comparison of other five algorithms in terms of the composite scores of F-score and Acc. Shades of the same color indicate different evaluating scores F-score and Acc. (Color figure online)

The function is as follows:

$$NA(p, b) = \frac{|V_p \bigcap V_b|^2}{|V_p| \times |V_b|} \tag{12}$$

where V_p is the set of proteins in the detected protein complex p and V_b is the set of proteins in the reference protein complex b. Following previous studies, p and b are considered to matched if $NA(p, b) > 0.25$.

Based on this parameter, we use four statistic measures for evaluating the performance of different methods: *Precision*, *Recall*, *F-score* and *Acc*. The *Precision* measures the proportion of predicted complexes that match at least one reference complex. The *Recall* measures the proportion of reference complexes that match at least one predicted complex. The *F-score* is the harmonic mean of *Precision*, and *Recall*, which can be used to evaluate the overall performance. The *Acc* is the geometric accuracy that consider the number of matched proteins.

4.3   Performance Comparison

We compared DANE with five protein complex detection methods: COACH [22], CMC [11], MCODE [1], ClusterOne [12] and IPCA [8] on six PPI datasets. The parameters of these methods are set to default values as mentioned in their original papers. All experimental results are listed in Table 3 and Fig. 2.

As shown in Table 3, our method achieved the highest *Precision* on four datasets except Gavin and Collins. Although MCODE achieved the highest *Precision* on these two datasets, the total number of predicted complexes was rather small since it only kept the dense complexes during its post-processing

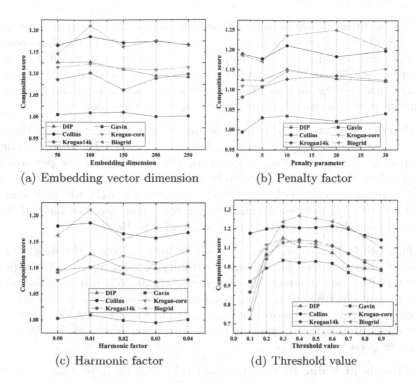

(a) Embedding vector dimension   (b) Penalty factor

(c) Harmonic factor   (d) Threshold value

Fig. 3. The performances of DANE based on four parameters.

step. Our method did not always achieve the highest *Recall*, probably because its number of predicted protein complexes was small. Overall, our method performed the best with respect to the overall evaluation metric *F-score* for all datasets. This showed that our method outperformed other methods in terms of comprehensive performance. Our method achieved the highest *Acc* on five datasets except Biogrid. IPCA performed the best on this dataset probably because it can form tighter cliques in a large network with many vertexes and edges, but its *F-score* was lower than DANE. We added the two indicators to get the composition scores, DANE had the highest value on all six datasets, which indicated our method had better performance than other algorithms on the protein prediction task.

4.4 Parameter Sensitivity

In this part, we examined the sensitivity of DANE with respect to four parameters: the embedding dimension d, the penalty factor β, the harmonic factor α and the threshold value θ. We conducted four experiments to adjust these four parameters separately. Before the experiments began, we set the default values of these four parameters to 100, 10, 0.01 and 0.3. In each experiment, we adjusted one parameter and fixed the other three parameters to the default values.

The Embedding Dimension d**.** The embedding dimension d affects the expression ability of vector representations, which is the basis of subsequent clustering processes. In the experiments, the embedding dimension d is varied between 50 and 250. As Fig. 3(a) shows, the effect of dimension d on DANE is not obvious. Although the best results on different datasets correspond to different dimensions d, we choose 100 as the default value of dimension.

The Penalty Factor β**.** The penalty factor β is aiming to emphasize the non-zero elements in the adjacency matrix. We vary β from 1 to 30 to investigate the impact of β, which is shown in Fig. 3(b). The method does not perform well when β is less than 10, this is probably because the β is too small to achieve enough emphasis, so the 0 elements in the adjacency matrix still play a major role. When the β is between 10 and 30, the performance of the method is not much different. Here, we set the default value of β to 10.

The Harmonic Factor α**.** The harmonic factor α balances the weight of the first-order, second-order and attribute proximity between vectors. We adjust α from 0 to 0.04 to get different vector representations. When $\alpha = 0$, the performance is only determined by the second-order proximity. As α increases, the first-order proximity plays more roles. As shown in Fig. 3(c), DANE achieves the best results on all datasets except Krogan-core when $\alpha = 0.01$. So, we choose 0.01 as the default value of α.

The Threshold Value θ**.** The threshold value θ is an important factor that influences the structure of protein complexes. This parameter measures the tightness of the connection between the protein and the subgraph. When the value θ is higher, it is harder for each neighbor protein to be added into this subgraph. In other words, internal connections of the resulting protein complex are tighter. As shown in Fig. 3(d), when θ is less than 0.3, the performance is relatively low. This is because when θ is small, most of the neighbors can be added into the subgraph. On the other hand, when θ is more than 0.7, the performance is relatively low on the contrary. The reason is that when θ is big, few neighbors can be added into the cluster. Although the best results on different datasets are achieved with different threshold values, 0.3 or 0.4 are relatively good choices in practice.

5 Conclusions

In this paper, we proposed a novel network embedding method for identifying protein complexes named DANE. Firstly, we used a deep autoencoder model to learn the vector representation for each protein from GO attributed PPI networks. We combined the first-order proximity and the second-order proximity, and added the biological attribute proximity into consideration. Secondly, we calculated the weighted adjacency matrix and used a clustering method based on the core-attachment principle to get the predicted protein complexes. Experiments on six different datasets showed that DANE outperforms five protein complex detection methods, which indicates our method can produce better prediction results. Our future work will focus on utilizing this new representation learning method to other biological networks such as gene-phenotype networks.

References

1. Bader, G.D., Hogue, C.W.: An automated method for finding molecular complexes in large protein interaction networks. BMC Bioinform. **4**(1), 2 (2003)
2. Collins, S.R., et al.: Toward a comprehensive atlas of the physical interactome of saccharomyces cerevisiae. Mol. Cell. Proteomics (MCP) **6**(3), 439 (2007)
3. Gene Ontology Consortium: The Gene Ontology (GO) project in 2006. Nucleic Acids Res **34**(Suppl. 1), D322–D326 (2006)
4. Gavin, A.C., et al.: Proteome survey reveals modularity of the yeast cell machinery. Nature **440**(7084), 631 (2006)
5. Grover, A., Leskovec, J.: node2vec: scalable feature learning for networks. In: Proceedings of the 22nd ACM SIGKDD International Conference on Knowledge Discovery and Data Mining, pp. 855–864. ACM (2016)
6. Hamp, T., Rost, B.: Evolutionary profiles improve protein-protein interaction prediction from sequence. Bioinformatics **31**(12), 1945–1950 (2015)
7. Krogan, N.J., et al.: Global landscape of protein complexes in the yeast saccharomyces cerevisiae. Nature **440**(7084), 637 (2006)
8. Li, M., Chen, J.E., Wang, J.X., Hu, B., Chen, G.: Modifying the DPClus algorithm for identifying protein complexes based on new topological structures. BMC Bioinform. **9**(1), 398 (2008)
9. Li, T., et al.: A scored human protein-protein interaction network to catalyze genomic interpretation. Nat. Methods **14**(1), 61 (2017)
10. Li, X.L., Foo, C.S., Ng, S.K.: Discovering protein complexes in dense reliable neighborhoods of protein interaction networks. Comput. Syst. Bioinform. **6**, 157–168 (2007)
11. Liu, G., Wong, L., Chua, H.N.: Complex discovery from weighted PPI networks. Bioinformatics **25**(15), 1891–1897 (2009)
12. Nepusz, T., Yu, H., Paccanaro, A.: Detecting overlapping protein complexes in protein-protein interaction networks. Nat. Methods **9**(5), 471 (2012)
13. Ozawa, Y., et al.: Protein complex prediction via verifying and reconstructing the topology of domain-domain interactions. BMC Bioinform. **11**(1), 350 (2010)
14. Perozzi, B., Al-Rfou, R., Skiena, S.: Deepwalk: online learning of social representations. In: Proceedings of the 20th ACM SIGKDD International Conference on Knowledge Discovery and Data Mining, pp. 701–710. ACM (2014)
15. Roweis, S.T., Saul, L.K.: Nonlinear dimensionality reduction by locally linear embedding. Science **290**(5500), 2323–2326 (2000)
16. Salakhutdinov, R., Hinton, G.: Semantic hashing. Int. J. Approx. Reason. **50**(7), 969–978 (2009)
17. Srihari, S., Leong, H.W.: A survey of computational methods for protein complex prediction from protein interaction networks. J. Bioinform. Comput. Biol. **11**(02), 1230002 (2013)
18. Stark, C., Breitkreutz, B.J., Reguly, T., Boucher, L., Breitkreutz, A., Tyers, M.: BioGRID: a general repository for interaction datasets. Nucleic Acids Res **34**(Suppl. 1), D535–D539 (2006)

19. Tomita, E., Tanaka, A., Takahashi, H.: The worst-case time complexity for generating all maximal cliques and computational experiments. Theoret. Comput. Sci. **363**(1), 28–42 (2006)
20. Wang, D., Cui, P., Zhu, W.: Structural deep network embedding. In: Proceedings of the 22nd ACM SIGKDD International Conference on Knowledge Discovery and Data Mining, pp. 1225–1234. ACM (2016)
21. Wang, J., Li, M., Deng, Y., Pan, Y.: Recent advances in clustering methods for protein interaction networks. BMC Genom. **11**(3), S10 (2010)
22. Wu, M., Li, X., Kwoh, C.K., Ng, S.K.: A core-attachment based method to detect protein complexes in PPI networks. BMC Bioinform. **10**(1), 169 (2009)
23. Xenarios, I., Salwinski, L., Duan, X.J., Higney, P., Kim, S.M., Eisenberg, D.: DIP, the database of interacting proteins: a research tool for studying cellular networks of protein interactions. Nucleic Acids Res. **30**(1), 303–305 (2002)
24. Zaki, N., Efimov, D., Berengueres, J.: Protein complex detection using interaction reliability assessment and weighted clustering coefficient. BMC Bioinform. **14**(1), 163 (2013)

Predicting Auction Price of Vehicle License Plate with Deep Residual Learning

Vinci Chow(✉) (iD)

The Chinese University of Hong Kong, Shatin, Hong Kong
vincichow@cuhk.edu.hk

Abstract. Due to superstition, license plates with desirable combinations of characters are highly sought after in China, fetching prices that can reach into the millions in government-held auctions. Despite the high stakes involved, there has been essentially no attempt to provide price estimates for license plates. We present an end-to-end neural network model that simultaneously predict the auction price, gives the distribution of prices and produces latent feature vectors. While both types of neural network architectures we consider outperform simpler machine learning methods, convolutional networks outperform recurrent networks for comparable training time or model complexity. The resulting model powers our online price estimator and search engine.

Keywords: Price estimate · Residual learning · License plate

1 Introduction

Chinese society place great importance on numerological superstition. Numbers such as 2 and 8 are often used solely because of the desirable qualities they represent (easiness and prosperity, respectively). For example, the Beijing Olympic opening ceremony occurred on 2008/8/8 at 8 p.m., while the Bank of China (Hong Kong) opened for business on 1988/8/8. Because license plates represent one of the most public displays of numbers for many people, people are willing an enormous amount of money for license plates with desirable combinations of characters. Local governments often auction off such license plates to generate public revenue. The five most expensive plates ever auctioned in Hong Kong have each sold for over US$1 million.

Unlike the auctioning of other valuable items, license plates generally do not come with a price estimate, even though price estimates have been shown to be a significant factor affecting the sale price [1,8]. It is also very difficult to discover which plates are available at what price because auction outcomes are not always available online. In Hong Kong, the government only provides the results of the three most recent auctions online, despite having hosted monthly auctions for several decades.

© Springer Nature Switzerland AG 2019
L. H. U and H. W. Lauw (Eds.): PAKDD 2019 Workshops, LNAI 11607, pp. 179–188, 2019.
https://doi.org/10.1007/978-3-030-26142-9_16

We build an online service, *markprice.ai*, that seeks to fill this gap by providing three specific features:

1. *Price estimation.* The primary service we provide is price estimation for license plate auctions in Hong Kong. In Hong Kong, license plates consist of either a two-letter prefix or no prefix, followed by up to four digits (e.g., HK 1, BC 6554, or 138). Plates can be desirable because the characters rhythm with auspicious Chinese phrases (e.g. "168" rhythms with "all the way to prosperity", while "186" does not rhythm with anything) or have visual appeal (e.g. "2112" is symmetric, while "2113" is not.) The model learns which features are valuable from the training data, allowing it to then generate a predicted price for the specific combination of characters a user enters.
2. *Distribution of predicted prices.* Even with a perfect model, there will be variation in the realized price due to factors we cannot capture. We do not wish to give the site's users a false impression of certainty, so it is important that we provide a distribution of possible outcomes. The main challenge here is providing a distribution for extremely expensive plates, for which there are very few examples to use to generate a distribution from.
3. *Search engine.* Beyond the price estimation service, we also want to help users research and discover plates that they might be interested in. We provide a way for users to not only search for past records of a specific plate, but also to discover other plates that are reasonably similar.

We detail our solution to the above three targets in this study. Our model is powered by a neural network that generates a distribution of predicted prices for a given license plate. In the process, the network also learns to generate latent feature vectors, which we extract to construct our search engine. With both linguistic and visual factors affecting a plate's value, our primary focus is on the relative performance of recurrent networks versus convolutional networks for our task. Our results suggest that while both neural network architecture outperform simpler machine learning methods, convolutional networks outperform recurrent networks for comparable training time or model complexity.

2   Related Studies

Early studies into the prices of license plates use hedonic regressions with a larger number of handpicked features [9–11]. Due to their reliance on ad-hoc features, these models adapt poorly to new data, such as when plates with new combinations of characters are auctioned off for the first time. In contrast, our model is able to learn the value of license plates from their prices. As we demonstrate below, no handpicked feature is needed to achieve high prediction accuracy, although the presence of such features does improve the consistency of the learned features. [3] attempts to model prices with a standard character-level recurrent neural network, achieving higher accuracy than that previous studies. In this paper we utilizes more advanced neural network designs, resulting in significant improvement in accuracy and consistency.

3   Model

3.1   Overall Structure

Figure 1 illustrates the structure of our model. Each plate is represented by a vector of numerically-encoded characters, padded to the same length. The characters are fed into a feature extraction unit, either one at a time for RNN or as a vector for CNN. The feature extraction unit outputs a feature vector, which is used in three ways. First, it is concatenated with auxiliary inputs and fed into a set of fully-connected layers to generate the predicted price. Second, it is fed into another set of fully-connected layers to generate a number of auxiliary targets. Third, it is fed into a k-nearest-neighbor clustering model to produce a list of similar plates. Finally, the predicted price is fed into another fully-connected layer responsible for producing distributional parameters for the final price distribution.

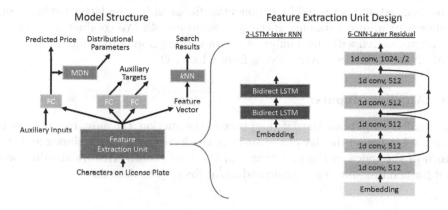

Fig. 1. Model structure and examples of feature extraction unit design

3.2   Feature Extraction Unit

The feature extraction unit begins with an embedding layer $g(s)$, which converts each character s to a vector representation $\boldsymbol{h}_s$: $g(s) = \boldsymbol{h}_s \equiv [h_s^1, ..., h_s^n]$. The dimension of the character embedding, n, is a hyperparameter. The values $h_s^1, ..., h_s^n$ are initialized with random values and learned through training. We experiment with n ranging from 8 to 24 as well as replacing learned embedding with one-hot encoding.

The embedding layer is followed by one or more layers of neurons. We explore three architectures:

1. *RNN.* As a baseline, we adopt the model used in [3], which has been shown to perform significantly better than simpler models such as hedonic regressions and n-grams. Specifically, the unit consists of one or more bi-directional,

batch-normalized recurrent layers with rectified linear units as activations. The feature vector is the sum of the last layer's recurrent output. We conduct a hyperparameter search with the number of layers varying from 1 to 7 and the number of neurons varying from 128 to 1024.

2. *LSTM.* It is widely recognized that Long Short-Term Memory (LSTM) networks performs better than simple recurrent networks in deep learning [6]. Following standard practice, we implement the unit as one or more bidirectional, batch-normalized LSTM layers with logistic and tanh activations. The feature vector is the output from the last time step of the last LSTM layer. As with RNN, we conduct a hyperparameter search with the number of layers varying from 1 to 5 and the number of neurons varying from 128 to 1600.

3. *Residual CNN.* Our implementation is a 1-demensional version of ResNet [5], illustrated in the rightmost panel of Fig. 1. The main features are residuals being added to the output after every two layers, and the number of filters being doubled whenever there is a 50% down sampling. All layers are batch normalized and activated by exponential linear units, the latter having been shown to improve training speed and accuracy [4]. We conduct a hyperparameter search with the number of layers varying from 1 to 7 and the number of filters in the first layer varying from 64 to 1024.

3.3   Auxiliary Inputs

The auxiliary inputs are the date and time of the auction, the most-recent general price level index on the day of the auction, the local stock market index and the return of the index in the past year and the past month. Historical values are used for training, but our actual product utilizes up-to-date data.

3.4   Auxiliary Targets

Previous versions of our model did not have auxiliary targets, but we discovered that while prediction accuracy was similar across different runs of the same model, the feature vectors generated were not. In hindsight, this should be expected: neural network training is non-convex problem, and there is no reason why the model should settle on a particular set of latent factors after every training run. This imposed a significant problem for our search engine, as search results would at times change noticeably after retraining the model with new data.

Our solution to these problem is to train the model with auxiliary targets, based on the handpicked features of [9]. The idea is to provide guidance on what features users might be looking for, while still allowing the model enough flexibility to learn additional features. Table 1 lists the 32 objective measures of plate characteristics, which include the number of characters, the number of repeated characters and whether the combination of characters is symmetric or sequential.

Table 1. Auxiliary targets

Letters	Numbers					
Repeated letters	x00	abab	aab	aaa	# of 0's	# of 5's
No letters	x000	aaab	abb	aaaa	# of 1's	# of 6's
= HK	symmetric	abbb	abcd	aabb	# of 2's	# of 7's
= XX	contains "13"	aaba	dcba		# of 3's	# of 8's
	= 911	abaa	aa		# of 4's	# of 9's

3.5   Mixture Density Network

A common way of estimating a distribution is to divide the target samples into mutually-exclusive categories and train a classifier. That does not work in our case because during inference the target is not bounded from above—there is always the possibility of a record-breaking price. Instead, we uses a mixture density network (MDN) to generate the distributional parameters [2].

The estimated probability density function of the realized price p for a given predicted price $\hat{p}$ is modelled as a Gaussian mixture:

$$P(p \mid \hat{p}) = \sum_{k=1}^{24} \frac{e^{z_k(\hat{p})}}{\sum_{i=1}^{24} e^{z_i(\hat{p})}} \phi(p \mid \mu_k(\hat{p}), \sigma_k(\hat{p})), \tag{1}$$

where ϕ represents the standard normal probability density distribution and $[z_1(\hat{p}), ..., z_i(\hat{p}), \mu_1(\hat{p}), ..., \mu_i(\hat{p}), \sigma_1(\hat{p}), ..., \sigma_i(\hat{p})]$ the output vector from a single fully-connected layer with a single input $\hat{p}$. σ's have exponential activations to ensure that they take on positive values, while μ's and z's have linear activations. We conduct a hyperparameter search with n varying from 3 to 24 and the number of hidden neurons varying from 64 to 256.

4   Experiment Setup

4.1   Data

The data used in this study come from the Hong Kong government and are an extension of the dataset used in [9–11]. They cover Hong Kong traditional license plate auctions from January 1997 to February 2017. To include as many ultra-expensive plates as possible, we also include the 10 most expensive plates since auctions commenced in 1973, information that is publicly available online. The data consist of 104,994 auction entries, almost twice that of previous studies. Each entry includes i. the characters on the plate, ii. the sale price (or a specific symbol if the plate was unsold), and iii. the auction date.

The distribution of prices is highly skewed—while the median sale price is $641, the mean sale price is $2,064. The most expensive plate in the data is "28," which sold for $2.3 million in February 2016. Following previous studies, we compensate for this skewness by using log prices within the model. The use of log price also means that the loss function depends on relative error rather absolute error.

Plates start at a reserve price of at least HK\$1,000 (\$128.2). The existence of reserve prices means that not every plate is sold, and 12.6% of the plates in our data were unsold. Because these plates do not possess a price, we follow previous studies and drop them from the dataset, leaving us 91,784 entries. The finalized data are randomly divided into three parts: 64% for training, 16% for validation and 20% for the final hold-out test.

4.2   Training

Continuous outputs of the model are trained using mean-squared error as the loss function, while binary outputs are trained using cross entropy as the loss function. The primary target and the whole of the auxiliary targets carry equal weight in the overall loss function. To allow ourselves the flexibility to use a different training duration for the mixture density network, the latter is trained separately after other parts of the model have been trained, using the negative log-likelihood of the Gaussian mixture as the loss function.

To compensate for the scarcity of expensive plates, we weight samples according to their log price. We further overweight the most expensive plates, specifically those with a log price above 12.5 (approximately \$34402), by a factor of 40. These weights are used in all training runs without further experimentation.

Drop out is applied to each layer in the feature extraction unit except the embedding layer. When we started our experiment with RNN we experimented with drop out rate ranging from 0 to 30%, but by the time we reached CNN we have decided to settle on 15% since there is little variation between different positive drop out rates—excluding a rate of zero, the correlation between validation RMSE and drop out rate is only 0.003.

In total, we have 432 sets of hyperparameters for both RNN and CNN. Due to time constraint, we had to do a relatively sparse search for LSTM with only 95 sets of hyperparameters. We train the model under each design and each set of hyperparameters three times, with early stopping and reloading of the best state. Recurrent versions of the model are trained for 120 epochs while convolutional versions are trained for 800. We pick these numbers because trials we conducted before this study suggest these values are large enough for the model to almost certainly stop early. The mixture density network is trained for 5000 epochs with reloading of the best state.

The Adam optimizer with a learning rate of 0.001 is used throughout [7]. Training is conducted with NVIDIA GTX 1080 s with mini-batch size of 2,048.

5   Experiment Results

5.1   Predicted Price

Table 2 lists the performance figures of the best model in each category and that of a number of simpler models for comparison. The best performing model is

Table 2. Model performance

Configuration	Train RMSE	Valid RMSE	Test RMSE	Train R^2	Valid R^2	Test R^2
Residual CNN	.3859	.4692	.4721	.8985	.8499	.8471
LSTM	.4747	.5007	.4982	.8464	.8290	.8297
RNN	.5069	.5443	.5492	.8473	.8197	.8237
Woo et al. [10]	.6739	.6808	.6769	..6905	.6840	.6857
Ng et al. [9]	.6817	.6880	.6856	.6833	.6773	.6775
unigram kNN-10	.8924	1.174	1.165	.4572	.0599	.0690

For Residual CNN, LSTM and RNN, the average numbers from the best-performing set of hyperparameters are reported

a 6-layer ResNet, with 512 filters per layer in the first five layers and a 50% down-sampling in the last, paired with a single 256-neuron fully-connected layer and an 8-channel embedding. This is followed by the bi-directional LSTM with 3200 neurons and two fully-connected layers of 512 neurons each. The basic bidirectional RNN comes in last among the three, performing at its best when there is a single recurrent layer of 1024 neurons and three fully-connected layers of 1024 neurons each. Both recurrent models perform best with one-hot encoding. The best convolutional model is able to explain a significantly higher fraction of variation in prices than the best recurrent models, both in sample (5.2% as measured by R-squared) and out of sample (1.7%).

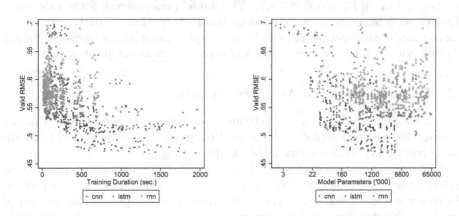

Fig. 2. Performance as a function of training time and model complexity

Figure 2 plots validation RMSE against training duration and complexity of a given model. Each marker is the average from all runs of a particular set of hyperparameters. Except for the smallest convolutional networks, it is clear from the plots that convolutional networks outperform recurrent networks for either comparable training time or comparable model complexity. One interesting observation from Fig. 2 is that the relationship between a model's performance and its training time/complexity is much more pronounced for convolutional networks than recurrent networks.

To better understand why convolutional networks perform better than recurrent networks, Table 3 lists the error figures for three linguistic patterns and three visual patterns. Although the best LSTM model is behind the best CNN model in all cases, the differences in error are much smaller for linguistic patterns than for visual patterns.

Table 3. Model performance by plate characteristics

	CNN RMSE	LSTM RMSE	Absolute diff.	Relative diff.
Linguistic patterns				
168 (all the way to prosperity)	.3284	.4156	.0872	27%
28 (easy prosperity)	.3662	.4291	.0629	17%
1314 (together forever)	.6049	.7000	.0951	16%
Visual patterns				
abba	.4466	.6139	.1673	37%
abcd	.4753	.6479	.1726	36%
aabb	.4934	.8006	.3071	62%

Figure 3 plots predicted prices against actual prices from the best model, grouped in bins of HK$1000 ($128.2). The model performs well for a wide range of prices, with bins tightly clustered along the 45-degree line. In particular, due to a better model architecture and weights of samples, the systemic underestimation of prices for the most expensive plates observed in [3] is not present here.

5.2   Feature Vector and Auxiliary Targets

To demonstrate the effectiveness of the auxiliary targets, we rerun the best model with auxiliary targets replaced by the auction price. This allows us to maintain model complexity while removing the auxiliary targets.

Table 4 lists the top-three search results for three representative plates from each of three runs of the best CNN model, with and without the auxiliary targets. We measure the consistency of the search results by computing the fraction of search results that appeared in all three runs. The examples listed in the table illustrate how training the model with auxiliary targets significantly increase consistency across different runs of the same model.

To evaluate consistency systemically, we generate 1000 random new plates and feed them through all six runs of the model. Figure 4 plots the consistency measure's distribution with and without the auxiliary targets. The search results are much more inconsistent when there is no auxiliary target (Mann-Whitney $z = -11.2$, $p = 0.0000$), with significantly more cases of zero consistency.

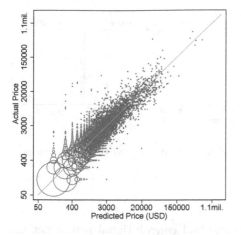

Fig. 3. Actual vs predicted price in test set

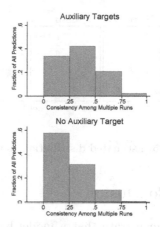

Fig. 4. Consistency of recommendations

Table 4. Auxiliary targets and search consistency

RMSE		Search results								
With auxiliary targets		2112			BB239			LZ3360		
Run 1	0.4750	2012	1812	2121	CC239	AA239	AL239	HV3360	BG3360	HC3360
Run 2	0.4681	1012	2012	1812	CC239	AA239	LL239	HV3360	BG3360	HC3360
Run 3	0.4671	1812	1012	2113	AA239	CC239	PP239	HV3360	BG3360	ND6330
Consistency		0.33			0.67			0.67		
Without auxiliary targets		2112			BB239			LZ3360		
Run 1	0.4884	9912	2223	8182	AA239	CC199	AA3298	HV3360	HC3360	JA6602
Run 2	0.4771	1212	1812	2012	CC239	BB989	AA239	KE9960	FE9960	JR6360
Run 3	0.4749	2832	8122	8182	CC239	AA239	AA269	HV3360	FM6369	JR6360
Consistency		0			0.33			0		

Each row is one training run. For each run, the top three search results are listed for each plate queried (2112, BB239, LZ3360). Consistency is calculated as the fraction of matches that appear in all three runs

5.3   Estimated Price Distribution

Figure 5 plots the estimated price distribution at selected prices from a mixture density network of 256 hidden neurons and a mixture of six Gaussian densities, approximately evenly spaced on a log scale. The red lines represent the estimated density for a given predicted price, while the bars represent the actual distribution of prices for the predicted price. The estimated density closely resembles the actual distribution for common, relatively low-value plates. For very expensive plates, the model is able to produce a density even if there is only a single sample at a given price. As can be seen from the examples, the spread of the estimated density generally covers any actual price that deviates from the predicted price.

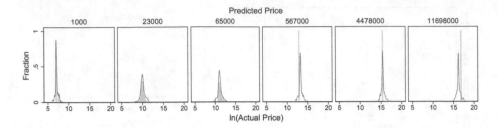

Fig. 5. Estimated distribution vs actual distribution at selected predicted prices

6   Conclusion

We demonstrate that a model based on residual convolutional neural net can generate accurate predicted prices and produce stable feature vectors for use in a search engine. We demonstrate that for comparable training time or model complexity, a model base on convolutional neural nets outperforms one that is base on recurrent neural network.

References

1. Ashenfelter, O.: How auctions work for wine and art. J. Econ. Perspect. **3**(3), 23–36 (1989). https://doi.org/10.1257/jep.3.3.23. http://www.aeaweb.org/articles?id=10.1257/jep.3.3.23
2. Bishop, C.: Mixture density networks. Technical report (1994)
3. Chow, V.: Predicting auction price of vehicle license plate with deep recurrent neural network. CoRR arXiv:1701.08711 (2017)
4. Clevert, D., Unterthiner, T., Hochreiter, S.: Fast and accurate deep network learning by exponential linear units (ELUs). CoRR arXiv:1511.07289 (2015)
5. He, K., Zhang, X., Ren, S., Sun, J.: Deep residual learning for image recognition. arXiv e-prints, December 2015
6. Hochreiter, S., Schmidhuber, J.: Long short-term memory. Neural Comput. **9**, 1735–1780 (1997)
7. Kingma, D.P., Ba, J.: Adam: A method for stochastic optimization. CoRR arXiv:1412.6980 (2014)
8. Milgrom, P.R., Weber, R.J.: A theory of auctions and competitive bidding. Econometrica **50**(5), 1089–1122 (1982). http://www.jstor.org/stable/1911865
9. Ng, T., Chong, T., Du, X.: The value of superstitions. J. Econ. Psychol. **31**(3), 293–309 (2010). https://doi.org/10.1016/j.joep.2009.12.002. http://www.sciencedirect.com/science/article/pii/S0167487009001275
10. Woo, C.K., Horowitz, I., Luk, S., Lai, A.: Willingness to pay and nuanced cultural cues: evidence from Hong Kong's license-plate auction market. J. Econ. Psychol. **29**(1), 35–53 (2008). https://doi.org/10.1016/j.joep.2007.03.002. http://www.sciencedirect.com/science/article/pii/S016748700700027X
11. Woo, C.K., Kwok, R.H.: Vanity, superstition and auction price. Econ. Lett. **44**(4), 389–395 (1994). https://doi.org/10.1016/0165-1765(94)90109-0. http://www.sciencedirect.com/science/article/pii/0165176594901090

Mining Multispectral Aerial Images for Automatic Detection of Strategic Bridge Locations for Disaster Relief Missions

Hafiz Suliman Munawar[1], Ji Zhang[2(✉)], Hongzhou Li[3], Deqing Mo[3], and Liang Chang[3]

[1] University of New South Wales (UNSW), Kensington, Australia
[2] The University of Southern Queensland, Toowoomba, Australia
ji.zhang@usq.edu.au
[3] Guilin University of Electronic Technology, Guilin, China

Abstract. We propose in this paper an image mining technique based on multispectral aerial images for automatic detection of strategic bridge locations for disaster relief missions. Bridge detection from aerial images is a key landmark that has vital importance in disaster management and relief missions. UAVs have been increasingly used in recent years for various relief missions during the natural disasters such as floods and earthquakes and a huge amount of multispectral aerial images are generated by UAVs in the missions. Being a multi- stage technique, our method utilizes these multispectral aerial images for identifying patterns for effective mining of bridge locations. Experimental results on real-world and synthetic images are conducted to demonstrate the effectiveness of our proposed method, showing that it is 40% faster than the existing Automatic Target Recognition (ATR) systems and can achieve a 95% accuracy. Our technique is believed to be able to help accelerate and enhance the effectiveness of the relief missions carried out during disasters.

Keywords: Image mining · Disaster management ·
Isotropic surround suppression · Image processing · Object recognition ·
Linear object detection · Bridge recognition · Road recognition

1 Introduction

Many countries are frequently hit by various natural disasters such as floods. The loss of lives proves to be extremely devastating for them from an economic perspective [1]. It is hoped that the rise in technology will provide more effective ways in which the disasters could be managed, and the system could be developed for dealing with the task of analyzing the weak zones and areas that are majorly hit by floods. As the natural disaster happens, the existing system of aerial imaging in many countries is quite weak in that they rely on the approach of manual analysis by personnel involved in recognizing the real-time static images [2]. With the help of the modern systems, the detection and mining results can be achieved in the times of disaster where the matter is not the cost but technology for saving lives. Automatic recognition and mining of bridges from aerial images are very important [4] as the bridges over water are one of

© Springer Nature Switzerland AG 2019
L. H. U and H. W. Lauw (Eds.): PAKDD 2019 Workshops, LNAI 11607, pp. 189–200, 2019.
https://doi.org/10.1007/978-3-030-26142-9_17

the most important spatial objects for recognizing the access to the flooded areas. The modern approach of using the image processing techniques for data and information mining through the existing digital images could be used. Once a pattern has been identified, the images can be processed, and feature extraction can be carried out followed by thorough analysis to interpret the meaning and knowledge gained through these images [5]. Extraction of features from images has been widely studied since the early days of remote sensing and aerial imaging. The research studies so far illustrate recursive scanning, clustering regions and searching features for extraction of objects from images [7]. For the cause of estimating positions and navigating UAVs, these applications have a wide usage holding advancements in future detection techniques by making use of multispectral images (Thermal, Satellite, visible, SAR, Infrared etc.) [8]. This study aims to develop an algorithm based on the latest technology for aiding the humanitarian purposes and ensuring that technology could come into use for the purpose of saving lives.

2 Literature Review

A profound effort is made to gather the resources deployed around the globe in the past to detect linear targets such as bridges and plausible avenues that can be marched on from this point [9]. From the perspective of relief work, bridges over water are extremely significant and are the most crucial for detection and understanding of an area; hence their detection is required for several applications including relief applications, updating geographical data- bases and in case of identifying the extent of destruction caused by natural calamities. Over the last few years, extensive research has been carried out on detection of bridges in infrared images, SAR (Synthetic Aperture Radar) images, and visible images [10]. The methods for bridge recognition can be broadly classified into five categories, namely knowledge-based methods, machine learning based methods, composite (feature & knowledge-based) methods, dynamic programming-based methods and context-based methods [11]. Currently, the knowledge-based methods are used for identifying the difference between the object and background but also use their spatial relationship. Trias-Sanz et al. [14] studied the difference of backscattering intensity between targets and background and used it along with the spatial relation and context dependency between bridges and rivers. A study conducted in this regard [15] proposed a method for automatically detecting and tracking bridges over water in IR images. First, they used the OTSU algorithm (using a global threshold) to separate the foreground and the background. Detection involves the detection of a river, detection of a bridge's edge/arch (using edge enhancement) and detection of a bridge's piers i.e., the bridge's horizontal and vertical extent is determined. Yuan et al. [16] gave a two-step knowledge-based strategy to recognize large bridges over water having a complicated background. Images are pre-processed to remove noise. Context-based information about bridges over water is considered and used in the process of marking the ROIs in the field of view. However, the confidence level for the detected ROIs is low [17]. Image mining is a diverse field that combines the tasks of information mining, image processing and computer vision. It is not simply the harvesting of data or information from the images, rather it requires detailed analysis and extraction of information from the images and requires the use of different techniques to ensure that the patterns and information present in the images are truly under- stood [18]. The essence of image mining is that this technique does not only identify the image patterns but also divulges into extracting the knowledge pre- sent in the image sets through the low-level pixel s of an image. The process is complex and requires various steps including analysis, feature recognition and extraction, classification, image indexing, retrieval and finally the data management [19].

2.1   Research Gap

The development of a new bridge detection and mining approach is motivated by our identification of the following research gaps existing in literature:

(1)   The traditional method uses a simple edge detection scheme, which is not enough for gathering concrete information about the areas being analyzed.
(2)   Another problem that is encountered in the literature is that the existing approaches search in the horizontal, vertical and diagonal directions. It is unnecessary and too expensive for detecting Bridge targets in all directions in complex situations.

Here effective edge detection and a large Gaussian kernel [20] are used. Further ROI regions are segmented using Hough transform and analysis of the ROI for being a possible target has been done. To evaluate the detection and recognition limits along with the presence of weather and camera effects, we propose a fully automatic object recognition system that involves a real-time cognitive approach utilizing sur- round suppression-based Hough voting scheme.

3   Our Method

Our method takes multiple stages to complete the bridge mining process, as elaborated as follows:

- The multispectral images are firstly grouped into eight land-cover types by means of a majority-must-be-granted logic based on the multi-seed super- vised classification technique. The eight classes include ice/snow, shrubs, concrete, water, sand, forest, soil and rock. For bridge mining, concrete and water are the most important classes. These three classes are thus selected. After this step, each and every pixel in the image is labeled accordingly as concrete, water or background;
- Bridges are then recognized in this tri-level image by using a knowledge- based approach that exploits the spatial arrangement of bridges and their respective sur- roundings by means of a five-step approach. The possible bridge pixel s are identified by using a neighborhood operator and the information of the spatial dimensions of a typical bridge.
- Lastly, these bridge segments are subjected to verification on the basis of directional water index in the line of different directions and their association with the segments of the road. All surrounding directions are covered by water except for the direction of the bridge.

Therefore, the water densities in different directions and connectivity between the bridge and road are used as the basis for confirmation. Signed directional water index along 9 different directions are computed. After smoothing the image, segmentation is carried out via a region-correlating. Using the segmentation outcome, geometrical knowledge of bridge and the water region, a bridge is detected. Our proposed method assumes the following prior knowledge in the mining process:

- Grey levels of a bridge and the land are higher as compared to water;
- The area encompassed by the river is larger than the bridge's area;
- The edges of a bridge are considered to be two near-parallel lines;
- The river is divided into two homogeneous regions by the bridge situated across it;
- The length of a bridge is greater than its width;
- When a bridge is close to the image acquisition system, it can be clearly seen as the divider between the two sides of a river. However, when seen from afar, the bridge appears as a thick dark line.

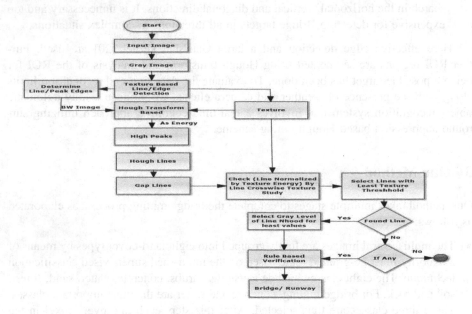

Fig. 1. Architecture of the bridge detection and mining method

The proposed algorithm is not specific to any orientation of the target nor rely on any structural formation of the target. Even with low image contrast, the proposed algorithm offers a good viability and precision in recognition of linear targets such as bridges. Figure 1 present below explains the overall methodology which has been used for the implementation of the proposed method.

3.1 Edge/Cornerness

Edge is the gradient magnitude in either direction whereas corner detection is an approach used within computer vision systems to extract certain kinds of features and infer the contents of an image. Corner detection is frequently used in motion detection, image registration, video tracking, image mosaicking, panorama stitching, 3D modeling and object recognition. A corner can be defined as the intersection of two edges while an edge is a sharp change in image brightness. Edges are a location with high gradient need smoothing to reduce noise prior to taking derivative, in x and y-direction. In the start image is converted to a gray level using rgb2gray, using the gray image. It is hard to find the gradient by using the Eq. 1 as shown below.

$$G(x) = e - x2/2\sigma2 \tag{1}$$

In order to simplify the computation, we adopt another equation equal to the Eq. 1. The Eq. 2 present below is a first-order derivative function of a Gaussian function.

$$G(x) = \left(-\frac{x}{\sigma^2}\right)e\frac{-x^2}{2\sigma^2} \tag{2}$$

Because the computation of 2D convolution is complex and large, find the gradient by convolving x-direction and y-direction individually. Basic Idea of corner detection algorithm is always finding a point where two edges meet, i.e. high gradient in two directions, this is why it is undefined at a single point as there would be only one gradient per point [16]. Edges always show strong brightness change in single direction while Corner shows strong brightness changes in orthogonal direction. The edges and cornerness features which can be used for applying Hough transform are represented in Fig. 3 while the Fig. 4 which follows shows the creation of masks and the implementation of the proposed algorithm.

Fig. 2. (Left to Right) Applied Harris on a gray image, R1 (Edges but not corner) and Applied Peaks

As shown in Fig. 2, we have edges, cornerness, and non-cornerness features, using these to find line after applying Hough transform and the sum of cornerness crosswise that line (considering no apparent texture across bridges).

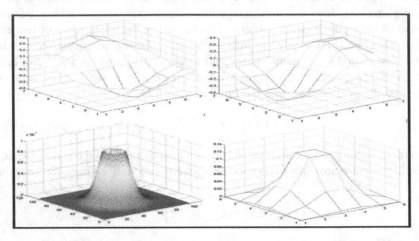

Fig. 3. (Left to Right) Gx Smoothing Mask, Gy Smoothing Mask, Inhibitor Kernel and Corner

3.2 Implementation

In order to determine the gradient image in the x-direction measuring the horizontal change in intensity is calculated by using this mask. Similarly, to determine the gradient image in the Y direction measuring the horizontal change in intensity is calculated by using this mask. The surround inhibition step is meant to suppress texture edges while leaving relatively unaffected the contours of objects and region boundaries. This biologically motivated mechanism introduced in Fig. 3 is particularly useful for contour-based object recognition.

In that case, texture edges play the role of noise that obscures object contours and region boundaries and should preferably be eliminated [21]. Different types of super-position can be used: L1, L2, and L-infinity norms. Alpha(α) This parameter controls the strength of surround suppression - the higher the value of Alpha (α), the more the strength (gradient magnitude) of an edge surrounded by other edges will be reduced. The default is 1 but one may need larger values in order to completely suppress texture edges K1 and K2. The size of the annual surround which has a substantial contribution to the suppression increases with K2. Default values are K1 = 1 and K2 = 4.

In order to compute corner, first, we have to smooth the integral I_x and I_y component found above by using this mask. Figure 4 represents the I_x and I_y components on a grayscale image after it has been converted from a color image.

The output image I is of the same class as the input image. If I is a color map, the input and output color maps [11] are both of class double. A gradient image in the x-direction measures the horizontal change in intensity.

Fig. 4. (Top Left to Right) Original, RGB to Gray, lx, ly, Grad

A gradient image in the y-direction measures the vertical change in intensity. Gradient magnitude of Gray images uses $_{Ix}$ and Iy. Simplified edge is computed using threshold.

3.3 Finding Edge, Energy and Texture

I_y^2 is the square of a smoothed version of integral I_y for corner detection and I_{xy} is the integral of a smoothed version of after being convolution. I_x^2 is the square of a smoothed version and smoothed version of integral $_{Ix}$ for Texture detection. I_y^2 is the square of a smoothed version of integral I_y for Texture detection. I_x is the integral of a smoothed version of after being convolution.

R2 is positive Texture density computed using Harris corner and edge detection. R1 is negative with large magnitude for an edge, computed using Harris corner and edge detection. To find R1_Edge, R1 is threshold using method, its value varies for visible and thermal imagery. The parameters of edge detection and texture determination have been represented in Fig. 5.

To find R2_Corner, R2 is threshold using its maximum value percentage; its value varies for visible and thermal imagery. The Fig. 6 shows the R2 corner and the peal results. To find Texture, R2 is threshold using its maximum value percentage; its value varies for visible and thermal imagery. To find Edge, R1 is threshold using its maximum value percentage; its value varies for visible and thermal imagery. The input image contains the minimum texture and the mean value which have been rep resented in the Fig. 7 present below.

Fig. 5. (Top Left to Right) I_x^2, I_y^2, I_{xy}, R2, R1 and R1 edge.

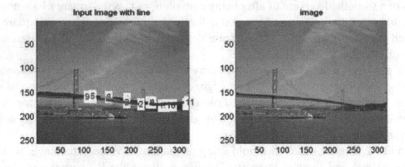

Fig. 6. (Top Left to Right) R2 Corner, R1 surround suppress edge, Peak Result, Hough Transform and Crosswise Image study of lines

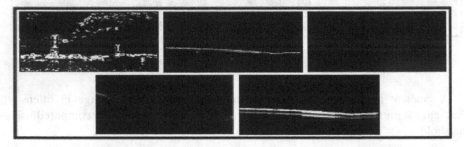

Fig. 7. The minimum Texture and mean value are our feature, so we plot the lines to express the Bridge as a final result.

4 Experimental Results

Projected data set is of 70–80 images including Google images. All parameters used in this algorithm depends upon the camera and image characteristics; fov, focal length, aov, Threshold for controlling Parameter R1 (cornerness) and R2 (edges), it depends upon sigma values provided for edges and corners and that also depends upon the parameters set. The yellow lines are not qualified as rejected by GAP analysis, green lines are showing least cornerness (almost zero) and therefore not considered in the respective target, red lines show Single Target qualified and the blue lines show other lines processed but not qualified to be a bridge. Considering the orientation of the bridge and it's across area characteristics i.e.

a. Self-resemblance of features (similar apparent texture) on both side of the bridge
b. Lower gray value of river or canal. Crosswise corner sum; Perfect corner sum across lines using Nhood, omitting corner pixel of the line and Ratio; Ratio of both side of the line from the Corners (Corner ratio) and Mean value (Mean ratio). These features crosswise corner sum, Ratios, slice (length of the line) and landmark check can be used for creating clustering classifier, to train the data set and neural networks.

In the following figures we have Ratio of both side of the line, Corners (Corner ratio) and Mean value (Mean ratio). Corner ratio is 1 and means the ratio is greater than 1. We can build a role for any candidate line to have some reasonable ratio values. While the Table 1 following the image represents the values obtained for the image analysis and represent the single target qualified based on minimum corner density. The same features are further represented in Fig. 9 and explained through Table 2 which represent a bridge based in New York, which contain the analysis of a bridge in China. These values with other features are helpful for creating clustering classifier to improve the performance and elimination and approval of candidate line to be the targeted bridge.

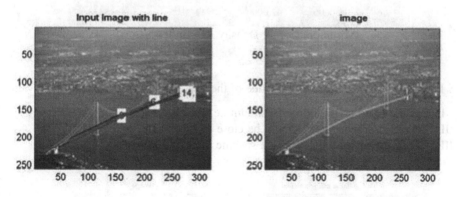

Fig. 8. The analysis of bridge image, determining the edges, ratio of sides and the mean ratio

Table 1. Single target qualified based on minimum corner density

Results...				
K	Length	Corner Sum	Corner Ratio	Mean Sum
1	188.0000	7.0000	1.3333	104.7900
2	98.0000	8.0000	1.6667	108.8500
3	137.0000	15.0000	1.5000	110.8367
4	162.0000	6.0000	1.0000	107.4900
5	188.0000	14.0000	1.3333	111.8100

Minimum Corner Density	Index
6	4

Table 2. Single target qualified based on minimum corner density

Results...				
K	Length	Corner Sum	Corner Ratio	Mean Sum
1	52.0000	32.0000	1.4615	60.3500
2	41.0000	28.0000	1.5455	40.8500
3	39.0000	2.0000	1.0000	64.0233
4	60.0000	35.0000	1.0588	49.3733
5	62.0000	2.0000	1.0000	65.9000
6	45.0000	42.0000	1.2105	96.5433

Minimum Corner Density	Index
2	3 5

In Fig. 8, we have Ratio of both side of the line, Texture density and Mean value.

   I. Texture Density should be minimum i.e. 6
  II. Texture crosswise ratio should be close to 1 i.e. 1.00
 III. Mean sum should not exceed a specific threshold.

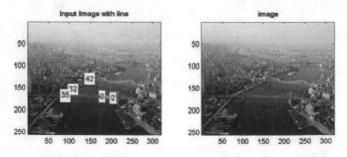

Fig. 9. Aerial image of a New York bridge

5   Conclusion and Future Work

The overall study was aimed at developing an algorithm that could be used for aid- ing the humanitarian relief purposes. The proposed algorithm works on corner and edge detection. The Hough transform is an image processing technique that is used to extract or detect features of a particular shape in the image and best works with large images where the effects of noise and undesired features are minimal. The result shows that this approach consists of the basic and effective method with as low as 0.8 s to recognize the target bridge from the image. The time is low indeed; real- time DSP/FPGA parallel processing machines perform operations much faster than normal PC. Making it suitable to operate and incorporate the speed with the standard time/frequency i.e. 4 images per second with lowest possible time expected to be as 0.20 s. The increase in the image databases have made it possible to carry out database wide search for the purpose of image mining. The use of the proposed algorithm can help in ensuring that the image-based data is utilized for the purpose of information mining and that the humanitarian purpose of saving lives through technology could truly be fulfilled.

The future enhancement may be carried out on hardware as well as software basis. The key goal of the algorithm is to ensure that the geographical database could be enhanced, and the image processing technique could be utilized for extracting infor- mation about the disaster-prone areas and analyze the images of these areas to identify the roads and bridges which could be used for making aid reach respective areas. The enhancements may be based on upgrading the existing hardware as well as the software to further increase the workability in the real-time environment and decrease the overall time required for mining information through the images.

Acknowledgement. The authors would like to thank the support from Guangxi Key Laboratory of Trusted Software (No. kx201615), Capacity Building Project for Young University Staff in Guangxi Province, Department of Education of Guangxi Province (No. ky2016YB149) and Major research project, Bureau of Science and Technology of Guilin (No.: 20180101-3).

References

1. Memon, N.: Malevolent Floods of Pakistan, 2010–2012 (2013)
2. Khan, S.I., Hong, Y., Gourley, J.J., Khattak, M.U., De Groeve, T.: Multi-sensor imaging and space-ground cross-validation for 2010 flood along Indus River, Pakistan. Remote Sens. **6**(3), 2393–2407 (2014)
3. Zhen, Z., Tianxu, Z., Guoyou, W.: Research on bridge recognition in long-range infrared images. ACTA Electronica Sinica **26**(11) (1998)
4. Zaihua, Y., Shuqian, Y.: Automatic recognition and tracking techniques for infrared bridge image. Infrared Laser Eng. (1998)
5. Hsu, W., Lee, M.L., Zhang, J.: Image mining: trends and developments. J. Intell. Inf. Syst. **19**(1), 7–23 (2002)
6. Gedika, E., Çinara, U., Karamana, E., Yardımcıa, Y., Halıcıb, U., Pakinc, K.: A new robust method for bridge detection from high resolution electro-optic satellite images. In: Proceedings of IEEE International GeoSciene and Remote Sensing Symposium, vol. 6 (2003)

7. Sui, H., Gong, J., Xiao, J., Li, M.: Automatic recognition of bridges over water and registration in remotely sensed images with GIS data. In: Proceedings of the ISPRS Commission VII Mid-term Symp, vol. 36 (2006)
8. Li, M., Yandong, T., Zelin, S.: Segmentation and recognition of bridge over water based on Mumford-Shah model. Infrared Laser Eng. 35(4), 499 (2006)
9. Abraham, L., Sasikumar, M.: A fully automatic bridge extraction technique for satellite images. Int. J. Inf. Process. 6(3), 89–97 (2012)
10. Chaudhuri, D., Samal, A.: An automatic bridge detection technique for multispectral images. In: CSE Conference and Workshop Papers (2008)
11. Fu, Y., Xing, K., Huang, Y., Xiao, Y.: Recognition of bridge over water in high-resolution remote sensing images. In: 2009 WRI World Congress on Computer Science and Information Engineering (2009)
12. Beuliga, S., von Schonermarka, M., Hubera, F.: A FPGA-based automatic bridge over water recognition in high-resolution satellite images. In: Image and Signal Processing for Remote Sensing XVIII, Proceedings of SPIE, vol. 8537 (2012)
13. Soyman, Y.: Robust automatic target recognition in FLIR imagery. In: Proceedings of SPIE, vol. 8391 (2012)
14. Trias-Sanz, R., Loménie, N.: Automatic bridge detection in high-resolution satellite images. In: Crowley, J.L., Piater, J.H., Vincze, M., Paletta, L. (eds.) ICVS 2003. LNCS, vol. 2626, pp. 172–181. Springer, Heidelberg (2003). https://doi.org/10.1007/3-540-36592-3_17
15. Cao, Z., Zhang, X.: Forward-looking infrared target recognition based on histograms of oriented gradients. In: Proceedings of SPIE, vol. 8003 (2011)
16. Yuan, X.-H., Jin, L.-Z., Li, J.-X., Xia, L.-Z.: Recognition of bridges over water through detecting and analyzing regions of interest. J. Infrared Millimeter Waves 22(5), 331–336 (2003)
17. Xu, W., Xu, L., Hu, Y.: Adaptive edge detection using a half neighborhood algorithm. J. Shanghai Univ. (Nat. Sci.) 12(2), 146–149 (2006)
18. Ji, Z., Hsu, W., Lee, M.L.: Image mining: issues, frameworks and techniques. In: Proceedings of the Second International Conference on Multimedia Data Mining. Springer, Heidelberg (2001)
19. Zahradnikova, B., Duchovicova, S., Schreiber, P.: Image mining: review and new challenges. Int. J. Adv. Comput. Sci. Appl. 6(7), 242–246 (2015)
20. Zhang, J.: Advancements of outlier detection: a survey. ICST Trans. Scalable Inf. Syst. 13 (1), 1–26 (2013)
21. Zhao, C., Shi, W., Deng, Y.: Novel edge detection method based on gradient. J. Opto-Electron. Eng. (2005)

Spike Sorting with Locally Weighted Co-association Matrix-Based Spectral Clustering

Wei Ji[1], Zhenbin Li[1], and Yun Li[2(✉)]

[1] College of Telecommunication and Information Engineering,
Nanjing University of Posts and Telecommunications, Nanjing, Jiangsu, China
jiwei@njupt.edu.cn
[2] College of Computer Science and Technology,
Nanjing University of Posts and Telecommunications, Nanjing, Jiangsu, China
liyun@njupt.edu.cn

Abstract. Spike sorting for neuron recordings is one of the core tasks in brain function studies. Spike sorting always consists of spike detection, feature extraction and clustering. Most of the clustering algorithms adopted in spike sorting schemes are subject to the shapes and structures of the signal except the spectral clustering algorithm. To improve the performance of spectral clustering algorithm for spike sorting, in this paper, a locally weighted co-association matrix is employed as the similarity matrix and the Shannon entropy is also introduced to measure the dependability of clustering. Experimental results show that the performance of spike sorting with the improved spectral clustering algorithm is superior to that of spike sorting with other classic clustering algorithms.

Keywords: Spike sorting · Co-association matrix · Local weighting · Spectral clustering

1 Introduction

Spike sorting for neuron recordings is very important in brain function studies. Neuron recordings picked up by extracellular recordings are a mixture of several neuron signals from the same brain area. Spike sorting is to separate the neuron recordings and group them into different clusters [1]. Then spike sorting always consists of spike detection, feature extraction and clustering. Spike detection is to detect the spikes from the collected neuron recordings. The simplest spike detection method is amplitude threshold detection [2, 3]. However, a wrong threshold will lead to a high mistake rate. To solve this problem, ovonic threshold method and peak detection method have been proposed [4]. After the spike detection, the feature extraction is implemented to capture features from the spikes. Principal component analysis (PCA) [5] and wavelet transform [6, 7] are widely used feature extraction methods. The final step of spike sorting is to cluster the spikes and identify the single neuron signal. The frequently used clustering algorithms in spike sorting is Gaussian mixed model (GMM) [8]. However, it is subject to the shapes and structures of spike. Spectral clustering algorithm always works well on complicated data [9], so it is a good choice for spike sorting.

© Springer Nature Switzerland AG 2019
L. H. U and H. W. Lauw (Eds.): PAKDD 2019 Workshops, LNAI 11607, pp. 201–213, 2019.
https://doi.org/10.1007/978-3-030-26142-9_18

The classic spectral clustering algorithm NJW always uses Gaussian kernel function as the similarity matrix [10]. To set the parameter σ in Gaussian kernel function, Zelnik-Manor and Perona adopt K-nearest neighbor algorithm to optimize the parameter σ [11]. In order to furtherly improve the performance of spectral clustering algorithm on spike sorting, in this paper, a spectral clustering algorithm based on locally weighted co-association matrix (LWCA-NJW) is proposed to cluster the spikes. In LWCA-NJW, locally weighted co-association matrix is employed as the similarity matrix, ensemble idea is embedded in locally weighted co-association matrix to improve the performance of clustering.

To evaluate the clustering performance, Davies-Bouldin Index (DBI) and Dunn Index (DI) are utilized. The spike sorting performance is compared among the improved spectral clustering algorithm LWCA-NJW, the classical spectral clustering algorithm NJW, GMM and some ensemble clustering algorithms. Experimental results show that the performance of spike sorting with the LWCA-NJW is superior to that of spike sorting with other clustering algorithms.

The remainder of this paper is organized as follows. Feature extraction for spike sorting with discrete wavelet transform is introduced in Sect. 2. Section 3 presents the classic spectral clustering algorithm and the proposed improvement. The procedure of spike sorting with the improved spectral clustering algorithm is summarized in Sect. 4. Section 5 provides the experimental results and discussion. The paper ends with conclusion in Sect. 6.

2 Feature Extraction for Spike Sorting

Wavelet transform is widely used to extract the features for spikes [12, 13]. The wavelet transform is briefly introduced by the following equation,

$$WT_x(a, b) = \frac{1}{\sqrt{|a|}} \int_{-\infty}^{+\infty} \Psi\left(\frac{t - b}{a}\right) x(t) dt \tag{1}$$

where $\Psi(t)$ is mother wavelet, a is the scale parameter and b is the translation parameter. In real application, discrete wavelet transform (DWT) is commonly used. The multiresolution decomposition of discrete wavelet transform can separate the signal into detailed components and approximation components at different scales [14].

In this paper, because the frequency range of spikes is from 0.4 Hz to 80 Hz, discrete wavelet transform with four level decomposition is utilized to extract features for spikes and each spike is ordinarily decomposed into five sub-bands: delta (0.5– 4 Hz), theta (4–8 Hz), alpha (8–12 Hz), beta (13–30 Hz) and gamma (30–60 Hz) [15]. In the first level decomposition, the spikes are filtered by the low pass filter and high pass filter, respectively. Then the filtered spikes are sampled to fetch approximation components (A1) and detailed components (D1) of the first level. The same procedure can be duplicated for the first level approximation components (A1) to get its approximation component (A2) and detailed component (D2) of the second level, and so on. The above mentioned procedure is shown in Fig. 1.

If the vector composed of A4, D4, D3, D2 and D1 is treated as a feature vector directly, the clustering performance could be decreased. One possible reason is that some wavelet coefficients in the feature vector fail to meet the requirement of distinguishing the different spikes. It's necessary to pick out the desirable wavelet coefficients or extract feature furtherly. In this paper, the mean, standard deviation, skewness, energy and entropy of wavelet coefficients [16] are calculated to represent the spikes.

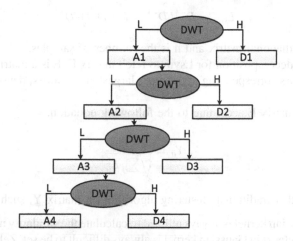

Fig. 1. Four level wavelet decomposition

3 Spectral Clustering Algorithm

Spectral clustering algorithm always works well on complicated data [9], so it is still a good choice for spike sorting.

3.1 Classic Spectral Clustering Algorithm – NJW

The main idea of NJW is to construct a graph, denoted as G = (V, E), where V is the set of vertexes in the graph and E represents the set of the edges in graph G [17]. The vertex represents a sample and edge represents the similarity between samples in the graph. Based on the graph, classic spectral clustering algorithm NJW is performed as follows.

1. Constructing the similarity matrix W based on the following equation,

$$W_{ij} = e^{-\frac{d^2(x_i, x_j)}{2\sigma^2}} \tag{2}$$

where W_{ij} and $d(x_i, x_j)$ are the similarity and distance between the samples x_i and x_j respectively.

2. Constructing the Laplace matrix Lsym based on the following equation,

$$D_{ii} = \sum_{j=1}^{n} W_{ij} \tag{3}$$

$$L = D - W \tag{4}$$

$$L_{sym} = D^{-\frac{1}{2}} L D^{-\frac{1}{2}} = I - D^{-\frac{1}{2}} W D^{-\frac{1}{2}} \tag{5}$$

where D is a diagonal matrix and n is the number of samples.

3. Making eigendecomposition for Lsym to fetch matrix E. E is a matrix composed of the eigenvectors corresponding to the top g largest eigenvalues, denoted as E = [e1, e2, ..., e.g.].

4. Normalizing matrix E according to the following equation,

$$Y_{ij} = \frac{E_{ij}}{\sqrt{\sum_j E_{ij}^2}} \tag{6}$$

5. Performing other traditional clustering algorithm on matrix Y, such as k-means.

In NJW, Gaussian kernel is always utilized to calculate the similarity matrix as shown in (2). The parameter σ in Gaussian kernel is always difficult to be set. Zelnik-Manor and Perona optimized the parameter σ by adopting K-nearest neighbors algorithm. The optimized similarity matrix is given by the following equation,

$$W_{ij} = e^{-\frac{d^2(x_i, x_j)}{\sigma_i \sigma_j}} \tag{7}$$

where $\sigma_i = d(x_i, x_i^K)$ is the distance between the sample x_i and x_i^K. x_i^K represents K-th nearest neighbor sample of x_i.

3.2 Spectral Clustering Based on Co-association Matrix (CA-NJW)

To alleviate the issue of parameter setting in Gaussian kernel function and to improve the clustering performance, co-association matrix [18] is utilized in spectral clustering. Given a data set DS, B partitions are obtained by clustering B times on DS. And l-th partition is denoted as π^l. B partitions is represented as an ensemble $\Pi = \{\pi^1, \pi^2, \ldots, \pi^B\}$. For example, given a data set DS = $\{x_1, x_2, \ldots, x_5\}$, three partitions, π^1, π^2 and π^3 are obtained by clustering three times on it. Five samples are assigned into three clusters for each partition, which is shown in Table 1. The elements are cluster labels of each sample in π^1, π^2 and π^3.

With the above definition, the co-association matrix CA is defined as follows,

$$CA = \{ca_{ij}\}_{n\times n} \tag{8}$$

$$ca_{ij} = \frac{1}{B}\sum_{l=1}^{B} \theta_{ij}^{l} \tag{9}$$

$$\theta_{ij}^{l} = \begin{cases} 1, & \pi_i^l = \pi_j^l \\ 0, & otherwise \end{cases} \tag{10}$$

where B is the number of partitions in ensemble Π and π_i^l is the cluster label of i-th sample in partition π^l. Co-association matrix evaluates the similarity among samples by counting the times that different samples are located in the same cluster. The more times that samples are grouped into the same cluster, the more similar they are.

Table 1. A clustering ensemble with three partitions.

Sample	Π		
	π^1	π^2	π^3
x_1	2	1	3
x_2	2	1	3
x_3	1	2	3
x_4	3	3	1
x_5	3	3	2

3.3 Spectral Clustering Based on Locally Weighted Co-association Matrix (LWCA-NJW)

The co-association matrix treats the clusters of B partitions equally without considering the dependability of clusters. In other words, the same weight is assigned to the clusters with different dependability, which will degrade the clustering performance. So we like to introduce the Shannon entropy [19] to measure the unreliability of each cluster and different weights are assigned to the clusters according to their dependability.

Before giving the definition of dependability, the cluster representation (CR) will be presented. Given a partition π^l, the cluster representation of π^l is defined by the following equation,

$$CR(\pi^l) = [\tau_1^l, \tau_2^l, \ldots, \tau_{k^l}^l] \tag{11}$$

$$\tau_c^l(i) = \begin{cases} 1, & \pi_i^l = c \\ 0, & otherwise \end{cases} \tag{12}$$

where k^l is the number of clusters in partition π^l and τ_c^l is the cluster representation of c-th cluster in partition π^l. If the i-th sample in π^l is grouped into c-th cluster, the i-th element in the τ_c^l is set as 1. Then τ_c^l is a column vector consists of 0 and 1. The cluster representation of ensemble Π is defined as (13).

$$T = \left[CR(\pi^1), CR(\pi^2), \ldots, CR(\pi^B) \right] \tag{13}$$

where T is a n $\times$ β matrix. n is the number of samples and β is the total number of clusters in ensemble Π, denoted as $\beta = \sum_{l=1}^{B} k^l$. For example, given the data set shown in Table 1, the cluster representation of π^1, π^2 and π^3 are shown in Table 2.

Table 2. Cluster representation for the three partitions.

Sample	CR (π^1)			CR (π^2)			CR (π^3)		
	τ_1^1	τ_2^1	τ_3^1	τ_1^2	τ_2^2	τ_3^2	τ_1^3	τ_2^3	τ_3^3
x_1	0	1	0	1	0	0	0	0	1
x_2	0	1	0	1	0	0	0	0	1
x_3	1	0	0	0	1	0	0	0	1
x_4	0	0	1	0	0	1	1	0	0
x_5	0	0	1	0	0	1	0	1	0

The unreliability of cluster T_f corresponding to π^l is defined by the following equation,

$$u(T_f, \pi^l) = -\sum_{c=1}^{k^l} log_2 \frac{[T_f, \tau_c^l]}{[T_f, T_f]}^{\frac{[T_f, \tau_c^l]}{[T_f, T_f]}} \tag{14}$$

where T_f is the f-th column of matrix T and τ_c^l is the c-th column of $CR(\pi^l)$. $[T_f, \tau_c^l]$ represents the inner product of T_f and τ_c^l. Obviously, $[T_f, \tau_c^l]/[T_f, T_f] \in [0, 1]$. If all the samples in T_f belong to τ_c^l, $[T_f, \tau_c^l]/[T_f, T_f]$ equals to 1 and the unreliability $u(T_f, \pi^l)$ equals to 0. If all the samples in T_f belong to several clusters in π^l, the value of $u(T_f, \pi^l)$ will increase in general. Specially, if all the samples in T_f belong to k^l clusters in π^l evenly, $[T_f, \tau_c^l]/[T_f, T_f]$ equals to $\frac{1}{k^l}$ and $u(T_f, \pi^l)$ obtains the maximum value $log_2^{k^l}$. Then $u(T_f, \pi^l) \in \left[0, log_2^{k^l} \right]$.

Suppose that all the partitions in ensemble Π are independent, the unreliability of cluster T_f with respect to the ensemble Π is defined as follows,

$$U(T_f, \Pi) = \frac{1}{B} \sum_{l=1}^{B} u(T_f, \pi^l) \tag{15}$$

If all the samples in T_f belong to one cluster in each partition, which means that the samples in T_f should be grouped into the same cluster in all the partitions. In this case,

$U(T_f, \Pi)$ equals to 0. If all the samples in T_f belong to different clusters in each partition, indicating that the samples in T_f should not be assigned to the same cluster, then the value of $U(T_f, \Pi)$ increases.

The normalized unreliability is defined as (16),

$$NU(T_f, \Pi) = \frac{1}{B} \sum_{l=1}^{B} nu(T_f, \pi^l) \tag{16}$$

$$nu(T_f, \pi^l) = \frac{u(T_f, \pi^l)}{\log_2^{k^l}} \tag{17}$$

The normalized dependability of T_f with respect to the ensemble Π is defined as (18),

$$ND(T_f, \Pi) = 1 - NU(T_f, \Pi) \tag{18}$$

With the definition of cluster dependability, it is possible to weight the cluster according to its dependability. Then the locally weighted co-association matrix is defined as the following equation,

$$LWCA = \{lwca_{ij}\}_{n \times n} \tag{19}$$

$$lwca_{ij} = \frac{1}{B} \sum_{l=1}^{B} \xi_{ij}^l \tag{20}$$

$$\xi_{ij}^l = \begin{cases} ND(\tau_{\pi_i^l}^l, \Pi), & \pi_i^l = \pi_j^l \\ 0, & otherwise \end{cases} \tag{21}$$

where $ND(\tau_{\pi_i^l}^l, \Pi)$ is the normalized dependability of the cluster where sample x_i is located in l-th partition.

4   Spike Sorting with LWCA-NJW

The improved spectral clustering algorithm above is utilized for spike sorting. To construct the locally weighted co-association matrix in LWCA-NJW, the spikes data set is sampled randomly with the rate r, and the k-means algorithm is performed on sampled data set. The procedure of spike sorting with LWCA-NJW is summarized in Algorithm 1.

Algorithm 1 Spike Sorting with LWCA-NJW

Input: Spikes, k (the number of clusters), B (the number of partitions), r (sampling rate).

1:    Extracting the features for spikes using discrete wavelet transform.
2:    Generating the ensemble Π with the size of B.
       For l=1, 2, ..., B
           Sample spikes data randomly.
           Perform k-means clustering on the sampled spikes to obtain one partition π^l.
       End
3:    Constructing the LWCA matrix with Π according to (19), (20) and (21).
4:    Considering the LWCA matrix as a similarity matrix to construct the Laplace matrix Lsym according to (5).
5:    Making eigendecomposition for Lsym to obtain matrix E and normalizing matrix E according to (6) to obtain matrix Y.
6:    Performing k-means clustering on the matrix Y to obtain the consensus partition π^*.

Output: π^*.

5 Spike Sorting Experiments

In this section, we like to evaluate the spike sorting performance of the proposed LWCA-NJW and other clustering algorithms on spikes data set.

5.1 Data Set and Evaluation Methods

The data set used in the experiments is from the University of Leicester. It consists of simulated spike signals constructed using a database of 594 different average spike shapes compiled from recordings in the neocortex and basal ganglia. To construct the signals, spikes randomly selected from the database were superimposed at random times and amplitudes to generate background noise. Then, a train of three distinct spike shapes were superimposed on the noise signal at random times. The data set is available on-line at https://vis.caltech.edu/~rodri/Wave_clus/Simulator.zip. The partial data are shown in Fig. 2.

Two frequently-used clustering performance evaluation index, Davies-Bouldin Index (DBI) and Dunn Index (DI), are used to evaluate the clustering quality. The DBI and DI are independent. The lower DBI or the higher DI indicate the better clustering performance. Given a partition with k clusters, i.e.. $C_1, C_2, \ldots, C_k$, the DBI index is defined as follow,

$$DBI = \frac{1}{k} \sum_{\lambda,\mu=1}^{k} \max_{\lambda \neq \mu} \left(\frac{avg(C_\lambda) + avg(C_\mu)}{d_{cen}(C_\lambda, C_\mu)} \right) \tag{22}$$

spikes ✕					
⊞ 3061x64 double					
	1	2	3	4	5	6
1	0.1243	0.1043	0.1066	0.1276	0.1459	0.1462
2	-0.0149	-7.8958e-05	0.0049	0.0043	-0.0022	-0.0209
3	0.1123	0.1295	0.1535	0.1809	0.2073	0.2238
4	-0.0117	-0.0207	-0.0279	-0.0211	0.0063	0.0496
5	-0.0159	0.0020	0.0222	0.0414	0.0624	0.0867
6	-0.1508	-0.0867	-0.0360	-0.0154	-0.0190	-0.0289
7	0.0840	0.0664	0.0403	-0.0029	-0.0523	-0.0892
8	0.1230	0.1017	0.0802	0.0657	0.0615	0.0664
9	0.0560	0.0613	0.0656	0.0635	0.0476	0.0102

Fig. 2. The partial data used in the experiments.

where $avg(C)$ is the average intra-distance among the samples in cluster C and $d_{cen}(C_\lambda, C_\mu)$ is the inter-distance between the center of cluster C_λ and C_μ.

The DI index is defined by the following equation,

$$DI = \frac{\min\limits_{1 \leq \lambda \neq \mu \leq k} d_{min}(C_\lambda, C_\mu)}{\max\limits_{1 \leq \alpha \leq k} diam(C_\alpha)} \tag{23}$$

where $diam(C)$ is the maximum intra-distance among the samples in cluster C and $d_{min}(C_\lambda, C_\mu)$ is the minimum inter-distance between the C_λ and C_μ, which is calculated by (24).

$$d_{min}(C_\lambda, C_\mu) = \min\limits_{x_i \in C_\lambda, x_j \in C_\mu} dist(x_i, x_j) \tag{24}$$

5.2   Experiments on Parameter k

Since the k-means clustering is used to generate the ensemble Π, it is important to determine the k value for the number of clusters in each partition. We like to draw a curve about a proper clustering performance index along with the number of clusters k. The common clustering performance index includes the radius and diameter of cluster, which are defined as follows respectively,

$$radius = \max\limits_{1 \leq i \leq |C|} dist(x_i, \Omega) \tag{25}$$

$$diameter = \max\limits_{1 \leq i < j \leq |C|} dist(x_i, x_j) \tag{26}$$

where $|C|$ is the total number of samples in cluster C and Ω is the center of cluster C. $dist(x_i, x_j)$ is the distance between the sample x_i and x_j.

In our case, the diameter of cluster is employed to study the effect of the parameter k. The experimental result is shown in Fig. 3. We can observe that the inflection point of the curve is located in k = 4. As a result, the number of clusters for the experimental spikes data is chosen as 4.

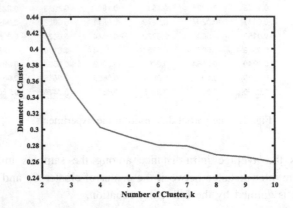

Fig. 3. Effect of the number of clusters on cluster diameter.

5.3 Experiments on Parameters *B* and *r*

To evaluate the effect of parameter B on spike sorting performance, different values are chosen from the set {10, 20, 30, 40, 50, 60, 70, 80, 90, 100} in experiments. The DBI index is computed 20 times for each B value, and the average DBI index with regard to the different B is shown in Fig. 4.

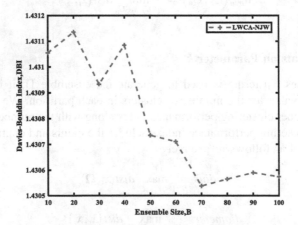

Fig. 4. Effect of the ensemble size on DBI

We can observe that the DBI index decreases along with the B value generally. The reason lies in that the original k-means algorithm for each partition tends to get local optimum, however, it can be alleviated through performing k-means repeatedly to obtain multiple partitions. The clusters in partition that get the local optimum are assigned a low weight. Of course, due to the randomness of sampling and the application of k-means, the DBI does not decrease dramatically along with the increasing of B. Based on the experimental results, the parameter B is set as 70 in our case.

For the parameter r, it is chosen from the set {0.1, 0.2, 0.3, 0.4, 0.5, 0.6, 0.7, 0.8, 0.9, 1.0}. The DBI index is also computed 20 times for each r value. The DBI indices with regard to the different sampling rate r are shown in Fig. 5.

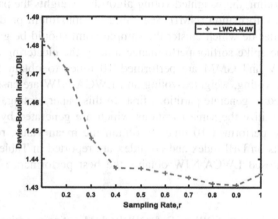

Fig. 5. Effect of the sampling rate on DBI.

As shown in the figure, the DBI index decreases along with the growth of r and it is obvious that the parameter r is better to be set as 0.9.

With the given k, B and r, the consensus partition obtained by the improved algorithm is shown in Fig. 6. The left figure represents the mixed spike signals and the clusters correspond to different neurons are shown on the right.

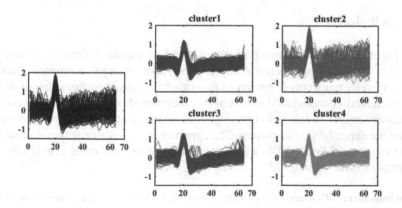

Fig. 6. The result of spike sorting using LCWA-NJW

5.4   Comparison with Other Clustering Algorithms on Spike Sorting

LWCA-NJW is compared with NJW and GMM first. Considering that the ensemble idea is embedded into the locally weighted co-association matrix, LWCA-NJW is also an ensemble clustering algorithm. So we like to compare LWCA-NJW with other ensemble clustering algorithms such as voting and weighted-voting [20]. The main idea of voting algorithm is to generate several partitions first, then counting the times that the sample points are grouped into some cluster in these partitions. If the sample point is assigned to a cluster for the most times, it is grouped into this cluster finally. Based on the voting algorithm, the weighted-voting algorithm weights the partitions with the normalized mutual information (NMI). Searching the maximum product of the times and the weight to decide which cluster the sample point should be grouped into.

To compare the spike sorting performance among the clustering algorithms mentioned above, NJW and GMM are performed 10 times to obtain the mean value, respectively. Since voting, weighted-voting and LWCA-NJW are ensemble clustering algorithm, it is need to generate partitions first. In this paper, voting, weighted-voting and LWCA-NJW utilize the same partitions which are generated by k-means. Similarly, they are also performed 10 times to obtain the mean value, respectively. The experimental results on DBI index and DI index are reported in Table 3. As shown in the table, the proposed LWCA-NJW obtains the best performance among them on spike sorting.

Table 3. Experimental results of DBI and DI for NJW, GMM, voting, w-voting and LWCA-NJW.

Algorithm	Davies-Bouldin Index, DBI	Dunn Index, DI
NJW	2.3880	0.0257
GMM	2.5863	0.0139
Voting	1.5905	0.0259
w-voting	1.5773	0.0268
LWCA-NJW	1.4287	0.0289

6   Conclusion

In this paper, a spectral clustering algorithm based on locally weighted co-association matrix (LWCA-NJW) is proposed to improve the spike sorting performance efficiently. Discrete wavelet transform is employed to extract the spikes features. Especially, the locally weighted co-association matrix with ensemble idea is introduced to construct the similarity matrix in spectral clustering. And Shannon entropy is also introduced to calculate the dependability of clusters. The experimental results on spikes data set show that the proposed LWCA-NJW always performs better than other classic clustering algorithms in spike sorting task.

Acknowledgement. This work was partially supported by Natural Science Foundation of China (No. 61603197, 61772284, 61876091).

References

1. Quian Quiroga, R.: Spike sorting. Curr. Biol. **22**(2), R45 (2012)
2. Takahashi, S., Anzai, Y., Sakurai, Y.: Automatic sorting for multi-neuronal activity recorded with tetrodes in the presence of overlapping spikes. J. Neurophysiol. **89**(4), 2245–2258 (2003)
3. Zhang, P.M., et al.: Spike sorting based on automatic template reconstruction with a partial solution to the overlapping problem. J. Neurosci. Methods **135**(1–2), 55–65 (2004)
4. Borghi, T., et al.: A simple method for efficient spike detection in multiunit recordings. J. Neurosci. Methods **163**(1), 176–180 (2007)
5. Shoham, S., Fellows, M.R., Normann, R.A.: Robust, automatic spike sorting using mixtures of multivariate t-distributions. J. Neurosci. Methods **127**(2), 111–122 (2003)
6. Hulata, E., Segev, R., Ben-Jacob, E.: A method for spike sorting and detection based on wavelet packets and Shannon's mutual information. J. Neurosci. Methods **117**(1), 1–12 (2002)
7. Takekawa, T., Isomura, Y., Fukai, T.: Accurate spike sorting for multi-unit recordings. Eur. J. Neurosci. **31**(2), 263–272 (2010)
8. Wood, F., Black, M.J.: A nonparametric Bayesian alternative to spike sorting. J. Neurosci. Methods **173**(1), 1–12 (2008)
9. Jia, H.J., et al.: The latest research progress on spectral clustering. Neural Comput. Appl. **24**(7–8), 1477–1486 (2014)
10. Ng, A.Y., Jordan, M.I., Weiss, Y.: On spectral clustering: analysis and an algorithm. Adv. Neural. Inf. Process. Syst. **2**, 849–856 (2001)
11. Zelnik-Manor, L., Perona, P.: Self-tuning spectral clustering. In: Advances in Neural Information Processing Systems (2004)
12. Übeyli, E.D.: Wavelet/mixture of experts network structure for EEG signals classification. Expert Syst. Appl. **34**(3), 1954–1962 (2008)
13. Quiroga, R.Q., Nadasdy, Z., Ben-Shaul, Y.: Unsupervised spike detection and sorting with wavelets and superparamagnetic clustering. Neural Comput. **16**(8), 1661–1687 (2004)
14. Quian, Q.R., et al.: Wavelet transform in the analysis of the frequency composition of evoked potentials. Brain Res. Protoc. **8**(1), 16–24 (2001)
15. Adeli, H., Ghosh-Dastidar, S., Dadmehr, N.: A wavelet-chaos methodology for analysis of EEGs and EEG subbands to detect seizure and epilepsy. IEEE Trans. Biomed. Eng. **54**(2), 205–211 (2007)
16. Fahmy, A.A., et al.: Feature extraction of epilepsy EEG using discrete wavelet transform. In: Computer Engineering Conference (2017)
17. Shi, J., Malik, J.: Normalized cuts and image segmentation. IEEE Trans. Pattern Anal. Mach. Intell. **22**(8), 888–905 (2000)
18. Fred, A.L., Jain, A.K.: Combining multiple clustering using evidence accumulation. IEEE Trans. Pattern Anal. Mach. Intell. **27**(6), 835–850 (2005)
19. Huang, D., Wang, C.D., Lai, J.H.: Locally weighted ensemble clustering. IEEE Trans. Cybern. **48**(5), 1460–1473 (2018)
20. Zhou, Z.H., Tang, W.: Cluster ensemble. Knowl.-Based Syst. **19**(1), 77–83 (2006)

Label Distribution Learning Based Age-Invariant Face Recognition

Hai Huang[✉], Senlin Cheng[✉], Zhong Hong, and Liutong Xu

School of Computer Science,
Beijing University of Posts and Telecommunications, Beijing, China
hhuang@bupt.edu.cn, 972705994@qq.com,
hertz0725@hotmail.com

Abstract. Face recognition is an important application of computer vision. Although the accuracy of face recognition is high, face recognition and retrieval across age is still challenging. Faces across age can be very different caused by the aging process over time. The problem is that the images are not too similar, but with the same label. To reduce the intraclass discrepancy, in this paper we pro-pose a new method called Label Distribution learning for the end-to-end neural network to learn more discriminative features. Extensive experiments conducted on the three public domain face aging datasets (MORPH Album 2, CACD-VS and LFW) have shown the effectiveness of the proposed approach.

Keywords: Label distribution · Face recognition · Intra-class

1 Introduction

Recent years, Face recognition is developing very fast. There are many new methods appearing and have near perfect results in the face database, especially in deep-learning [1–8]. As a major challenge in face recognition, age-invariant face recognition is extremely valuable on various application scenarios. For example, we can match face images in different ages and find missing persons and child. it is very important and age-invariant face recognition has attracted much attention from both academic and industry for decades.

For face recognition, some researchers begin to design different loss functions to learn discriminative features so that it can have a better performance. So contrastive loss [20], triplet loss [21], were proposed to have better intra-class compactness and inter-class separability. However, we need to select some training pairs and triplets when using contrastive loss and triplet loss although they do really improve the performance and extract more discriminative features. The selection of samples is important for the training result, if all training samples are selected, the complexity can go up to $O(N^2)$ where N is the total number of training samples. The L-softmax loss [22] will enhance the angle margin between different classes. But the angle margin is hard to train. A-softmax [23] was developed to explicitly enforce the angle margin with two limitations based on the L-softmax. The center loss [24] learns different centers of different classes, it was combined with the softmax loss and the intra-class compactness, but center loss did not consider the inter-class separability. A deep convolutional

L. H. U and H. W. Lauw (Eds.): PAKDD 2019 Workshops, LNAI 11607, pp. 214–222, 2019.
https://doi.org/10.1007/978-3-030-26142-9_19

neural network can learn a good feature if their intra-class compactness and inter-class separability are well maximized. Actually, the feature of age-invariant face recognition has the same question.

For age-invariant face recognition, some people construct 2D or 3D aging models to reduce the age variation in face matching [9–11]. These models usually rely on strong parametric assumptions, accurate age estimation, as well as clean training data, and therefore they do not work well in unconstrainted environments. After some years, some scholars propose to separate the feature into identity and age components [12] using hidden factor analysis. sometimes the inter-class variation is much smaller than the intra-class variation because of the age. To solve this problem, there are always two technical schemes: generative scheme and discriminative scheme. The generative scheme can make faces one or more fixed age category then do recognition with the face representations. But the weakness of generative scheme is obvious, firstly, generation models sometimes generate noise that make results bad. Secondly, the generation models has two steps to do face recognition, it is not easy for us to optimize models. The discriminative scheme can construct the sophisticated discriminative model. Recently, The discriminative scheme develop rapidly, some researchers combine the deep learning to have a better performance than before. For example, they proposed a new coding framework called CARC that leveraged a reference image set (available from Internet) for age-invariant face recognition and retrieval. They also introduced a new large-scale face dataset, CACD. They used the linear combination of jointly learned deep features to represent identity and age information. The LF-CNNs (latent factor guided convolutional neural networks) achieved the state-of-the-art recognition accuracy in age-invariant face recognition.

The problem is that dissimilar images, but the same label. In this paper, we aim at designing a new deep learning approach called label distribution learning to effectively make the distance of dissimilar images for same label closer. Specifically, we use the siamese neural network trained with a label distribution loss function that attempts to bring class conditional probability distributions closer to each other.

2 Proposed Method

In this section, we introduce the our method called label distribution learning. Consider the conditional probability distributions for two input images x_1 and x_2, which can be given by $p(y|x_1; \Theta)$ and $p(y|x_2; \Theta)$. For a classification problem with N output classes, each of these distributions is an N-dimensional vector, with each element i denoting the belief of the classifier in class y_i given input x. we should learn parameters Θ to make the image x_1 and x_2 of their label distributions closer under some distance metric, that is, make the feature x_1 and x_2 closer.

2.1 Label Distribution Distance

In the method above, we can use some ways like metric learning to measure the distance between two conditional probability distribution. The Euclidean distance and Kullback-Leibler (KL) divergence are defined as the loss function measuring the similarity. Firstly, We consider the Euclidean Distance. With two images x_1 and x_2, N is the number of classes, it can be expressed as:

$$L_{Eu} = \sum_{i=1}^{N} (p(y_i|x_1; \Theta) - p(y_i|x_2; \Theta))^2$$

$$= ||p(y|x_1; \Theta) - p(y|x_2; \Theta))||_2^2 \quad (1)$$

The Kullback-Leibler (KL) divergence can be expressed as

$$L_{KL} = \sum_{i=1}^{N} (p(y_i|x_1; \Theta) ln \frac{p(y_i|x_1; \Theta)}{p(y_i|x_2; \Theta)}) \quad (2)$$

2.2   The Proposed Deep Learning Model

We train a Siamese-like neural network with shared weights, training each network individually using softmax and add the label distribution loss between the conditional probability distributions obtained from each network, as shown in Fig. 1.

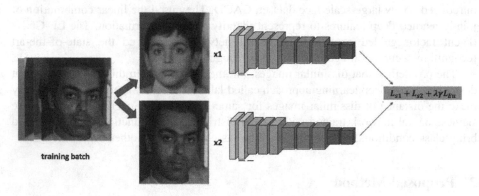

Fig. 1. The CNN training model, We use a Siamese-like architecture, with softmax loss in each network, followed by a label distribution loss when two samples are the same class. We split each incoming batch of samples into two mini-batches, and feed the network in pairs.

When training, we split an incoming batch of training samples into two parts, and evaluating softmax loss on each sub-batch identically, followed by a distribution loss function term when two samples are the same class. Because they are dissimilar images, but the same label, We make class conditional probability distributions closer to each other. When testing, we use only one branch of the network, and calculate the highest probability to predict which category it is.

The joint loss function can be expressed as:

$$L = L_{s1} + L_{s2} + \lambda\gamma L_{Eu} \tag{3}$$

The L_{s1} and L_{s2} are the softmax loss function, L_s can be expressed as:

$$L_s = -\log\left(\frac{e^{w_y^T x_i + b_y}}{\sum_{j=1}^m e^{w_j^T x_i + b_j}}\right) \tag{4}$$

Where λ denotes the scalar used for balancing the joint loss functions, γ is a constant 1 or 0, when two images are the same class, we set γ as 1, the other is 0. When γ is 0, this total loss becomes original softmax loss. The finally term represents the label distribution loss function, we tried two methods, but the Euclidean Distance is better than Kullback-Leibler (KL) divergence, we use L_{Eu}.

Taking Eqs. (1) and (4) into Eq. (3), we get:

$$L = -\log\left(\frac{e^{w_y^T x_1 + b_y}}{\sum_{j=1}^m e^{w_j^T x_1 + b_j}}\right) - \log\left(\frac{e^{w_y^T x_2 + b_y}}{\sum_{j=1}^m e^{w_j^T x_2 + b_j}}\right)$$
$$+ \lambda\gamma\|p(y|x_1; \Theta) - p(y|x_2; \Theta))\|_2^2 \tag{5}$$

Then the above parameters can be computed with backward propagation algorithm with two stages, we can optimize the w and the label distribution model parameters are implemented together using the stochastic gradient descent (SGD) method with the backpropagation algorithm.

Actually, the first two terms can be written as entropy loss between $p(y_i|x; \Theta)$ and y, where y denotes the predict label, N is the number of classes. The formula can be expressed as:

$$L_{CE}(p(y|x), y^{\sim}; \Theta) = -\sum_{i=1}^N y_i^{\sim} \log\left(\frac{p(y_i|x; \Theta)}{y_i^{\sim}}\right) > 0 \tag{6}$$

The total loss function can be expressed as:

$$L_{total} = \sum_{i=1}^2 \left(L_{CE}(p(y|x_i), y^{\sim}; \Theta)\right)$$
$$+ \lambda\gamma\|p(y|x_1; \Theta) - p(y|x_2; \Theta))\|_2^2 \tag{7}$$

Then the above parameter θ can be computed with only one stage, using the stochastic gradient descent (SGD) method with the backpropagation algorithm.

3 Experiment

To improve our work is effective, we evaluate our approach on existing public-domain cross-age face benchmark datasets MORPH Album2, CACD-VS and FG-NET.

3.1 Experiments on the MORPH Album 2 Dataset

There are 78,000 face images of 20,000 identities in the MORPH Album2 dataset. The data has been split into training and testing set. The training set contains 10,000 identities. The rest of 10,000 identities belong to testing set where each identity has 2 photos with a large age gap. We evaluate the rank 1 identification rates of our algorithm. We compare different baseline model: (1) Softmax: the CNN-baseline model trained by the original Softmax loss, (2) center loss: the CNNbaseline model guided by the center loss(3) L-Softmax: the CNNbaseline model guided by the L-Softmax loss, (4) A-softmax: the CNNbaseline model guided by the A-Softmax loss, (5) the proposed approach. Table 1 compares the rank 1 identification rates testing on 10,000 subjects of Morph Album 2 over Softmax, L-softmax, center loss, A-softmax, and Label distribution Learning.

The result is shown in Table 1. We can observe that:

(1) A-softmax makes the rank 1 identification rates increased by 1.43% compared with the nets only used softmax loss.
(2) Our proposed Label distribution Learning can achieve a better performance and the rank 1 identification rates is increased by 0.83% compared with the net used A-softmax loss.

Table 1. Rank-1 Identification Rates of different baselines on Morph Album 2.

Method	Rank-1 Identification Rates
Softmax	94.84%
center loss	95.94%
L-softmax	95.90%
A-softmax	96.27%
Our Label distribution learning	**97.10%**

To prove our method is effective and stable, we also test our performance with other methods. Table 2 shows the Rank-1 Identification Rates of different methods. We split two different schemes on Morph Album 2: testing on 10,000 subjects or 3,000 subjects. For fairly comparing against other methods, we evaluate the proposed methods on both schemes.

The result is shown in Table 2. We can observe that the performance of our method is better than other methods.

3.2   Experiments on the CACD-VS Dataset

CACD dataset has 163,446 images from 2,000 distinct celebrities. The age ranges from 10 to 62 years old. This dataset collects from the celebrity. The images include various illumination condition, different poses and makeup with the effect of various illumination condition, different poses and makeup, which can effectively reflect the robustness of the age-invariant face recognition algorithm. As shown in Table 3, We compare the accuracy of different models and the above baselines on the CACD-VS Dataset.

Table 2. The Rank-1 Identification Rates of different methods on Morph Album 2 dataset

Method	Test subjects	Rank 1 Rates (%)
HFA [12]	10000	91.14
CARC [13]	10000	92.80
MEFA [14]	10000	93.80
MEFA+SIFT+MLBP [14]	10000	94.59
LPS+HFA [15]	10000	94.87
LF-CNNs [16]	10000	97.51
Our label distribution	10000	98.10
GSM [17]	3000	94.40
AE-CNNs [18]	3000	98.13
Our label distribution	3000	98.56

We can observe that our proposed Label distribution Learning can achieve a better performance and the accuracy is increased by 0.4% compared with the net used A-softmax loss.

Table 3. The accuracy of different baselines and methods on the CACD-VS Dataset.

Method	Accuracy
High-Dimensional LBP [19]	81.6%
HFA	84.4%
CARC	87.6%
LF-CNNs	98.5%
Human, Average [25]	85.7%
Human, Voting [25]	94.2%
Softmax	98.4%
center loss	98.62%
L-softmax	98.58%
A-softmax	98.7%
Our Label distribution learning	**99.10%**

3.3 Experiments on the LFW Dataset

LFW [9] is a very famous benchmark for general face recognition. The dataset has 13,233 face images from 5,749 subjects. We test our model on 6,000 face pairs. The training data are disjoint from the testing data.

As can been seen in Table 4, we can observe that:

(1) The performance of our proposed label distribution learning is better than LF-CNNs, a method used in age-Invariant face recognition.
(2) When the dataset is the same order of magnitude, our method can achieve the best performance in these methods.

Table 4. The accuracy of different methods on LFW dataset

Method	Images	accuracy (%)
DeepFace [26]	4M	97.35
Facenet [27]	200M	99.65
DeepId2+ [28]		99.47
center loss	0.7M	99.28
A-softmax	0.5M	99.42
LF-CNNs	0.7M	99.10
Our label distribution	0.5M	99.32
Our label distribution	1.7M	99.46

4 Conclusion

Age-Invariant FaceRecognition is a remained challenging computer vision task, The problem is that the images are not too similar, but with the same label. To reduce the intraclass discrepancy, in this paper we propose a new method called Label Distribution learning for the end-to-end neural network to learn more discriminative features. We believe that there are some more effective methods solving this problem.

References

1. Ahonen, T., Hadid, A., Pietikainen, M.: Face description with local binary patterns: application to face recognition. IEEE Trans. Pattern Anal. Mach. Intell. (T-PAMI) **12**, 2037–2041 (2006)
2. Belhumeur, P., Hespanha, J.P., Kriegman, D.: Eigenfaces vs. fisherfaces: recognition using class specific linear projection. IEEE Trans. Pattern Anal. Mach. Intell. (T-PAMI) **7**, 711–720 (1997)
3. Zhou, E., Cao, Z., Yin, Q.: Naive-deep face recognition: touching the limit of LFW benchmark or not? In: CVPR (2015)
4. Taigman, Y., Yang, M., Ranzato, M., Wolf, L.: DeepFace: closing the gap to human-level performance in face verification. In: CVPR (2014)

5. Samal, A., Iyengar, P.A.: Automatic recognition and analysis of human faces and facial expressions: a survey. Pattern Recogn. **25**(1), 65–67 (1992)
6. Zhang, K., Zhang, Z., Li, Z., Qiao, Y.: Joint face detection and alignment using multi-task cascaded convolutional networks. In: ECCV (2016)
7. Schroff, F., Kalenichenko, D., Philbin, J.: Facenet: a unified embedding for face recognition and clustering. In: Proceedings of CVPR (2015)
8. Huang, G.B., Ramesh, M., Berg, T., Learned-Miller, E.: Labeled faces in the wild: a database for studying face recognition in unconstrained environments. Technical Report 07-49, University of Massachusetts, Amherst (2007)
9. Geng, X., Zhou, Z.H., Smith-Miles, K.: Automatic age estimation based on facial aging patterns. IEEE Trans. Pattern Anal. Mach. Intell. **29**(12), 22342240 (2007)
10. Lanitis, A., Taylor, C.J., Cootes, T.F.: Toward automatic simulation of aging effects on face images. IEEE Trans. Pattern Anal. Mach. Intell. **24**(4), 442–455 (2002)
11. Park, U., Tong, Y., Jain, A.K.: Age-invariant face recognition. IEEE Trans. Pattern Anal. Mach. Intell. **32**(5), 947954 (2010)
12. Gong, D., Li, Z., Lin, D., Liu, J., Tang, X.: Hidden factor analysis for age invariant face recognition. In: 2013 IEEE 14th International Conference on Computer Vision (ICCV). IEEE (2013)
13. Chen, B.-C., Chen, C.-S., Hsu, W.H.: Cross-age reference coding for age-invariant face recognition and retrieval. In: Fleet, D., Pajdla, T., Schiele, B., Tuytelaars, T. (eds.) ECCV 2014. LNCS, vol. 8694, pp. 768–783. Springer, Cham (2014). https://doi.org/10.1007/978-3-319-10599-4_49
14. Gong, D., Li, Z., Tao, D., Liu, J., Li, X.: A maximum entropy feature descriptor for age invariant face recognition. In: IEEE Conference on Computer Vision and Pattern Recognition (CVPR), pp. 5289–5297 (2015)
15. Li, Z., Gong, D., Li, X., Tao, D.: Aging face recognition: a hierarchical learning model based on local patterns selection. IEEE Trans. Image Process. (TIP) **25**(5), 21462154 (2016)
16. Wen, Y., Li, Z., Qiao, Y.: Latent factor guided convolutional neural networks for age-invariant face recognition. In: IEEE Conference on Computer Vision and Pattern Recognition (CVPR) (2016)
17. Lin, L., Wang, G., Zuo, W., Feng, X., Zhang, L.: Cross-domain visual matching via generalized similarity measure and feature learning. IEEE Trans. Pattern Anal. Mach. Intell. (TPAMI) **39**(6), 10891102 (2017)
18. Zheng, T., Deng, W., Hu, J.: Age estimation guided convolutional neural network for age-invariant face recognition. In: IEEE Conference on Computer Vision and Pattern Recognition Workshops (CVPRW) (2017)
19. Chen, D., Cao, X., Wen, F., Sun, J.: Blessing of dimensionality: high-dimensional feature and its efficient compression for face verification. In: IEEE Conference on Computer Vision and Pattern Recognition (CVPR), pp. 3025–3032 (2013)
20. Sun, Y., Chen, Y., Wang, X., Tang, X.: Deep learning face representation by joint identification-verification. In: Advances in Neural Information Processing Systems, pp. 1988–1996 (2014)
21. Schroff, F., Kalenichenko, D., Philbin, J.: Facenet: a unified embedding for face recognition and clustering. In: Proceedings of the IEEE Conference on Computer Vision and Pattern Recognition, pp. 815–823 (2015)
22. Liu, W., Wen, Y., Yu, Z.: Large-margin softmax loss for convolutional neural networks. In: ICML (2016)
23. Liu, W., Wen, Y., Yu, Z., Li, M., Raj, B., Song, L.: Sphereface: deep hypersphere embedding for face recognition. In: CVPR (2017)

24. Wen, Y., Zhang, K., Li, Z., Qiao, Yu.: A discriminative feature learning approach for deep face recognition. In: Leibe, B., Matas, J., Sebe, N., Welling, M. (eds.) ECCV 2016. LNCS, vol. 9911, pp. 499–515. Springer, Cham (2016). https://doi.org/10.1007/978-3-319-46478-7_31
25. Chen, B.C., Chen, C.S., Hsu, W.H.: Face recognition and retrieval using cross-age reference coding with cross-age celebrity dataset. IEEE Trans. Multimedia 17(6), 804815 (2015)
26. Taigman, Y., Yang, M., Ranzato, M., Wolf, L.: Deepface: closing the gap to human-level performance in face verification. In: IEEE Conference on Computer Vision and Pattern Recognition (CVPR) (2014)
27. Schroff, F., Kalenichenko, D., Philbin, J.: Facenet: a unified embedding for face recognition and clustering. In: IEEE Conference on Computer Vision and Pattern Recognition (CVPR) (2015)
28. Sun, Y., Wang, X., Tang, X.: Deeply learned face representations are sparse, selective, and robust. In: IEEE Conference on Computer Vision and Pattern Recognition (CVPR) (2015)

Overall Loss for Deep Neural Networks

Hai Huang$^{(\boxtimes)}$, Senlin Cheng$^{(\boxtimes)}$, and Liutong Xu

School of Computer Science, Beijing University of Posts and
Telecommunications, Beijing, China
hhuang@bupt.edu.cn, 972705994@qq.com

Abstract. Convolutional Neural Network (CNN) have been widely used for
image classification and computer vision tasks such as face recognition, target
detection. Softmax loss is one of the most commonly used components to train
CNN, which only penalizes the classification loss. So we consider how to train
intra-class compactness and inter-class separability better. In this paper, we
proposed an Overall Loss to make inter-class having a better separability, which
means that Overall loss penalizes the difference between each center of classes.
With Overall loss, we trained a robust CNN to achieve a better performance.
Extensive experiments on MNIST, CIFAR10, LFW (face datasets for face
recognition) demonstrate the effectiveness of the Overall loss. We have tried
different models, visualized the experimental results and showed the effectiveness of our proposed Overall loss.

Keywords: Overall loss · Intra-class · Deep learning

1 Introduction

Image classification and recognition are still difficult problem in machine learning and
computer vision tasks. Recent years, the deep convolutional neural network
(CNN) have a state-of-art performance in many image classification tasks such as hand-
written digit recognition [1], visual object classification [2–6], object recognition [7],
and face recognition such as DeepFace [8], FaceNet [9], Deep-Id [10–12], OpenFace
[13]. In order to train a CNN model, there are some tricks when training CNN such as
Regularization [14], Dropout [15], new activations [16], different CNN network
structure [7], different pooling methods [5, 17] and so on. On the other hand, some
researchers also optimize the task from loss function by using softmax in the last layer
to classify images to different classes.

Recently, some researchers begin to design different loss functions which could
learn discriminative features so that it can have a better performance. A deep convolutional neural network can learn a good feature if its intra-class compactness and inter-
class separability are well maximized. So contrastive loss [11], triplet loss [9], were
proposed to have better intra-class compactness and inter-class separability. However,
we need to select some training pairs and triplets when using contrastive loss and triplet
loss although they do really improve the performance and extract more discriminative
features. The selection of samples is important for the training result, if all training
samples are selected, the complexity can go up to $O(N^2)$ where N is the total number of

L. H. U and H. W. Lauw (Eds.): PAKDD 2019 Workshops, LNAI 11607, pp. 223–231, 2019.
https://doi.org/10.1007/978-3-030-26142-9_20

training samples. The L-softmax loss [18] will enhance the angle margin between different classes. But the angle margin is hard to train. A-softmax [19] was developed to explicitly enforce the angle margin with two limitations based on the L-softmax. The center loss [20] learns different centers of different classes, it was combined with the softmax loss and the intra-class compactness, but center loss did not consider the inter-class separability.

In this paper, we propose the Overall loss, which penalize the inter-class separability and align the center of each class to have a general representation for all classes. Overall loss considers the difference between the L2 norm and angle norm of each center of classes and the other centers. When training, we combine multiple losses (softmax loss, center loss, and Overall loss) so that we can train a more robust CNN to have a better representation for all classes.

Experiments and visualizations show the effectiveness of our method. The experiments on MNIST, CIFAR10, LFW (face datasets for face recognition) and compare different CNN models, our method have a better performance.

2 Related Work

In order to learn better discriminative feature, there are two ways to improve the performance. One way is to improve the deep CNN structures, another way is to design better loss functions.

CNN Structures: VGGNet explored the relationship between the depth of the convolutional neural network and its performance through stacking 3 * 3 small convolutional kernals and 2 * 2 maximum pooling layer. GooleNet controls the amount of computation and the number of parameters, and its classification performance is very good. ResNet Solves the missing information problem in information transmission of traditional CNN, the entire network only need to learn the differences between the input and output, simplifies the learning difficulty and goal. The R-CNN resorted to a recurrent CNN for visual classification by incorporating recurrent connections into each convolutional layer. Although existing CNN structures have achieved promising results for classification, they still have limitation because of softmax-loss.

Loss Functions: Recent years, There are many related works about loss functions such as contrastive loss, triplet loss, center loss, L-softmax loss and A-softmax loss. Expected for the contrastive loss and triplet loss, they need pre-selected sample pairs or triples, the aim of other loss functions not only consider the intra-class compactness, but also the inter-class separability. Our work combined the center loss with softmax loss to learn discriminative features. The softmax loss differ from softmax which is widely used in CNN and it can be written as:

$$L_{softmax} = \frac{1}{N}\sum_i L_i = \frac{1}{N}\sum_i -\log\frac{e^{f_{y_i}}}{\sum_j e^{f_j}} \tag{1}$$

where x_i denotes the deep feature of the $i - th$ training sample. y_i is its corresponding label. f_j represents the $j - th$ element of the class output vector f of the final full connection layer, and N is the number of training samples. f is the output of the final full connection layer, so f_{y_i} can be expressed as $f_{y_i} = W_{y_i}^T x_i$.

3   Proposed Method

In this section, we introduce center loss and indicate its weakness. We introduce the Overall loss, present the formulation of the Overall loss, which considers the intra-class compactness and inter-class separability. And we will show how to optimize the Overall loss with stochastic gradients descent.

3.1   Center Loss

As shown in Fig. 1(a), the features extracted from the deep neural network trained based on softmax loss are distinguishable, but the discriminant ability is not enough because of the characteristics of the class changes, the distance between intra-class is greater than the distance between inter-class. So the center loss is developed to improve the ability of the features extracted from deep neural networks. As shown in Fig. 1(b), this loss can minimize the intra-class variations while keeping the features of different classes separable. The loss function can be expressed as:

$$L_c = \frac{1}{2} \sum_{i=1}^{m} \left\| x_i - c_{y_i} \right\|_2^2 \tag{2}$$

where m denotes the number of training samples, x_i denotes the ith training sample. y_i is its corresponding label. L_c denotes its center loss. c_{y_i} denotes the y_ith class center of deep features. Then the author combined softmax loss and center loss to train the network, it can be expressed as:

$$L = L_{softmax} + \lambda L_c \tag{3}$$

Where λ is a scalar to balance the center loss and softmax loss, $L_{softmax}$ denotes the softmax loss, L_c denotes the center loss and L denotes the sum of the two losses.

The Weakness of Center Loss: We believe that a good design loss function should simultaneously consider the intra-class compactness and inter-class separability. The center loss consider the intra-class and have a large margin. However, this loss does not consider the inter-class. This phenomenon makes the distance of different classes not far enough so that the feature learn from deep neural network will be not more discriminative.

3.2   Overall Loss

The overall loss denoted as L_o, it can be expressed as:

$$
L_o = \sum_{i,j=1,i\neq j}^{m} \frac{1}{q + \left\| x^i - c_{y_j} \right\|_2^2}
$$
$$
+ \sum_{c_j \in N} \sum_{c_k \in N, c_k \neq c_j} \left(\frac{c_k \cdot c_j}{\left\| c_k \right\|_2 \left\| c_j \right\|_2} + 1 \right) \tag{4}
$$

where N is the set of expression labels; c_k and c_j denotes the k_{th} and j_{th} center with the L_2 norm, . represents the dot product. The first term $\frac{1}{1 + \left\| x^i - c_{y_j} \right\|_2^2}$ considers the sum of the distances of training samples to their non-corresponding class centers, q is a constant used for preventing the denominator equal to 0. We set q equals to 1. The second term $\left(\frac{c_k \cdot c_j}{\left\| c_k \right\|_2 \left\| c_j \right\|_2} + 1 \right)$ penalizes the different class centers. But this term will not influence the x_i when doing backward propagation. By minimizing the overall loss, the samples of different centers will be more separated. Our Overall loss can improve the power of the deep features extracted from deep neural networks.

Here we add softmax loss and center loss to jointly supervise the training of the CNN model. Total loss can be written as: $L = L_{softmax} + \lambda L_c + \gamma L_o$.

Except for $L_{softmax}$ and L_c, the gradient $\left(\frac{\partial L_o}{\partial x} \right)$ of the L_o with respect to x_i can be computed as:

$$
\frac{\partial L_o}{\partial x} = \frac{\partial}{\partial x_i} \left(\frac{1}{1 + \left\| x_i - c_{y_j} \right\|_2^2} \right)
$$
$$
= \frac{-2x_i - c_{y_j}}{\left(1 + \left(x_i - c_{y_j} \right)^2 \right)^2} \tag{5}
$$

Then this term can be optimized by the standard Stochastic Gradient Descent (SGD). The second term aims to update the cluster center, the jth class center can be calculated as

$$
\Delta c_j = \frac{\sum_m^{i=1} \delta(y_i, j)(c_j - x_i)}{1 + \sum_m^{i=1} \delta(y_i, j)}
$$
$$
+ \frac{\gamma}{N-1} \sum_{c_k \in N, c_k \neq c_j} \frac{c_k}{\left\| c_k \right\|_2 \left\| c_j \right\|_2}
$$
$$
- \left(\frac{c_k \cdot c_j}{\left\| c_k \right\|_2 \left\| c_j \right\|_2^3} \right) c_j \tag{6}
$$

Then this term can be updated in each mini-batch.

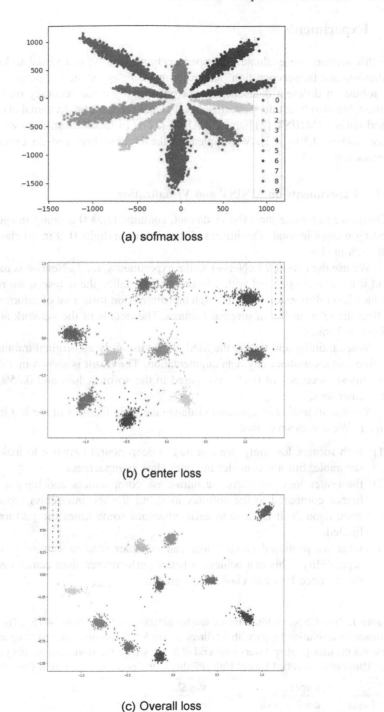

(a) sofmax loss

(b) Center loss

(c) Overall loss

Fig. 1. Visualization of MNIST. Note: The center loss can be observed that the distance between the classes is very small compared to softmax, and the Overall loss we proposed has a greater distance than the center loss. In our experiments, we select $\gamma = 0.0001$ to train the deep neural network.

4 Experiments

In this section, we evaluate the experiments on two typical visual tasks: visual classification and face recognition. The experiments demonstrate our proposed Overall loss is robust on different datasets, can not only improve the accuracy on visual classification, but also boost the performance on visual recognition. In visual classification, we used dataset (MNIST [21] and CIFAR10 [22]). In face recognition, we use the usual face datasets LFW [23]. We implement the Overall loss and do experiments using Tensorflow.

4.1 Experiments on MNIST and Visualization

Firstly, we introduce the MNIST dataset, contains 60,000 training images and 10,000 testing images in total. The images all hand-written digits 0–9 in 10 classes which are 28 * 28 in size.

We use the network LeNets++ in the experiments, the LeNets++ is based on LeNet and it is more deeper and wider. In order to visualize the datasets, we reduce the last hidden layer dimension to 2, although this dimension losts a lot of information, also can reflect the relationship of original features. The details of the network architecture are given in Table 1.

When training and testing the MNIST, we use only all original training images and testing images without any data augmentation. The result is shown in Table 2. Overall loss boosts accuracy of 0.41% compared to the softmax loss and 0.25% compared to the center loss.

We use all testing images and visualize the deep features of the last hidden layer in Fig. 1. We can observe that:

(1) with softmax loss only, we can train a deep neural network to make the features separable, but not consider the intra-class compactness.
(2) the center loss considers the intra-class compactness and have a better performance compared to the softmax loss, but the the inter-class separability is not good enough. It is close to each other and some times the performance will be limited.
(3) what we proposed Overall loss can consider a large margin of the inter-class separability. This can achieve a better performance than center loss. Intuitively, the distance between classes gets greater.

Table 1. The CNNs architecture we use for MNIST, called LeNets++. $(5, 32)_{/1,2} * 2$ denotes 2 cascaded convolution layers with 32 filters of size 5 * 5, the stride and padding are 1 and 2. $2/_{2,0}$ denotes the max-pooling layers with grid of 2 * 2, where the stride and padding are 2 and 0. We use Parametric Rectified Linear Unit (PRelu) as the nonlinear activation function.

	stage1	stage2	stage3	stage4
Layer	conv + pool	conv + pool	conv + pool	FC
LeNets	$(5, 20)_{/1,0} + 2_{/2,0}$	$(5, 50)_{/1,0} + 2_{/2,0}$		500
LeNets++	$(5, 32)_{/1,2} * 2 + 2_{/2,0}$	$(5, 64)_{/1,2} * 2 + 2_{/2,0}$	$(5, 128)_{/1,2} * 2 + 2_{/2,0}$	2

Table 2. Classification accuracy(%) on MNIST dataset

Method	Accuracy(%)
Softmax	98.8
Center loss	98.94
Our Overall loss	**99.21**

4.2   Experiments on CIFAR10

The CIFAR10 dataset includes 10 classes of natural images with 50,000 training images and 10,000 testing images. Each image is RGB image of size 32 * 32.

The deep neural network is 20-layer ResNet. In detail, we use Parametric Rectified Linear Unit (PRelu) as the nonlinear activation function to do data augmentation in training. In testing, we only use the original testing images. While training, we use Batch Normalization, set loss weight $\lambda = 0.1$.

The result is shown in Table 3. We can observe that:

(1) the center loss makes the accuracy increased by 0.85% compared with the nets only used softmax loss.

(2) our proposed Overall loss make the accuracy increased by 1.20% compared with the net only used softmax loss.

(3) our proposed Overall loss can achieve a better performance and the accuracy is increased by 0.43% compared with the net used center loss.

Table 3. Classification accuracy(%) on CIFAR10 dataset

Method	Accuracy(%)
20-layer ResNet based on softmax loss	91.23
20-layer ResNet based on center loss	91.98
20-layer ResNet based Overall loss	**92.43**

4.3   Experiments on LFW

The LFW dataset is established to study face recognition in unrestricted environments. This collection contains more than 13,233 face images from 5749 persons (all from the Internet). Each face was given a standard name. This set is widely used to evaluate the performance of face verification algorithm.

Firstly, we use MT-CNN to detect the faces and align them based on 5 points (eyes, nose and mouth), then we train on another dataset called CASIA-WebFace (490k labeled face images belonging to over 10,000 individuals) and test on 6,000 face pairs on LFW. We train a single network for feature extraction. For good comparision, we use the network FaceNet, the architecture is the Inception-ResNet. Based on the network publicly available, we achieve a better performance than many other models. The result is shown in Table 4. We can observe that train FaceNet in same dataset and use the Overall loss have a better performance than other softmax loss and center loss. Secondly, Overall loss can get a better performance than other models.

Table 4. Verification accuracy(%) on LFW dataset

Method	Images	Accuracy(%)
DeepFace	4M	97.35
Fusion	10M	98.37
DeepId-2+	0.5M	98.70
SeetaFce	0.5M	98.60
DeepFR	2.6M	98.95
FaceNet(softmax)	260M	98.41
FaceNet(center loss)	260M	99.53
FaceNet(Overall loss)	**260M**	**99.69**

5   Conclusion

In this paper, we proposed an Overall loss for deep neural networks. With softmax loss and center loss joint training, we not only consider the intra-class compactness but also consider the inter-class separability. We train a robust CNN to learn discriminative features so that it can have a better performance. The extensive experiments and visualizations of different datasets demonstrated the effectiveness of the proposed approach.

References

1. Wan, L., Zeiler, M., Zhang, S.: Regularization of neural networks using dropconnect. In: ICML (2013)
2. Krizhevsky, A., Sutskever, I., Hinton, G.E.: Imagenet classification with deep convolutional neural networks. In: Advances in Neural Information Processing Systems, pp. 1097–1105 (2012)
3. Simonyan, K., Zisserman, A.: Very deep convolutional networks for large-scale image recognition. arXiv preprint arXiv:1409.1556 (2014)
4. Szegedy, C., et al.: Going deeper with convolutions. In: Proceedings of the IEEE Conference on Computer Vision and Pattern Recognition, p. 19 (2015)
5. He, K., Zhang, X., Ren, S., Sun, J.: Delving deep into rectifiers: surpassing human-level performance on imagenet classification. In: Proceedings of the IEEE International Conference on Computer Vision, pp. 1026–1034 (2015)
6. He, K., Zhang, X., Ren, S., Sun, J.: Deep residual learning for image recognition, arXiv preprint arXiv:1512.03385 (2015)
7. Szegedy, C., Liu, W., Jia, Y.: Going deeper with convolutions. In: CVPR (2015)
8. Parkhi, O.M., Vedaldi, A., Zisserman, A.: Deep face recognition. In: British Machine Vision Conference, vol. 1, p. 6 (2015)
9. Schroff, F., Kalenichenko, D., Philbin, J.: Facenet: a unified embedding for face recognition and clustering. In: Proceedings of the IEEE Conference on Computer Vision and Pattern Recognition, pp. 815–823 (2015)
10. Sun, Y., Wang, X., Tang, X.: Deep learning face representation from predicting 10,000 classes. In: Proceedings of the IEEE Conference on Computer Vision and Pattern Recognition, pp. 1891–1898 (2014)

11. Sun, Y., Chen, Y., Wang, X., Tang, X.: Deep learning face representation by joint identification-verification. In: Advances in Neural Information Processing Systems, pp. 1988–1996 (2014)
12. Sun, Y., Wang, X., Tang, X.: Deeply learned face representations are sparse, selective, and robust. In: Proceedings of the IEEE Conference on Computer Vision and Pattern Recognition, pp. 2892–2900 (2015)
13. Liu, X., Kan, M., Wu, W., Shan, S., Chen, X.: VIPLFaceNet: an open source deep face recognition SDK, arXiv preprint arXiv:1609.03892 (2016)
14. Srivastava, N., Hinton, G.E., Krizhevsky, A.: Dropout: a simple way to prevent neural networks from overfitting. JMLR **15**, 1929–1958 (2014)
15. Krizhevsky, A., Sutskever, I., Hinton, G.E.: Imagenet classification with deep convolutional neural networks. In: NIPS (2012)
16. He, K., Zhang, X., Ren, S.: Delving deep into rectifiers: surpassing human-level performance on imagenet classification. In: CVPR (2015)
17. Goodfellow, I.J., Warde-Farley, D., Mirza, M., Courville, A.C., Bengio, Y.: Maxout networks. In: ICML, vol. 3, no, 28, pp. 1319–1327 (2013)
18. Liu, W., Wen, Y., Yu, Z.: Large-margin softmax loss for convolutional neural networks. In: ICML (2016)
19. Liu, W., Wen, Y., Yu, Z., Li, M., Raj, B., Song, L.: Sphereface: deep hypersphere embedding for face recognition. In: CVPR (2017)
20. Wen, Y., Zhang, K., Li, Z., Qiao, Yu.: A Discriminative feature learning approach for deep face recognition. In: Leibe, B., Matas, J., Sebe, N., Welling, M. (eds.) ECCV 2016. LNCS, vol. 9911, pp. 499–515. Springer, Cham (2016). https://doi.org/10.1007/978-3-319-46478-7_31
21. LeCun, Y., Cortes, C., Burges, C.J.C.: The MNIST database of handwritten digits (1998)
22. Krizhevsky, A., Hinton, G.: Learning multiple layers of features from tiny images (2009)
23. Huang, G.B., Ramesh, M., Berg, T., Learned-Miller, E.: Labeled faces in the wild: a database for studying face recognition in unconstrained environments. Technical report 07-49, University of Massachusetts, Amherst, October 2007

Sentiment Analysis Based on LSTM Architecture with Emoticon Attention

Changliang Li[1(✉)], Changsong Li[2(✉)], and Pengyuan Liu[3(✉)]

[1] Kingsoft AI Lab, No. 33 Xiaoyingxi, Haidian, Beijing, China
lichangliang@kingsoft.com
[2] Peking University, No. 5 Yiheyuan, Haidian, Beijing, China
lichangsong@pku.edu.cn
[3] Beijing Language and Culture University, No. 15 Xueyuan, Haidian, Beijing, China
liupengyuan@blcu.edu.cn

Abstract. Sentiment analysis is one of the most important research directions in natural language processing field. People increasingly use emoticons in text to express their sentiment. However, most existing algorithms for sentiment classification only focus on text information but don't full make use of the emoticon information. To address this issue, we propose a novel LSTM architecture with emoticon attention to incorporate emoticon information into sentiment analysis. Emoticon attention is employed to use emoticons to capture crucial semantic components. To evaluate the efficiency of our model, we build the first sentiment corpus with rich emoticons from movie review website and we use it as our experiment dataset. Experiments results show that our approach is able to better use emoticon information to improve the performance on sentiment analysis.

Keywords: Sentiment analysis · Emoticon · Attention

1 Introduction

Sentiment analysis has attracted increasing research interest in recent years. The objective is to classify the sentiment polarity of a text as positive, negative or neural. There has been a variety of approaches for this task. Representative approaches at present include machine learning algorithm and neural network models. [1] employ machine learning techniques to the sentiment analysis problem. Under this direction, most of studies [2] focus on designing effective features to obtain better classification performance. Feature engineering is important but labor-intensive. Neural network models [3–6] are popular for their capacity to learn text representation from data without careful engineering of features.

Sentiment analysis is a special case of text classification problem. For such tasks, neural network models take the text information as input and generate the semantic representations. Many models based on neural network have achieved excellent performance in sentiment analysis. However, these models only focus on the text content but ignore the crucial characteristics of emoticons. In recent years, people have become

L. H. U and H. W. Lauw (Eds.): PAKDD 2019 Workshops, LNAI 11607, pp. 232–242, 2019.
https://doi.org/10.1007/978-3-030-26142-9_21

more and more interested in using emoticons when chatting or commenting online. It is a common sense that the emoticon makes significant influence on the ratings and emoticon is an additional factor to help extract sentiments. For instance, a text containing ":)" is most likely to have a positive emotion but containing ":(" is most likely to have a negative emotion. Even though, there are some work have focused on using emoticons as noisy labels to learn the classifiers from the data [7, 8] and exploiting emoticons in lexicon-based polarity classification [9]. However, they only consider the word-level preference rather than semantic levels.

Attention has become an effective mechanism to obtain superior results in a variety of NLP tasks such as machine translation [10], sentence summarization [11], and read comprehension [12]. In this paper, we propose an attention mechanism based on emoticon information to enhance the sentiment representation and improve the classification performance of our model. We explore the potential correlation of emoticons and sentiment polarity in sentence-level sentiment analysis. In order to capture information in response to given sentences with emoticons, we design an attention based Bi-LSTM. We evaluate our approach on DouBan movie review dataset, which contains short movie reviews data and each of the movie review contains one or more emoticons.

To Summarize, our effort provide the following contributions:

(1) Most existing algorithms for sentiment analysis only focus on text information and don't full make use of the emoticon information. We propose a neural network model with emoticon information for sentiment analysis, and we consider the information conveyed by emoticons is assumed to affect the surrounding text on sentence level.
(2) We explore the attention mechanism based on emoticon information for sentiment analysis. Traditional attention-based neural network models only take the local text information into consideration. In contrast, our model puts forward the idea of emoticon attention by utilizing the emoticon information.
(3) We build corpus with rich emoticons from DouBan and we use it as our experiment dataset to verify the effectiveness of our model. The experimental results demonstrate that our model are able to better use emoticon information to improve the performance on sentiment analysis.

2 Related Work

2.1 Sentiment Analysis

Sentiment analysis is a long standing research topic. Readers can refer to [13] for a recent survey. In this section, we describe some related work about sentiment analysis.

The problem of sentiment analysis has been of great interest in the past decades because of its practical applicability. For example, a sentiment model could be employed to rank products and merchants [14]. In [15], Twitter sentiment was applied to predict election results. In [16], a method was reported for predicting comment volumes of political blogs. In [17], movie reviews and blogs were used to predict box-office revenues for movies. In [18], sentiment flow in social networks was investigated. In [19], expert investors in microblogs were identified and sentiment analysis of stocks was performed. In [20], sentiment analysis was used to characterize social relations. [21] used deep learning to predict movie reviews' sentiment polarity.

Existing approaches to sentiment analysis can be grouped into three main categories: knowledge-based techniques, statistical methods, and hybrid approaches. Knowledge-based techniques classify text by affect categories based on the presence of unambiguous affect words such as happy, sad, afraid, and bored. Statistical methods leverage on elements from machine learning such as latent semantic analysis, support vector machines, "bag of words" and Semantic Orientation—Pointwise Mutual Information [22]. Hybrid approaches leverage on both machine learning and elements from knowledge representation such as subtle manner, e.g., through the analysis of concepts that do not explicitly convey relevant information, but which are implicitly linked to other concepts that do so. With the trends of deep learning in computer vision, speech recognition and natural language processing, neural models are introduced into sentiment analysis field due to its ability of text representation learning.

2.2 Neural Network Models for Sentiment Analysis

Neural network models have achieved promising results for sentiment analysis due to its ability of text representation learning. There're three main neural network models for sentiment analysis, recursive neural network, convolution neural network and recurrent neural network. Socher conducts a series of recursive neural network models to learn representations based on the recursive tree structure of sentences, including Recursive Autoencoder (RAE) [23], Matrix-Vector Recursive Neural Network (MV-RNN) [4] and Recursive Neural Tensor Network (RNTN) [24]. [6] (Kim et al. 2014) and [25] adopt convolution neural network (CNN) to learn sentence representations and achieve outstanding performance in sentiment analysis. [26] investigate tree-structured long-short term memory (LSTM) networks on sentiment analysis.

2.3 Emoticons

People's facial expression like laughing and weeping are often considered to be involuntary ways of expressing oneself in face-to-face communication, whereas the use of their respective equivalents emoticons like ":-)" and ":-(" in computer mediated communication is intentional [27].

Even though some works has taken into account the information conveyed by emoticons, most of existing sentiment analysis models just consider emoticons as noisy labels. For instance, [7, 8, 28] uses the emoticons like "O(∩_∩)O" and "(︶︿︶)" as noisy labels to construct training data for polarity classification. The basic assumption is that a text contains "O(∩_∩)O" is most likely to have a positive emotion and that containing "(︶︿︶)" to be negative. Provided that emotions are important cues for sentiment in text, the key to harvesting information from emoticons lies in understanding how they related to a text's overall polarity. To address this issue, [9] exploit emoticons in lexicon-based polarity classification. Nevertheless, it only considers the word-level preference rather than semantic levels and ignores the interplay of emoticons and textual cues for sentiment, for instance, in cases when emoticons are used to intensify sentiment that is already conveyed by the text. In contrast, we propose an efficient neural sentiment analysis model with emoticons which serve as attention to take the interplay of emoticons and textual into consideration.

3   Method

We describe the proposed sentiment analysis model with emoticon attention in this section. Figure 1 gives the overall architecture of our model. First, we use Bidirectional Long Short-Term Memory (Bi-LSTM) network to learn the representation of input sentences, due to its ability in capturing both past and future information. Furthermore, all emoticons in sentence are extracted to enhance sentence semantic representations. Finally, the enhanced sentence representation is used as input of sentiment analysis model.

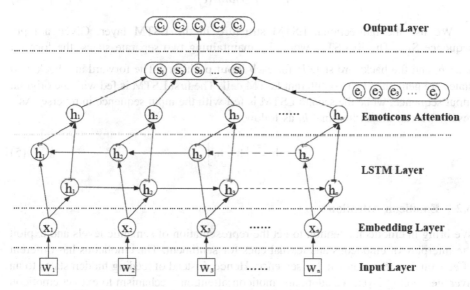

Fig. 1. Neural sentiment analysis model with emoticon attention.

3.1   Bi-directional Long Short-Term Memory Network

In this subsection, we describe the Bidirectional Long Short-Term Memory (Bi-LSTM) network for sentiment analysis. Recurrent neural network (RNNs) is a very useful model in dealing with language data. RNNs, particularly ones with gated architectures such as the LSTM, are very powerful at capturing statistical regularities in sequential inputs. To learn the semantic representation of a sentence, we adopt Bi-LSTM network as our sentiment analysis model.

Given an input sentence, we represent this sentence as $S = \{w_1, w_2, \ldots w_n\}$. In which w_j is the j-th word in sentence S and n is the length of sentence. In the embedding layer, we embed each word in a sentence into a low dimensional semantic space. That means each word w_j is mapped to its embedding $x_j \in R^d$. Then we obtain a $\{x_1, x_2 \ldots x_n\}$ represent the word vector in a sentence. At time step j, we use the word vector x_j as LSTM cell's input. More formally, given an input word vector x_j, the current cell state c_j and hidden state h_j can be update with previous cell state c_{j-1} and hidden state h_{j-1}:

$$\begin{bmatrix} i_j \\ f_j \\ o_j \end{bmatrix} = \begin{bmatrix} \sigma \\ \sigma \\ \sigma \end{bmatrix} (W \cdot [h_{j-1}, x_j] + b) \tag{1}$$

$$\hat{c}_j = tanh(W \cdot [h_{j-1}, x_j] + b) \tag{2}$$

$$c_j = f_j \otimes c_{j-1} + i_j \otimes \hat{c}_j \tag{3}$$

$$h_j = o_j \otimes tanh(c_j) \tag{4}$$

We use the bidirectional LSTM structure in the LSTM layer. Given a input sequence $S_{1:n}$. The Bi-LSTM works by maintaining two separate states, the forward state $\vec{h}_j$ and the backward state $\overleftarrow{h}_j$ for each input position j. The forward and backward states are generated by two different LSTM cell. The first LSTM is fed with the original input sequence, while the second LSTM is fed with the input sequence in reverse. As a result, we can obtain the final h_j as follow:

$$h_j = \left[\vec{h}_j, \overleftarrow{h}_j\right] \tag{5}$$

3.2 Emoticon Attention

We bring in emoticon attention to get the representation of semantic levels and exploit the interplay of emoticons and textual cues for sentiment. The emoticons has different effects on different words in the sentence. Hence, instead of feeding hidden states to an average pooling layer, we adopt an emoticon attention mechanism to extract emoticon specific words that are important to the meaning of sentence. Finally, we aggregate the representations of those informative words to form the sentence representation. Formally, the enhanced sentence representation is a weighted sum of hidden states as:

$$s = \sum_{j=1}^{n} \alpha_j h_j \tag{6}$$

Where α_j measures the importance of the j-th word for current emoticons. For each input sentence, there are k emoticons $\{e_1, e_2, \ldots, e_k\}$. Here, we embed each emoticon as a continuous and real-valued vector $e_j \in R^{d_e}$, where d_e is the dimension of emoticon embeddings respectively. Then the vector of all the related words are made an average operation. After that, we get a emoticon representation $e \in R^{d_e}$. Thus, the attention weight α_j for each hidden state can be defined as:

$$\alpha_j = \frac{\exp(att(h_j, e))}{\sum_{k=1}^{n} \exp(att(h_k, e))} \tag{7}$$

Where att is an attention function which scores the importance of words for composing sentence representation. The attention function att is defined as:

$$att(h_j, e) = v^T \tanh(W_H h_j + W_E e + b) \tag{8}$$

Where W_H and W_F are weight matrices, v is vector and v^T denotes its transpose.

3.3 Sentiment Classification

Since sentence representation s is hierarchically extracted from the words in the sentences, it is a high level representation of the sentence. Hence, we regard it as a features for sentence sentiment classification. We use a non-linear layer to project sentence representation s into the target space of C classes:

$$\hat{s} = \tanh(W_c s + b_c) \tag{9}$$

Afterwards, we use a softmax layer to obtain the sentence sentiment distribution:

$$p_c = \frac{\exp(\hat{s}_c)}{\sum_{k=1}^{C} \exp(\hat{s}_k)} \tag{10}$$

where C is the number of sentiment classes, p_c is the predicted probability of sentiment class c. In our model, cross-entropy error between gold sentiment distribution and predicted sentiment distribution is defined as loss function for optimization while training

$$L = -\sum_{s \in S} \sum_{c=1}^{C} p_c^g(s) \cdot \log(p_c(s)) \tag{11}$$

4 Experiments

In this section, we describe the experiment settings and give empirical results and analysis.

4.1 Data

In order to study the role of emoticon in the sentiment analysis of text, we constructed a dataset of DouBan movie reviews (DBMR), which is the first Chinese sentiment corpus with rich emoticons.

DBMR is collected from DouBan site, which is the largest movie review website in China. We crawled about 9 million raw data from douban.com. Besides, we have built a common emoticon set that contains about 100 commonly used emoticons. We cleaned the data and kept all the emoticons in the text according to a series of strict rules such as removing reviews with less than 3 words and so on. Then we extracted 47250 samples containing emoticons from the original data as our experimental data sets (DBMR).

In our emoticon set, each emoticon contains different emotional meanings such as laughing emoticon ':)' and weeping ': ('. Some of these emoticons contain more complex meanings, such as '= =' which means awkward, indifferent or satiric. Each of the movie reviews contains one or more emoticons. There are some samples shown in Table 1. In order for non-Chinese readers to better understand the examples, we translate the reviews into English besides. In Table 1, proportion gives the distribution of different emoticons.

Table 1. DBMR samples.

Emoticon	Meaning	Proportion	Sample	Label
= =	awkward, indifferent, atiric	0.402	= =唯一亮点是那首歌了吧 (= =The only highlight is that song...)	2
(~ ̄~)	Angry,bored	0.003	纯粹打发时间 ⌣ ̄⌣ 唉(Just to pass the time ..o(~ ̄~)o alas)	2
o(∩_∩)o	Happy, Delightful, excitement	0.019	没有传说中的看不懂。就是一群疯子的故事 o(∩_∩)o 哈哈 (It is not beyond comprehension, It's the story of a bunch of crazy people o(∩_∩)o haha.	5

The sentiment label of DBMR dataset ranges from 1 to 5, which is scored by the reviewer. 1 means very negative, 2 means negative, 3 means neural, 4 means positive and 5 means very positive. Our task aims to identify the sentiment score of a movie review.

We split train, validation and test sets in the proportion of 8:1:1(80% for training, 10% for validation, and 10% for test). Validation set is used to find the optimal parameters for model. The statistical information of the dataset is shown in Table 2.

Table 2. DBMR Samples

Label	Train-set	Validation-set	Test-set
1	2191	310	298
2	6318	769	829
3	10562	1271	1277
4	9348	1206	1182
5	9381	1169	1139
Total	37800	4725	4725

4.2   Metrics

We employ standards accuracy rate [29] to measure the overall sentiment analysis performance. The higher accuracy is, the better performance is. The accuracy rate is defined as following equation:

$$Accuracy = \frac{T}{N} \tag{12}$$

T is the number of predicted sentiment label equal to ground truth label, N is the overall number of text.

4.3   Baseline

In order to fully assess the efficiency of our model, we compare with following methods, which are widely used as baselines in other sentiment analysis work.

Majority Method: It is a basic baseline method, which assigns the majority sentiment label in train set to each instance in the test set. This method has been widely used as a baseline in sentiment analysis task [30].

SVM: We implement the standard SVM method and adopt tf-idf as word representation as baseline methods in our task. This method has been widely used as a baseline in sentiment analysis task [31].

NB: We implement the standard Naive Bayes method and adopt tf-idf as word representation as baseline methods in our task.

LSTM/Bi-LSTM: Long Short-Term Memory [32] and the bidirectional variant as introduced previously, the Bidirectional LSTM can capture both past and future information.

CNN: Convolutional Neural Network [33] generates sentence representation by convolution and pooling operations.

4.4   Model Comparison and Analysis

Table 3. Sentiment analysis result

Models	Without emoticon information in text	With emoticon information in text
Majority	0.270	0.270
SVM	0.349	0.373
NB	0.357	0.385
CNN	0.409	0.421
CNN-kmax	0.408	0.422
LSTM-1	0.399	0.420
LSTM-2	0.401	0.422
Bi-LSTM	0.413	0.422
Bi-LSTM + EA	0.421	0.433

As shown the experimental results in Table 3, we divided the results into two parts, the left one of which only consider the text information and the right one consider the text with emoticons.

Majority is just a simple statistics method without any semantic analysis. So it's no wonder that it performs worst. Morever, from the results, we observe that neural network models, the basic implementation of our model, significantly outperforms traditional machine learning algorithms. It indicates the efficiency of deep neural networks.

Besides, in the right part of Table 3, we show the performance of models with emoticon information. From this part, we can see that the emoticon information is helpful for sentiment analysis. For example, with the consideration of such information in DouBan, LSTM achieves 2.1% improvement, Bi-LSTM achieves 0.9% improvement and CNN achieves 1.2% improvement.

In our experiment, most of the baseline models achieved 42% accuracy, but we consider that relying on base network structures such as CNN or RNN is difficult to make full use of emoticons in a sentence, and it's hard to fully extract the emotional information in the sentence. We consider the importance of emoticons and adopt an attention mechanism to extract emoticon specific words that are important to the emotional attributes of sentence. Our proposed Bi-LSTM model with emoticon attention (Bi-LSTM + EA) outperforms all the other baseline methods. It indicates our model incorporates emoticon information in an effective and efficient way.

5   Conclusion and Future Work

In this paper, we propose a neural network which incorporates emoticon information via semantic level attentions. With the emoticon attention, our model can take account of the emoticons in semantic level. In experiments, we evaluate our model on sentiment analysis task. The experimental results show that our model achieves significant and consistent improvements compared to other popular models. In this paper, we only consider the emoticon information. In fact, most movies usually have some background information such as director and actor. We will take ad-vantages of that information in sentiment analysis in future.

References

1. Pang, B., Lee, L., Vaithyanathan, S.: Thumbs up?: sentiment classification using machine learning techniques. In: Empirical Methods in Natural Language Processing, pp. 79–86 (2002)
2. Mohammad, S.M., Kiritchenko, S., Zhu, X.: NRC-Canada: building the state-of-the-art in sentiment analysis of tweets. arXiv preprint arXiv:1308.6242 (2013)
3. Socher, R., Bauer, J., Manning, C.D.: Parsing with compositional vector grammars. In: Proceedings of the 51st Annual Meeting of the Association for Computational Linguistics, vol. 1: Long Papers, pp. 455–465 (2013)
4. Socher, R., Huval, B., Manning, C.D., Ng, A.Y.: Semantic compositionality through recursive matrix-vector spaces. In: Empirical Methods in Natural Language Processing, pp. 1201–1211 (2012)
5. Socher, R., Lin, C.C., Manning, C.D., Ng, A.Y.: Parsing natural scenes and natural language with recursive neural networks. In: Proceedings of the 28th International Conference on Machine Learning, ICML 2011, pp. 129–136 (2011)
6. Kim, Y.: Convolutional neural networks for sentence classification. In: EMNLP (2014)
7. Go, A., Bhayani, R., Huang, L.: Twitter sentiment classification using distant supervision. CS224 N Project Report, Stanford, vol. 1 (2009)
8. Pak, A., Paroubek, P.: Twitter as a corpus for sentiment analysis and opinion mining. In: LREc (2010)
9. Hogenboom, A., Bal, D., Frasincar, F., Bal, M., de Jong, F., Kaymak, U.: Exploiting emoticons in sentiment analysis. In: Proceedings of the 28th Annual ACM Symposium on Applied Computing, pp. 703–710 (2013)
10. Bahdanau, D., Cho, K., Bengio, Y.: Neural machine translation by jointly learning to align and translate. In: ICLR (2015)
11. Rush, A.M., Chopra, S., Weston, J.: A neural attention model for abstractive sentence summarization. arXiv preprint arXiv:1509.00685 (2015)
12. Hermann, K.M., Kocisky, T., Grefenstette, E., Espeholt, L., Kay, W., Suleyman, M., et al.: Teaching machines to read and comprehend. In: Advances in Neural Information Processing Systems, pp. 1693–1701 (2015)
13. Liu, B.: Sentiment Analysis: Mining Opinions, Sentiments, and Emotions. Cambridge University Press, Cambridge (2015)
14. McGlohon, M., Glance, N.S., Reiter, Z.: Star quality: aggregating reviews to rank products and merchants. In: ICWSM (2010)
15. Tumasjan, A., Sprenger, T.O., Sandner, P.G., Welpe, I.M.: Predicting elections with twitter: what 140 characters reveal about political sentiment. In: ICWSM, vol. 10, pp. 178–185 (2010)
16. Yano, T., Smith, N.A.: What's worthy of comment? Content and comment volume in political blogs. In: ICWSM (2010)
17. Sadikov, E., Parameswaran, A.G., Venetis, P.: Blogs as predictors of movie success. In: ICWSM (2009)

18. Miller, M., Sathi, C., Wiesenthal, D., Leskovec, J., Potts, C.: Sentiment flow through hyperlink networks. In: ICWSM (2011)
19. Feldman, R., Rosenfeld, B., Bar-Haim, R., Fresko, M.: The stock sonar—sentiment analysis of stocks based on a hybrid approach. In: Twenty-Third IAAI Conference (2011)
20. Groh, G., Hauffa, J.: Characterizing social relations via NLP-based sentiment analysis. In: ICWSM (2011)
21. Li, C., Xu, B., Wu, G., He, S., Tian, G., Zhou, Y.: Parallel recursive deep model for sentiment analysis. In: Cao, T., Lim, E.-P., Zhou, Z.-H., Ho, T.-B., Cheung, D., Motoda, H. (eds.) PAKDD 2015. LNCS (LNAI), vol. 9078, pp. 15–26. Springer, Cham (2015). https://doi.org/10.1007/978-3-319-18032-8_2
22. Turney, P.D.: Thumbs up or thumbs down?: semantic orientation applied to unsupervised classification of reviews. In: Proceedings of the 40th Annual Meeting on Association for Computational Linguistics, pp. 417–424 (2002)
23. Socher, R., Pennington, J., Huang, E., Ng, A.Y., Manning, C.D.: Semi-supervised recursive autoencoders for predicting sentiment distributions. In: Empirical Methods in Natural Language Processing, pp. 151–161 (2011)
24. Socher, R., Perelygin, A., Wu, J.Y., Chuang, J., Manning, C.D., Ng, A.Y., et al.: Recursive deep models for semantic compositionality over a sentiment treebank. In: Empirical Methods in Natural Language Processing, pp. 1631–1642 (2013)
25. Santos, C.N.D., Gatti, M.A.D.C.: Deep convolutional neural networks for sentiment analysis of short texts. In: International Conference on Computational Linguistics, pp. 69–78 (2014)
26. Tai, K.S., Socher, R., Manning, C.D.: Improved semantic representations from tree-structured long short-term memory networks. In: ACL (2015)
27. Kendon, A.: On gesture: its complementary relationship with speech. In: Nonverbal Behavior and Communication, pp. 65–97 (1987)
28. Rao, D., Ravichandran, D.: Semi-supervised polarity lexicon induction. In: Proceedings of the 12th Conference of the European Chapter of the Association for Computational Linguistics, pp. 675–682 (2009)
29. Jurafsky, D.: Speech & Language Processing. Pearson Education India, Noida (2000)
30. Tang, D., Qin, B., Feng, X., Liu, T.: Target-dependent sentiment classification with long short term memory. CoRR, abs/1512.01100 (2015)
31. Chen, H., Sun, M., Tu, C., Lin, Y., Liu, Z.: Neural sentiment classification with user and product attention
32. Cho, K., Van Merrienboer, B., Gulcehre, C., Bahdanau, D., Bougares, F., Schwenk, H., et al.: Learning phrase representations using RNN encoder-decoder for statistical machine translation. In: Empirical Methods in Natural Language Processing, pp. 1724–1734 (2014)
33. Kalchbrenner, N., Grefenstette, E., Blunsom, P.: A convolutional neural network for modelling sentences. arXiv preprint arXiv:1404.2188 (2014)

Aspect Level Sentiment Analysis with Aspect Attention

Changliang Li[1(✉)], Hailiang Wang[2(✉)], and Saike He[2(✉)]

[1] Kingsoft AI Lab, Beijing 100085, China
lichangliang@kingsoft.com
[2] Institute Automation, Chinese Academy of Sciences, Beijing, China
{hailiang.wang,saike.he}@ia.ac.cn

Abstract. Aspect level sentiment classification is a fundamental task in the field of sentiment analysis, which goal is to inferring sentiment on entities mentioned within texts or aspects of them. Since it performs finer-grained analysis, aspect level sentiment classification is more challenging. Recently, neural network approaches, such as LSTMs, have achieved much progress in sentiment analysis. However, most neural models capture little aspect information in sentences. Aspect level sentiment of a sentence is determined not only by the content but also by the concerned aspect. In this paper, we propose a novel LSTM with Aspect Attention model (LSTM_AA) for aspect level sentiment classification. Our model introduces aspect attention to relate the aspect level sentiment of a sentence closely to the concerned aspect, as well as to explore the connection between an aspect and the content of a sentence. We experiment on the SemEval 2014 datasets and results show that our model performs comparable to state-of-the-art deep memory network, and substantially better than other neural network approaches. Besides, our approach is more robust than deep memory network which performance heavily depends on the hops.

Keywords: Aspect attention · Sentiment analysis ·
Natural language processing

1 Introduction

Sentiment analysis, also called opinion mining, is the field of study that analyzes people's opinions, sentiments, evaluations, appraisals, attitudes, and emotions towards entities such as products, services, organizations, individuals, issues, events, topics, and their attributes [11]. It represents a large problem space. The field of sentiment analysis has seen a lot of attention in the last few years. The corresponding growth of the field has resulted in the emergence of various subareas, each addressing a different level of analysis or research question.

Aspect level sentiment classification is a fundamental task in the field of sentiment analysis [9,15,17]. The problem of aspect level sentiment analysis has

© Springer Nature Switzerland AG 2019
L. H. U and H. W. Lauw (Eds.): PAKDD 2019 Workshops, LNAI 11607, pp. 243–254, 2019.
https://doi.org/10.1007/978-3-030-26142-9_22

received much attention these years because of its practical applicability. The goal of aspect-level sentiment analysis is to inferring sentiment on aspects mentioned within texts. It is based on the idea that an opinion consists of a sentiment and an aspect. An opinion without its aspect being identified is of limited use. Realizing the importance of aspects also helps us understand the sentiment analysis problem better. For example, the sentence "The iPhone's call quality is good, but its battery life is short" evaluates two aspects, call quality and battery life, of iPhone. The sentiment on iPhone's call quality is positive, but the sentiment on its battery life is negative. Based on aspect level of analysis, a structured summary of opinions about entities and their aspects can be produced.

Aspect level sentiment classification is more challenging since it performs finer-grained analysis. Researchers typically use machine learning algorithms and build sentiment classifier in a supervised manner. Neural networks have achieved state-of-the-art performance in a variety of NLP tasks including sentiment analysis. However, neural network models are still in infancy to deal with aspect level sentiment classification. Conventional neural models like long short-term memory (LSTM) [5,24] cannot explicitly reveal the importance of aspect word. In some works, target dependent sentiment classification can be benefited from taking into account target information, such as in Target-Dependent LSTM (TD-LSTM) and Target-Connection LSTM (TC-LSTM) [26]. However, those models can only take into consideration the target but not aspect information.

We believe that aspect plays a key role in aspect level sentiment analysis, and the context words play different roles while inferring the sentiment towards different aspects of a sentence. In this paper, we propose a LSTM with aspect attention (LSMT-AA) model for aspect-level sentiment classification, which is inspired by the recent success of computational models with attention mechanism [1,3,22]. Attention has become an effective mechanism to obtain superior results. Our model can highlight aspect word's role and is able to attend different parts of a sentence when different aspects are concerned. Our model mainly consists of two parts. One is LSTM model to generate sentence representation. Another is aspect information as attentions over context words. Besides, our approach is data-driven, and does not rely on syntactic parser or sentiment lexicon.

We apply the proposed approach to laptop and restaurant datasets from SemEval 2014 task 4: Aspect Based Sentiment Analysis [17]. Experimental results show that our approach performs comparable to a top system using deep memory networks [26], and substantially better than other neural network approaches. Besides, our approach is more robust than deep memory network which performance heavily depends on the hops. Our model gives a new perspective to research on aspect level sentiment analysis.

The rest of our paper is structured as follows: Sect. 2 discusses related works, Sect. 3 gives a detailed description of our LSTM with Aspect Attention model, Sect. 4 presents experiments and analysis, and Sect. 5 summarizes this work and the future direction.

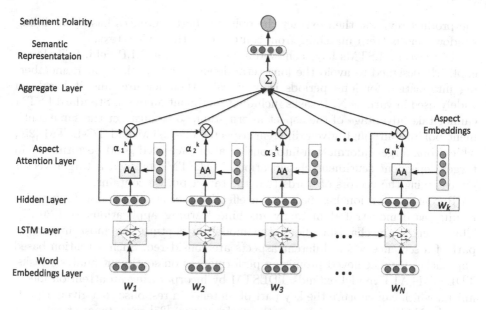

Fig. 1. LSTM_AA Model

2   Related Work

The problem of aspect specific sentiment analysis has been of great interest in the past decades because of its practical applicability. The majority of current approaches attempt to detect the polarity of the entire sentence, regardless of the entities mentioned or aspects. Traditional approaches to solve those problems are to manually design a set of features. With the abundance of sentiment lexicons [16,19], the lexicon-based features were built for sentiment analysis [13]. Most of these studies focus on building sentiment classifiers with features, which include bag-of-words and sentiment lexicons, using SVM [8,14]. However, the results highly depend on the quality of features. In addition, feature engineering is labor intensive.

[7] formulated this problem and proposed association mining based algorithm to extract product features. Wordnet was used to capture sentiment polarity of words. [18] approached this problem by proposing rule-based ontologies. [2] and [12] proposed models for uncovering parts of reviews which mention specific aspects. [10] proposed sentence level topic models to extract aspects and identifying the sentiment polarity. Though these models account for joint modeling of aspects and sentiments, they make assumptions about the syntax of the words and how the syntax governs if a particular word is an aspect or a sentiment. This is not ideal because there are words that we encounter in real world (such as "tasty") which play a dual role of representing both aspects and sentiments. [6]suggest an approach to Aspect-Based Sentiment Analysis that incorporates structural information of reviews by employing Rhetorical Structure. [23] apply the Latent Dirichlet Allocation model to discover multi aspect global topics of

the product reviews, then extract the opinion short sentences based on sliding windows and pattern matching from context over the review text.

These years, LSTMs have achieved a great success in NLP field. LSTMs are explicitly designed to avoid the long-term dependency problem [5]. Remembering information for long periods is practically their feature, and it has been widely used in various NLP tasks including sentiment analysis. Standard LSTM cannot take advantage of the aspect information, so it must get the same sentiment polarity although given different aspects. TD-LSTM and TC-LSTM [25], which took target information into consideration, achieved good performance in target-dependent sentiment classification. TC-LSTM obtained a target vector by averaging the vectors of words that the target phrase contains.

Recently, attention has become an effective mechanism to obtain superior results, as demonstrated in many machine learning applications [1,4,20,21]. The attention mechanism allows the model to capture the most important part of a sentence when different aspects are considered. Such attention-based approaches have achieved promising performances on sentiment analysis tasks. TDLSTM+ATT model extends TDLSTM by incorporating an attention mechanism, which can capture the key part of sentence in response to a given aspect, over the hidden vectors to improve the performance. [27] concatenate the aspect vector into the sentence hidden representations for computing attention weights for aspect level sentiment analysis.

Despite the effectiveness of those methods, it is still challenging to discriminate different sentiment polarities at a fine-grained aspect level. Therefore, we are motivated to design a powerful model, which can fully employ aspect information for aspect level sentiment classification.

3   LSTM with Aspect Attention Model

In this section, we will introduce our LSTM with Aspect Attention model (LSTM_AA) for aspect level sentiment classification in detail. First, we give the formalization of aspect level sentiment classification. Then, we discuss how to obtain semantic representation via the LSTM network. At last, we present our aspect attention, which highlights aspect word's role and reveal the connection between different semantic parts of a sentence with aspects concerned. An overall illustration of LSTM sentiment classification with aspect attention model (LSTM_AA) is shown in Fig. 1.

3.1   Formalization

Given a sentence $s = (w_1, w_2, \cdots, w_{a\_i}, \cdots, w_N)$ which consists of N words and one or more aspect words represented as w_{a_i}, aspect level sentiment classification aims at determining the sentiment of sentence s towards the aspect w_{a_i}. For example, the sentence "great food but the service was dreadful!" has two aspects w_{a_1} food and w_{a_2}: service. The sentiment polarity of sentence towards aspect "food" is positive, while the polarity towards aspect "service" is negative.

3.2   LSTM Sentiment Classification Model

In this work, to model the semantic representations of sentences, we adopt Long Short-Term Memory (LSTM) network because of its excellent ability to remember information for long periods.

A LSTM unit is composed of three multiplicative gates, which control the proportions of information to forget and to pass on to the next time step.

Given an input word w_i, we compute the values for i_t, the input gate, and C_t the candidate value for the states of the memory cells at time t:

$$i_t = \sigma(W_{in}w_i + U_{in}h_{t-1} + b_{in}) \tag{1}$$

$$\widehat{C} = tanh(W_c w_i + U_c h_{t-1} + b_c) \tag{2}$$

σ is the element-wise sigmoid function; W_{in} and W_c denote the weight matrices of input gate and cell respectively for input word w_i; U_{in} and U_c denotes the weight matrices of hidden state h_{t-1}; b_{in} and b_c denote the bias vectors.

Then we compute the value for f_t, the activation of the memory cells' forget gates at time t :

$$f_t = \sigma(W_f w_i + U_f h_{t-1} + b_f) \tag{3}$$

W_f denotes the weight matrices of forget gate for input word w_i; U_f denotes the weight matrices of hidden state h_{t-1}; b_f denotes the bias vectors.

Given the value of the input gate activation i_t, the forget gate activation f_t and the candidate state value $\widehat{C_t}$, we can compute C_t the memory cells' new state at time t:

$$C_t = i_t \odot \widehat{C_t} + f_t \odot C_{t-1} \tag{4}$$

$\odot$ is the element-wise product. With the new state of the memory cells, we can compute the value of their output gates and, subsequently, their outputs :

$$o_t = \sigma(W_o w_i + U_o h_{t-1} + b_o) \tag{5}$$

$$h_t = o_t \odot tanh(C_t) \tag{6}$$

W_o denotes the weight matrices of output gate for input word w_i; U_o denotes the weight matrices of hidden state h_{t-1}; b_o denotes the bias vectors.

Thus, from an input sequence $\{w_1, w_2, \cdots, w_N\}$, the memory cells in the LSTM layer will produce a representation sequence $\{h_1, h_2, h_3, \ldots, h_i, \cdots, h_N\}$.

3.3   Aspect Attention

Aspect information is vital when classifying the polarity of one sentence given aspect. We may get opposite polarities if different aspects are considered. To make the best use of aspect information, our model introduces aspect attention

Table 1. Examples in two datasets

Domain	Sentence	Aspect	Sentiment
Laptop	air has higher resolution but the fonts are small	resolution	Pos
		fonts	Neg
	usb3-peripherals are noticeably less expensive than the thunderbolt ones	thunderbolt	Neg
		usb3-peripherals	Pos
Restaurant	great food but the service was dreadful!	food	Pos
		service	Neg
	the appetizers are ok, but the service is slow	appetizers	Neu
		service	Neg

to both explore the connection between an aspect and the different semantic parts of a sentence, and highly relate the sentiment polarity of a sentence closely to the concerned aspect.

Since we have the semantic representations $\{h_1, h_2, h_3, \cdots, h_i, \cdots, h_N\}$ in hidden layer, then we aggregate the representations using aspect attention to form the sentence representation for the k-th aspect in aggregate layer, which is a weighted sum of hidden states as:

$$s_k = \sum_{i=1}^{N} a_i^k h_i \tag{7}$$

Where a_i^k is aspect attention factor, which reflects the importance of the i-th word for the k-th aspect word in the sentence.

In aspect attention layer, we represent each aspect asp as continuous and real valued vectors $asp \in R^{da}$, where da is the dimensions of aspect embeddings. As a result, the aspect attention factor α_i for each hidden state can be defined as:

$$a_i^k = \frac{exp(e(h_i, asp^k))}{\sum_{j=1}^{N} exp(e(h_j, asp^k))} \tag{8}$$

Where e is a score function which explores the connection between an aspect and the different semantic parts of a sentence. The score function e is defined as:

$$e(h_i, asp^k) = V tanh(W_H h_i + W_A asp^k + b) \tag{9}$$

Where W_H, V and W_A are weight matrices.

3.4 Sentiment Classification

Since s_k is a high-level representation of the sentence semantic representation for the k-th aspect. Hence, we regard it as features for aspect level sentiment classification. We use a non-linear layer to project sentence semantic representation s_k into the target space of C classes:

$$\widehat{s_k} = tanh(W_c s_k + b_c) \tag{10}$$

W_c is weight matrices. Afterwards, we use a softmax layer to obtain the aspect level sentiment distribution:

$$p_c = \frac{exp(\widehat{s_k})}{\sum_{k=1}^{c} exp(\widehat{s_k})} \tag{11}$$

where C is the number of sentiment classes; p_c is the predicted probability of aspect level sentiment class c. In our model, cross-entropy error between gold sentiment distribution and our model's aspect level sentiment distribution is defined as loss function for optimization when training:

$$L = -\sum_{s \in S} \sum_{c=1}^{C} p_c^t(s) \cdot log(p_c(s)) \tag{12}$$

p_c^t is the gold probability of sentiment class c with ground truth being 1 and others being 0, S represents the training sentences.

4 Experiment

In this section, we introduce the experimental settings and empirical results on the task of aspect level sentiment classification.

4.1 Experiment Settings

We conduct experiments on SemEval 2014 task 4: Aspect Based Sentiment Analysis (Pontiki, Galanis et al. 2014). The task aims to determine the polarity of each aspect given a set of pre-identified aspects. For example, given a sentence, "The restaurant was too expensive.", there is an aspect price whose polarity is negative. Two domain-specific datasets for laptops and restaurants, consisting of over 6000 sentences with fine-grained aspect-level human annotations have been provided for training.

Table 1 shows some examples in two datasets. Each sentence has two aspects, and for each aspect the sentiment polarity is different. Pos means positive; Neg means negative and Neu means neutral.

Statistics of the datasets are given in Table 2. It is worth noting that the original dataset contains the fourth category conflict, which means that a sentence expresses both positive and negative opinion towards an aspect. Like work of [26], we remove conflict category as the number of instances is very tiny, incorporating which will make the dataset extremely unbalanced. Evaluation metric is classification accuracy.

Table 2. Statistics of two datasets

Dataset	Pos.	Neg.	Neu.
Laptop-Train	994	870	464
Laptop-Test	341	128	169
Restaurant-Train	2164	807	637
Restaurant-Test	728	196	196

4.2   Baselines

In order to fully evaluate our LSMT-AA model, we compare with the following baseline methods on both datasets.

– Majority method
  It is a basic baseline method, which assigns the majority sentiment label in training set to each instance in the test set.
– LSTM models
  We compare with three LSTM models [25]. In LSTM, a LSTM based recurrent model is applied from the start to the end of a sentence, and the last hidden vector is used as the sentence representation. TDLSTM extends LSTM by taking into account of the aspect, and uses two LSTM networks, a forward one and a backward one, towards the aspect. TDLSTM+ATT extends TDLSTM by incorporating an attention mechanism [1] over the hidden vectors.
– ContextAVG
  We also compare with ContextAVG method [26]. In this approach, context word vectors are averaged and the result is added to the aspect vector. The output is fed to a softmax function.
– Deep memory networks (MemNet)
  The approach consists of multiple computational layers with shared parameters [26]. Each layer is content and location based attention model, which first learns the importance/weight of each context word and then utilizes this information to calculate continuous text representation. The text representation in the last layer is regarded as the feature for sentiment classification.

4.3   Experiment Results

Experimental results are given in Table 3. We can find that on Laptop dataset, our LSTM_AA model achieves state-of-the-art. While on Restaurant dataset, our model substantially outperforms all other baseline methods, except MemNet approach when it selects more than 5 hops. However, we can find that our LSTM_AA model is more robust than MemNet, which performance depends closely on the number of hops. On Laptop dataset, MemNet needs to select 7 hops to get its best performance. While on Restaurant dataset, MemNet needs to select 9 hops to get its best performance. The results demonstrate our model is more powerful and robust for aspect level sentiment classification.

Table 3. Experiment Results

Method	Laptop	Restaurant
Majority	53.45	65
LSTM	66.45	74.28
TDLSTM	68.13	75.63
TDLSTM+ATT	66.24	74.31
ContextAVG	61.22	71.33
MemNet(1)	67.66	76.10
MemNet(2)	71.14	78.61
MemNet(3)	71.74	79.06
MemNet(4)	72.21	79.87
MemNet(5)	71.89	80.14
MemNet(6)	72.21	80.05
MemNet(7)	72.37	80.32
MemNet(8)	72.05	80.14
MemNet(9)	72.21	**80.95**
LSTM_AA	**72.88**	80.00

Among three LSTM models, TDLSTM performs better than LSTM. Standard LSTM cannot capture any aspect information in sentence, so it must get the same sentiment polarity although given different aspects. Since it cannot take advantage of any aspect information, it is no wonder that it has poor performance on aspect level sentiment analysis.

TDLSTM can improve the performance of sentiment classifier by treating an aspect as a target. Since there is no attention mechanism in TDLSTM, it cannot reveal which words are important for a given aspect.

From the result of TDLSTM+ATT, we can see that it is useless to simply employ attention mechanism to TDLSTM architecture. This highlights the role and necessary of our proposed aspect attention.

We can also find that the performance of ContextAVG is very poor, which means that assigning the same weight/importance to all the context words is not an effective way.

4.4 Effects of Aspect Attention

In this section, we will further show the advantages of our LSTM_AA model through some typical examples. We visualize the aspect attention weight of each context word to get a better understanding of our LSTM_AA model.

Figure 2 gives one example, "the food is mediocre and the service was severely slow", to show effects of aspect attention of our LSTM_AA model. The aspects of this sentence are service and food respectively. The color depth expresses the

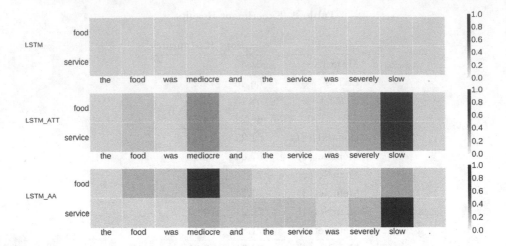

Fig. 2. Visualization of aspect attention

importance degree of the weight in aspect attention vector. Given aspect "service", aspect attention can detect the most important word from the whole sentence for aspect is "slow". As a result, the model correctly predicts the polarity towards aspect "service" as negative. While given aspect "food", aspect attention can detect the most important word from the whole sentence for aspect is "mediocre". As a result, the model correctly predicts the polarity towards aspect "food" as neutral. The predicted sentiment is closely related to the given aspect. Meanwhile, aspect attention fully explores the connection of different semantic parts and the given aspect. Therefore, it is helpful to predict the aspect level sentiment correctly.

However, in standard LSTM architecture, each word contributes equally to the sentence semantic representation. So it is not hard to understand that it performances badly. LSTM architecture with traditional attention mechanism, though it can explore the role of different parts of the sentence, it cannot capture the aspect information. Given different aspects, it can only get one sentiment polarity like standard LSTM architecture. It demonstrates that the traditional attention mechanism cannot address the aspect level sentiment analysis.

From this case, we can see that our LSTM_AA model is very fit to address the aspect level sentiment analysis.

5 Conclusion and Future Work

In this paper, we propose a LSTM with aspect attention (LSTM_AA) model for aspect level sentiment analysis. Sentiment polarity of a sentence is determined not only by the content but also highly by the concerned aspect. Therefore, it is worthwhile to consider the aspect information. Our model introduces aspect attention to relate the sentiment polarity of a sentence closely to the concerned aspect, as well as to explore the connection between an aspect and the different semantic parts of a sentence. We experiment on the SemEval 2014 datasets and results show that our model gives good performance and is more robust than state-of-the-art approach. Besides, our approach is data-driven, more robust, and does not rely on syntactic parser or sentiment lexicon. Our approach gives a new perspective to research on aspect level sentiment analysis.

The proposals have shown potentials for aspect level sentiment analysis. We will explore more in future. For example, in this paper, we only consider the sentence level. We will research paragraph or document for aspect level sentiment analysis in future.

References

1. Bahdanau, D., Cho, K., Bengio, Y.: Neural machine translation by jointly learning to align and translate. arXiv preprint arXiv:1409.0473 (2014)
2. Brody, S., Elhadad, N.: An unsupervised aspect-sentiment model for online reviews. In: Human Language Technologies: The 2010 Annual Conference of the North American Chapter of the Association for Computational Linguistics. Association for Computational Linguistics (2010)
3. Graves, A., Wayne, G., Danihelka, I.: Neural turing machines. arXiv preprint arXiv:1410.5401 (2014)
4. Hermann, K.M., et al.: Teaching machines to read and comprehend. In: Advances in Neural Information Processing Systems (2015)
5. Hochreiter, S., Schmidhuber, J.: Long short-term memory. Neural Comput. 9(8), 1735–1780 (1997)
6. Hoogervorst, R., et al.: Aspect-based sentiment analysis on the web using rhetorical structure theory. In: Bozzon, A., Cudre-Maroux, P., Pautasso, C. (eds.) ICWE 2016. LNCS, vol. 9671, pp. 317–334. Springer, Cham (2016). https://doi.org/10.1007/978-3-319-38791-8_18
7. Hu, M., Liu, B.: Mining and summarizing customer reviews. In: Proceedings of the Tenth ACM SIGKDD International Conference on Knowledge Discovery and Data Mining. ACM (2004)
8. Kiritchenko, S., et al.: NRC-Canada-2014: detecting aspects and sentiment in customer reviews. In: Proceedings of the 8th International Workshop on Semantic Evaluation (SemEval 2014) (2014)
9. Kumar, A., et al.: Iit-tuda at semeval-2016 task 5: beyond sentiment lexicon: combining domain dependency and distributional semantics features for aspect based sentiment analysis. In: Proceedings of the 10th International Workshop on Semantic Evaluation (SemEval-2016) (2016)
10. Lakkaraju, H., et al.: Exploiting coherence in reviews for discovering latent facets and associated sentiments. In: Proceedings of 2011 SIAM International Conference on Data Mining, April 2011
11. Liu, B.: Sentiment analysis and opinion mining. Synth. Lect. Hum. Lang. Technol. 5(1), 1–167 (2012)
12. McAuley, J., Leskovec, J., Jurafsky, D.: Learning attitudes and attributes from multi-aspect reviews. In: 2012 IEEE 12th International Conference on Data Mining (ICDM). IEEE (2012)
13. Mohammad, S.M., Turney, P.D.: Nrc emotion lexicon. National Research Council, Canada (2013)

14. Mullen, T., Collier, N.: Sentiment analysis using support vector machines with diverse information sources. In: Proceedings of the 2004 Conference on Empirical Methods in Natural Language Processing (2004)

15. Pang, B., Lee, L.: Opinion mining and sentiment analysis. Found. Trends® Inf. Retr. **2**(1–2), 1–135 (2008)

16. Perez-Rosas, V., Banea, C., Mihalcea, R.: Learning sentiment Lexicons in Spanish. In: LREC, vol. 12 (2012)

17. Pontiki, M., et al.: SemEval-2016 task 5: aspect based sentiment analysis. In: Proceedings of the 10th International Workshop on Semantic Evaluation (SemEval-2016) (2016)

18. Popescu, A.-M., Etzioni, O.: Extracting product features and opinions from reviews. In: Kao, A., Poteet, S.R. (eds.) Natural Language Processing and Text Mining, pp. 9–28. Springer, London (2007). https://doi.org/10.1007/978-1-84628-754-1_2

19. Rao, D., Ravichandran, D.: Semi-supervised polarity lexicon induction. In: Proceedings of the 12th Conference of the European Chapter of the Association for Computational Linguistics. Association for Computational Linguistics (2009)

20. Rocktäschel, T., et al.: Reasoning about entailment with neural attention. arXiv preprint arXiv:1509.06664 (2015)

21. Rush, A.M., Chopra, S., Weston, J.: A neural attention model for abstractive sentence summarization. arXiv preprint arXiv:1509.00685 (2015)

22. Sukhbaatar, S., Weston, J., Fergus, R.: End-to-end memory networks. In: Advances in Neural Information Processing Systems (2015)

23. Sun, Q., et al.: Research on semantic orientation classification of Chinese online product reviews based on multi-aspect sentiment analysis. In: Proceedings of the 3rd IEEE/ACM International Conference on Big Data Computing, Applications and Technologies. ACM (2016)

24. Tai, K.S., Socher, R., Manning, C.D.: Improved semantic representations from tree-structured long short-term memory networks. arXiv preprint arXiv:1503.00075 (2015)

25. Tang, D., et al.: Target-dependent sentiment classification with long short term memory. CoRR, abs/1512.01100 (2015)

26. Tang, D., Qin, B., Liu, T.: Aspect level sentiment classification with deep memory network. arXiv preprint arXiv:1605.08900 (2016)

27. Wang, Y., Huang, M., Zhao, L.: Attention-based LSTM for aspect-level sentiment classification. In: Proceedings of the 2016 Conference on Empirical Methods in Natural Language Processing (2016)

The 1st Pacific Asia Workshop on Deep Learning for Knowledge Transfer (DLKT 2019)

Transfer Channel Pruning
for Compressing Deep Domain
Adaptation Models

Chaohui Yu[1,2], Jindong Wang[1,2], Yiqiang Chen[1,2(✉)], and Zijing Wu[3]

[1] Beijing Key Laboratory of Mobile Computing and Pervasive Device,
Institute of Computing Technology, Chinese Academy of Sciences, Beijing, China
yqchen@ict.ac.cn
[2] University of Chinese Academy of Sciences, Beijing, China
[3] Columbia University, New York, NY 10024, USA

Abstract. Deep unsupervised domain adaptation has recently received increasing attention from researchers. However, existing methods are computationally intensive due to the computational cost of CNN (Convolutional Neural Networks) adopted by most work. There is no effective network compression method for such problem. In this paper, we propose a unified *Transfer Channel Pruning (TCP)* approach for accelerating deep unsupervised domain adaptation (UDA) models. TCP is capable of compressing the deep UDA model by pruning less important channels while simultaneously learning transferable features by reducing the cross-domain distribution divergence. Therefore, it reduces the impact of negative transfer and maintains competitive performance on the target task. To the best of our knowledge, TCP is the *first* approach that aims at accelerating deep unsupervised domain adaptation models. TCP is validated on two benchmark datasets – Office-31 and ImageCLEF-DA with two common backbone networks – VGG16 and ResNet50. Experimental results demonstrate that TCP achieves comparable or better classification accuracy than other comparison methods while significantly reducing the computational cost. To be more specific, in VGG16, we get even higher accuracy after pruning 26% floating point operations (FLOPs); in ResNet50, we also get higher accuracy on half of the tasks after pruning 12% FLOPs.

Keywords: Unsupervised domain adaptation ·
Transfer channel pruning · Accelerating

1 Introduction

Deep neural networks have significantly improved the performance of diverse machine learning applications. However, in order to avoid overfitting and achieve better performance, a large amount of labeled data is needed to train a deep network. Since the manual labeling of massive training data is usually expensive

© Springer Nature Switzerland AG 2019
L. H. U and H. W. Lauw (Eds.): PAKDD 2019 Workshops, LNAI 11607, pp. 257–273, 2019.
https://doi.org/10.1007/978-3-030-26142-9_23

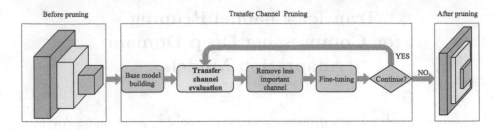

Fig. 1. The framework of the proposed Transfer Channel Pruning (TCP) approach.

in terms of money and time, it is urgent to develop effective algorithms to reduce the labeling workload on the domain to be learned (i.e. *target* domain).

A popular solution to solve the above problem is called *transfer learning*, or *domain adaptation* [27,40,41], which tries to transfer knowledge from well-labeled domains (i.e. *source* domains) to the target domain. Specifically, unsupervised domain adaptation (UDA) is considered more challenging since the target domain has no labels. The key is to learn a discriminative model to reduce the distribution divergence between domains. In recent years, deep domain adaptation methods have produced competitive performance in various tasks [19,39]. This is because that they take advantages of CNN (convolutional neural networks) to learn more transferable representations [19] compared to traditional methods. Popular CNN architectures such as AlexNet [17], VGGNet [32], and ResNet [11] are widely adopted as the backbone networks for deep unsupervised domain adaptation methods. Then, knowledge can be transferred to the target domain by reducing the cross-domain distance such as MMD (maximum mean discrepancy) [26] or KL divergence [26].

Unfortunately, it is still challenging to deploy these deep UDA models on resource constrained devices such as mobile phones since there is a huge computational cost required by these methods. In order to reduce resource requirement and accelerate the inference process, a common solution is *network compression*. Network compression methods mainly include network quantization [30,44], weight pruning [10,12,24], and low-rank approximation [3,9]. Especially channel pruning [12,24], which is a type of weight pruning and compared to other methods, it does not need special hardware or software implementations. And it can reduce negative transfer by pruning some redundant channels, so it is a good choice for compressing deep UDA models.

However, it is not feasible to apply the above network compression methods directly to the UDA problems. The reasons are two folds. Firstly, these compression methods are proposed to solve *supervised* learning problems, which is not suitable for the UDA settings since there are no labels in the target domain. Secondly, even if we can acquire some labels manually, applying these compression methods directly to UDA will result in *negative transfer* [27], since they fail to consider the distribution discrepancy between the source and target domains. Currently, there is no effective network compression method for UDA.

In this paper, we propose a unified network compression method called **Transfer Channel Pruning (TCP)** for accelerating deep unsupervised domain adaptation models. The general framework of our method TCP is shown in Fig. 1. Starting from a deep unsupervised domain adaptation base model, TCP iteratively evaluates the importance of channels with the transfer channel evaluation module and remove less important channels for both source and target domains. TCP is capable of compressing the deep UDA model by pruning less important channels while simultaneously learning transferable features by reducing the cross-domain distribution divergence. Experimental results demonstrate that TCP achieves better classification accuracy than other comparison pruning methods while significantly reducing the computational cost. To the best of our knowledge, TCP is the *first* approach to accelerate the deep UDA models.

To summarize, the contributions of this paper are as follows:

(1) We present TCP as a unified approach for accelerating deep unsupervised domain adaptation models. TCP is a generic, accurate, and efficient compression method that can be easily implemented by most deep learning libraries.
(2) TCP is able to reduce negative transfer by considering the cross-domain distribution discrepancy using the proposed *Transfer Channel Evaluation* module.
(3) Extensive experiments on two public UDA datasets demonstrate the significant superiority of our TCP method.

2 Related Work

Our work is mainly related to unsupervised domain adaptation and network compression.

2.1 Unsupervised Domain Adaptation

Unsupervised domain adaptation (UDA) is a specific area of transfer learning [27, 40], which is to learn a discriminative model in the presence of the domain-shifts between domains. The main problem of UDA is how to reduce the domain shift between the source and target domains. There are many methods to tackle this problem: traditional (shallow) learning and deep learning.

Traditional (shallow) learning methods have several aspects: (1) Subspace learning. Subspace Alignment (SA) [5] aligns the base vectors of both domains and Subspace Distribution Alignment (SDA) [34] extends SA by adding the subspace variance adaptation. Gong *et al.* proposed the Geodesic Flow Kernel (GFK) [8] to sample indefinite points along the geodesic flow between domains. CORAL [33] aligns subspaces in second-order statistics. (2) Distribution alignment. Pan *et al.* proposed the Transfer Component Analysis (TCA) method to align the marginal distributions between domains. Based on TCA, Joint Distribution Adaptation (JDA) [20] is proposed to match both marginal and conditional distributions. Later works extend JDA by adding regularization, structural consistency [14] and domain invariant clustering [36]. But these works treat the two distributions equally and fail to leverage the different importance of distributions. Recently, Wang *et al.* proposed the Manifold Embedded Distribution Alignment (MEDA) [40,41] approach to dynamically evaluate the different effect of marginal and conditional distributions and achieved the state-of-the-art results on domain adaptation.

As for deep learning methods, CNN can learn nonlinear deep representations and capture underlying factors of variation between different tasks [1]. These deep representations can disentangle the factors of variation, which enables the transfer of knowledge between tasks.

Recent works on deep UDA approaches can be mainly summarized into two cases [29]. The first case is the discrepancy-based deep UDA approach, which assumes that fine-tuning the deep network model with labeled or unlabeled target data can diminish the shift between the two domains. The most commonly used methods for aligning the distribution shift between the source and target domains are maximum mean discrepancy (MMD), correlation alignment (CORAL) [35], Kullback–Leibler (KL) divergence [45], among others. The second case can be referred to as an adversarial-based deep UDA approach [7]. In this case, a domain discriminator that classifies whether a data point is drawn from the source or target domain is used to encourage domain confusion through an adversarial objective to minimize the distance between the source and target distributions [6]. Recent works on deep UDA embed domain-adaptation modules into deep networks to improve transfer performance [7], where significant performance gains have been obtained. UDA has wide applications in computer vision [13,19] and natural language processing and is receiving increasing attention from researchers. In this paper, we only concentrate on accelerating the discrepancy-based deep UDA approaches, and we will discuss the adversarial-based deep UDA approaches in the future work.

As far as we know, no previous UDA approach has focused on the acceleration of the network.

2.2 Network Compression

These years, for better accuracy, designing deeper and wider CNN models has become a general trend, such as VGGNet [32] and ResNet [11]. However, as the CNN grow bigger, it is harder to deploy these deep models on resource constrained devices. Network compression becomes an efficient way to solve this problem. Network compression methods mainly include network quantization, low-rank approximation and weight pruning. Network quantization is good at decreasing the presentation precision of parameters so as to reduce the storage space. Low-rank approximation reduces the storage space by low-rank matrix techniques, which is not efficient for point-wise convolution [2]. Weight pruning mainly includes two methods, neural pruning [10,18] and channel pruning [12, 23,24].

Channel pruning methods prune the whole channel each time, so it is fast and efficient than neural pruning which removes a single neuron connection each time. It is a structured pruning method, compared to network quantization and low-rank approximation, it does not introduce sparsity to the original network structure and also does not require special software or hardware implementations. It has demonstrated superior performance compared to other methods and many works [12,23,24] have been proposed to perform channel pruning on pre-trained models with different kinds of criteria.

These above pruning methods mainly aim at supervised learning problems, by contrast, there have been few studies for compressing unsupervised domain adaptation models. As far as we know, we are the first to study how to do channel pruning for deep unsupervised domain adaptation.

TCP is primarily motivated by [24], while our work is different from it. TCP is presented for pruning unsupervised domain adaptation models. To be more specific, we take the discrepancy between the source and target domains into consideration so we can prune the less important channels not just for the source domain but also for the unlabeled target domain. We call this Transfer Channel Evaluation, which is highlighted in yellow in Fig. 1.

3   Transfer Channel Pruning

In this section, we introduce the proposed Transfer Channel Pruning (TCP) approach.

3.1   Problem Definition

In unsupervised domain adaptation, we are given a source domain $\mathcal{D}_s = \{(\mathbf{x}_i^s, y_i^s)\}_{i=1}^{n_s}$ of n_s labeled examples and a target domain $\mathcal{D}_t = \{\mathbf{x}_j^t\}_{j=1}^{n_t}$ of n_t unlabeled examples. $\mathcal{D}_s$ and $\mathcal{D}_t$ have the same label space, i.e. the marginal distributions between two domains are different, i.e. $P_s(\mathbf{x}_s) \neq P_t(\mathbf{x}_t)$. The goal of deep UDA is to design a deep neural network that enables learning of transfer classifiers $y = f_s(\mathbf{x})$ and $y = f_t(\mathbf{x})$ to close the source-target discrepancy and can achieve the best performance on the target dataset.

For a pre-trained deep UDA model, its parameters can be denoted as $\mathbf{W}$. Here we assume the l_{th} convolutional layer has an output activation tensor $\mathbf{a}_l$ of size of $h_l \times w_l \times k_l$, where k_l represents the number of output channels of the l_{th} layer, and h_l and w_l stand for the height and width of feature maps of the l_{th} layer, respectively. Therefore, the goal of TCP is to prune a UDA model in order to accelerate it with comparable or even better performance on the target domain. In this way, we can obtain smaller models that require less computation complexity and memory consumption, which can be deployed on resource constrained devices.

3.2   Motivation

We compress the deep UDA model using model pruning methods for their efficiency. A straightforward model pruning technique is a *two-stage* method, which

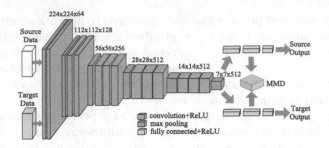

Fig. 2. The basic deep UDA architecture. (Color figure online)

first prunes the model on the source domain with supervised learning and then fine-tune the model on the target domain. However, negative transfer [27] is likely to happen during this pruning process since the discrepancy between the source and target domains is ignored.

In this work, we propose a unified **Transfer Channel Pruning (TCP)** approach to tackle such challenge. TCP is capable of compressing the deep UDA model by pruning less important channels while simultaneously learning transferable features by reducing the cross-domain distribution divergence. Therefore, TCP reduces the impact of negative transfer and maintains competitive performance on the target task. In short, TCP is a generic, accurate, and efficient compression method that can be easily implemented by most deep learning libraries.

To be more specific, Fig. 1 illustrates the main idea of TCP. There are mainly three steps. Firstly, TCP learns the base deep UDA model through *Base Model Building*. The base model is fine-tuned with the standard UDA criteria. Secondly, TCP evaluates the importance of channels of all layers with the *Transfer Channel Evaluation* and performs further fine-tuning. Specifically, the convolutional layers, which usually dominate the computation complexity, are pruned in this step. Thirdly, TCP *iteratively refines* the pruning results and stops after reaching the trade-off between accuracy and FLOPs (i.e. computational cost) or parameter size.

3.3   Base Model Building

In this step, we build the base UDA model with deep neural networks. Deep neural networks have been successfully used in UDA with state-of-the-art algorithms [7,19,37] in recent years. Previous studies [37,43] have shown that the features extracted by deep networks are general at lower layers, while specific at the higher layers since they are more task-specific. Therefore, more transferable representations can be learned by transferring the features at lower layers and then fine-tune the task-specific layers. During fine-tuning, the cross-domain discrepancy can be reduced by certain adaptation distance. Since our main contribution is not designing new deep UDA networks, we build the base model like most discrepancy-based deep UDA approaches [19,35,38]. In the following, we will briefly introduce the main idea of the base model, and more details can be found in the original paper of these discrepancy-based deep UDA approaches.

As shown in Fig. 2, we learn transferable features via several convolutional and pooling layers (the blue and purple blocks). Then, the classification task can be accomplished with the fully-connected layers (the yellow blocks). MMD (maximum mean discrepancy) [26] is adopted as the adaptation loss in order to reduce domain shift. MMD has been proposed to provide the distribution difference between the source and target datasets. And it has been widely utilized in many UDA methods [19,21,25]. The MMD loss between two domains can be computed as

$$L_{mmd} = \left\| \frac{1}{n_s} \sum_{\mathbf{x}_i \in \mathcal{D}_s} \phi(\mathbf{x}_i) - \frac{1}{n_t} \sum_{\mathbf{x}_j \in \mathcal{D}_t} \phi(\mathbf{x}_j) \right\|_{\mathcal{H}}^2 , \tag{1}$$

where $\mathcal{H}$ denotes RKHS (Reproducing Kernel Hilbert Space) and $\phi(\cdot)$ denotes some feature map to map the original samples to RKHS.

Several popular architectures can serve as the backbone network of the base model, such as AlexNet [17], VGGNet [32], and ResNet [11]. After obtaining the base model, we can perform channel pruning to accelerate the model.

3.4 Transfer Channel Evaluation

The goal of transfer channel evaluation is to iteratively evaluate the importance of output channels of layers in order to prune the $\mathcal{K}$ least important channels. Here $\mathcal{K}$ is controlled by users. In the pruning process, we want to preserve and refine a set of parameters $\mathbf{W}'$, which represents those important parameters for both source and target domains. Let $L(\mathcal{D}_s, \mathcal{D}_t, \mathbf{W})$ be the cost function for UDA and $\mathbf{W}' = \mathbf{W}$ at the starting time. For a better set of parameters $\mathbf{W}'$, we want to minimize the loss change after pruning a channel $\mathbf{a}_{l,i}$. This can be considered as an optimization problem. Here we introduce the absolute difference of loss:

$$|\Delta L(\mathbf{a}_{l,i})| = |L(\mathcal{D}_s, \mathcal{D}_t, \mathbf{a}_{l,i}) - L(\mathcal{D}_s, \mathcal{D}_t, \mathbf{a}_{l,i} = 0)| , \tag{2}$$

which means the loss change after pruning the i_{th} channel of the l_{th} convolutional layer. And we want to minimize $|\Delta L(\mathbf{a}_{l,i})|$ by selecting the appropriate channel $\mathbf{a}_{l,i}$. Pruning will stop until a trade-off between accuracy and pruning object (FLOPs or parameter size) has been achieved.

However, it is hard to find a set of optimal parameters in one go, because the search space is $2^{|\mathbf{W}|}$ which is too huge to compute and try every combination. Inspired by [24], our TCP solves this problem with a greedy algorithm by iteratively removing the $\mathcal{K}$ least important channels at each time.

Criteria. Criteria is the criterion for judging the importance of channels. Since the key to channel pruning is to select the least important channel, especially for UDA, we design the criteria of TCP carefully. There are many heuristic criteria, including the L_2-norm of filter weights, the activation statistics of feature maps, mutual information between activations and predictions and Taylor expansion, etc. Here we choose the *first-order Taylor expansion* as the base criteria since its efficiency and performance has been verified in [24] for pruning supervised learning models. Compared with our TCP, we also take pruning as an optimization problem, however, the objective we want to optimize is the final performance on the unlabeled target dataset. So we design our criteria in a different way which is better for pruning deep UDA models.

According to Taylor's theorem, the Taylor expansion at point $x = a$ can be computed as:

$$f(x) = \sum_{p=0}^{P} \frac{f^{(p)}(a)}{p!}(x-a)^p + R_p(x), \tag{3}$$

where p denotes the p_{th} derivative of $f(x)$ at point $x = a$ and the last item $R_p(x)$ represents the p_{th} remainder. To approximate $|\Delta L(\mathbf{a}_{l,i})|$, we can use the first-order Taylor expansion near $\mathbf{a}_{l,i} = 0$ which means the loss change after removing $\mathbf{a}_{l,i}$, then we can get:

$$f(\mathbf{a}_{l,i} = 0) = f(\mathbf{a}_{l,i}) - f'(\mathbf{a}_{l,i}) \cdot \mathbf{a}_{l,i} + \frac{|\mathbf{a}_{l,i}|^2}{2} \cdot f''(\xi), \tag{4}$$

where ξ is a value between 0 and $\mathbf{a}_{l,i}$, and $\frac{|\mathbf{a}_{l,i}|^2}{2} \cdot f''(\xi)$ is a Lagrange form remainder which requires too much computation, so we abandon this item for accelerating the pruning process. Then back to Eq. (2), we can get:

$$L(\mathcal{D}_s, \mathcal{D}_t, \mathbf{a}_{l,i} = 0) = L(\mathcal{D}_s, \mathcal{D}_t, \mathbf{a}_{l,i}) - \frac{\partial L}{\partial \mathbf{a}_{l,i}} \cdot \mathbf{a}_{l,i}. \tag{5}$$

Then, we combine Eqs. (2) and (5) and get the criteria G of TCP:

$$G(\mathbf{a}_{l,i}) = |\Delta L(\mathbf{a}_{l,i})| = |\frac{\partial L}{\partial \mathbf{a}_{l,i}} \cdot \mathbf{a}_{l,i}|, \tag{6}$$

which means the absolute value of product of the activation and the gradient of the cost function, and $\mathbf{a}_{l,i}$ can be calculated as:

$$\mathbf{a}_{l,i} = \frac{1}{N}\sum_{n=1}^{N} \frac{1}{h_l \times w_l} \sum_{p=1}^{h_l} \sum_{q=1}^{w_l} \mathbf{a}_{l,i}^{p,q}, \tag{7}$$

where N is the number of batch size.

Loss Function of TCP. To make TCP focus on pruning UDA models, we simultaneously take the source domain and the unlabeled target domain into consideration. The loss function of TCP consists of two parts, $L_{cls}(\mathcal{D}_s, \mathbf{W})$ and $L_{mmd}(\mathcal{D}_s, \mathcal{D}_t, \mathbf{W})$. Here, $L_{cls}(\mathcal{D}_s, \mathbf{W})$ is a cross-entropy loss which denotes the classification loss on source domain and can be computed as:

$$L_{cls}(\mathcal{D}_s, \mathbf{W}) = -\frac{1}{N}\sum_{i=1}^{N}\sum_{c=1}^{C} P_{i,c} log(h_c(x_i^s)) \tag{8}$$

where C is the number of classes of source dataset, $P_{i,c}$ is the probability of x_i^s belonging to class c, and $h_c(x_i^s)$ denotes the probability that the model predicts x_i^s as class c. And $L_{mmd}(\mathcal{D}_s, \mathcal{D}_t, \mathbf{W})$ denotes the MMD loss between the source and target domains that presented in Eq. (1). The total loss function can be computed as:

$$L(\mathcal{D}_s, \mathcal{D}_t, \mathbf{W}) = L_{cls}(\mathcal{D}_s, \mathbf{W}) + \beta L_{mmd}(\mathcal{D}_s, \mathcal{D}_t, \mathbf{W}), \qquad (9)$$

where

$$\beta = \frac{4}{1 + e^{-1 \cdot \frac{i}{ITER}}} - 2. \qquad (10)$$

Here, β is a dynamic value which takes values in $(0, 1)$. $i \in (0, ITER)$ where $ITER$ is the number of pruning iterations. We design β in this way for two main reasons, on the one hand, during the early stage of pruning, the weights have not converged and keep unstable so the L_{mmd} is too large and makes the pruned model hard to converge. On the other hand, in the rest of the pruning process, the L_{mmd} becomes more important that can guide the pruned model to focus more on the target domain. So the criteria of TCP can be computed as:

$$G(\mathbf{a}_{l,i}) = |\frac{\partial L_{cls}(\mathcal{D}_s, \mathbf{W})}{\mathbf{a}_{l,i}^s} \cdot \mathbf{a}_{l,i}^s + \beta \frac{\partial L_{mmd}(\mathcal{D}_s, \mathcal{D}_t, \mathbf{W})}{\mathbf{a}_{l,i}^t} \cdot \mathbf{a}_{l,i}^t|, \qquad (11)$$

where $\mathbf{a}_{l,i}^s$ and $\mathbf{a}_{l,i}^t$ denote the activation with source data and target data respectively.

3.5   Iterative Refinement

After the transfer channel evaluation, each channel is sorted according to $G(\mathbf{a}_{l,i})$ and the $\mathcal{K}$ least important channels are removed after each pruning iteration. Then, a short-term fine-tuning is adopted to the pruned model with 5 epochs to help the model to converge and the pruning is done after a trade-off between accuracy and FLOPs or parameter size has been achieved. This step is done iteratively to prune the network.

The learning procedure of TCP is described in Algorithm 1.

4   Experimental Analysis

In this section, we evaluate the performance of TCP via experiments on pruning deep unsupervised domain adaptation models. We evaluate our approaches for VGGNet [32] and ResNet [11] on two popular datasets – Office-31 [31] and ImageCLEF-DA[1]. All our methods are implemented based on the PyTorch [28] framework and the code will be released soon at [42].

---

[1] http://imageclef.org/2014/adaptation.

Algorithm 1. TCP: Transfer Channel Pruning

Input: Source domain $\mathcal{D}_s = \{(\mathbf{x}_i^s, y_i^s)\}_{i=1}^{n_s}$, target domain $\mathcal{D}_t = \{\mathbf{x}_j^t\}_{j=1}^{n_t}$, the baseline **W**.

Output: A pruned model **W**′ for deep unsupervised domain adaptation.

1: Fine-tune the unsupervised domain adaptation baseline until the best performance achieved on the unlabeled target dataset;
2: **for** *iteration i* **do**
3:    Sort the importance of channels by criteria Eq. (11) and identify less significant channels;
4:    Remove the $\mathcal{K}$ least important channels of the layers;
5:    Short-term fine-tune;
6:    **if** the trade-off between accuracy and FLOPs or parameter size has achieved **then**
7:       *break*
8:    **end if**
9: **end for**
10: Long-term fine-tune;
11: **return** pruned model **W**′.

4.1 Datasets

Office-31. This dataset is a standard and maybe the most popular benchmark for unsupervised domain adaptation. It consists of 4,110 images within 31 categories collected from everyday objects in an office environment. It consists of three domains: *Amazon* (**A**), which contains images downloaded from amazon.com, *Webcam* (**W**) and *DSLR* (**D**), which contain images respectively taken by web camera and digital SLR camera under different settings. We evaluate all our methods across six transfer tasks on all the three domains **A→W**, **W→A**, **A→D**, **D→A**, **D→W** and **W→D**.

ImageCLEF-DA. This dataset is a benchmark dataset for ImageCLEF 2014 domain adaptation challenge, and it is collected by selecting the 12 common categories shared by the following public datasets and each of them is considered as a domain: *Caltech* − 256 (**C**), *ImageNet ILSVRC* 2012 (**I**), *Pascal VOC* 2012 (**P**) and *Bing* (**B**). There are 50 images in each category and 600 images in each domain. We evaluate all methods across six transfer tasks following existing work [6,19,41]: **I→P**, **P→I**, **I→C**, **C→I**, **P→C** and **C→P**. Compared with Office-31, this dataset is more balanced and can be a good comparable dataset to Office-31.

4.2 Implementation Details

We mainly compare three methods: (1) **Two-stage**: which is the most straightforward method that applies channel pruning to the source domain task first, then fine-tune for the target domain task with the pruned model. (2) **TCP_w/o_DA**: Our TCP method without the MMD loss, here we call it domain adaptation (DA) loss. Which also means $\beta = 0$ all the time in Eq. (9). (3) **TCP**: Our full TCP method with DA loss.

We evaluate all the methods on two popular backbone networks: VGG16 [32] and ResNet50 [11]. As baselines, VGG16-based and ResNet50-based are the original models that are not pruned. As for VGG16-based model, it has 13 convolutional layers and 3 fully-connected layers. We prune all the convolutional layers and the first fully-connected layer and we only use the activations of the second fully-connected layer as image representation and build the MMD loss which is shown in Fig. 2. And as for ResNet50-based model, we use similar settings as VGG16-based model with a few differences. Because of the shortcut and residual branch structure, we only prune the inside convolutional layers of each bottleneck block. The MMD loss is built with the only fully-connected layer. Moreover, we also take the Batch Normalization (BN) [16] layers into consideration and reconstruct the whole model during pruning.

In practice, all the input images are cropped to a fixed size 224×224 and randomly sampled from the resized image with horizontal flip and mean-std normalization. At first, we fine-tune all the UDA models on each unsupervised domain adaptation tasks for 200 epochs with learning rate from 0.01 to 0.0001 and the batch size $= 32$. During pruning, we set $\mathcal{K} = 64$ which means 64 channels will be removed after each pruning iteration. Here we set $\mathcal{K} = 64$ because we want to speed up the pruning process. In fact, this is a tradeoff between accuracy and time. If we set $\mathcal{K}$ to a smaller value, we may achieve better performance on the target domain, but we also need more time to do the pruning and fine-tuning procedure to get the same FLOPs reduction. And if we set $\mathcal{K}$ to a larger value, the pruning process can be accelerated while the accuracy on the target domain can not be guaranteed.

After that, extra 5 epochs are adopted to help the pruned model to converge. And we follow [15, 22] to prune the baseline with different compression rate. The VGG16-based baseline is pruned with 26% and 70% FLOPs reduced while the ResNet50-based baseline is pruned with 12% and 46% FLOPs reduced. ResNet50 has lower compression rate since the bottleneck structure stops some layers from being pruned. As for the VGG16-based and ResNet50-based baselines, they are the original well fine-tuned UDA models without pruning. To be more specific, as for the VGG16-based baseline, we build the UDA model as Fig. 2. The backbone network is VGG16 and MMD loss is built on the second fully-connected layer to compare and align the distribution between source and target domains. Then we fine-tune the UDA model for 200 epochs with learning rate from 0.01 to 0.0001 and batch size $= 32$. And as for the ResNet50-based baseline, the structure is similar to the VGG16-based baseline, we use ResNet50 as the backbone network and MMD loss is built on the only fully-connected layer.

We follow standard evaluation protocol for UDA and use all source examples with labels and all target examples without labels [8]. The labels for the target domain are only used for evaluation. We adopt *classification accuracy* on the target domain and *parameter reduction* as the evaluation metrics: higher accuracy and fewer parameters indicate better performance.

Table 1. The performance on Office-31 dataset (VGG16-based and ResNet50-based). Here, $FLOPs \downarrow$ and $Param \downarrow$ denote the decrement of FLOPs and parameter size compared with the baseline, Acc means the accuracy on target domain.

Models	FLOPs↓	A→W		D→W		W→D		A→D		D→A		W→A		Average	
		Acc	Param↓	Acc	Param↓	Acc	Param↓	Acc	Param↓	Acc	Param↓	Acc	Param↓	Acc	Param↓
VGG16-base		74.0%		94.0%		97.5%		72.3%		54.1%		55.2%		74.5%	
Two_stage		69.4%	29.4%	94.5%	31.2%	99.0%	30.6%	69.8%	29.3%	42.7%	32.5%	47.2%	28.3%	70.4%	30.2%
TCP_w/o_DA	26%	73.0%	29.7%	95.5%	32.7%	99.3%	29.2%	75.8%	25.8%	45.9%	29.3%	50.4%	29.6%	73.3%	29.4%
TCP		76.1%	36.8%	96.1%	36.2%	99.8%	32.6%	76.2%	35.9%	47.9%	37.1%	51.2%	39.8%	74.5%	36.4%
Two_stage		57.1%	63.3%	88.1%	68.0%	96.3%	64.5%	55.0%	61.0%	31.8%	66.2%	32.7%	63.1%	60.2%	64.3%
TCP_w/o_DA	70%	53.5%	62.5%	89.2%	61.3%	97.9%	57.5%	61.8%	54.8%	35.3%	62.6%	34.5%	58.8%	62.0%	59.6%
TCP		74.1%	69.3%	89.5%	69.8%	98.8%	65.2%	65.9%	66.2%	35.6%	68.5%	38.5%	68.2%	67.1%	67.9%
ResNet50-base		80.3%		97.1%		99.2%		78.9%		64.3%		62.3%		80.3%	
Two_stage		75.8%	32.4%	96.7%	31.4%	99.5%	35.5%	76.0%	30.4%	48.0%	28.3%	50.1%	29.4%	74.4%	31.2%
TCP_w/o_DA	12%	79.8%	33.5%	97.0%	35.5%	100%	34.5%	77.1%	36.2%	47.8%	34.5%	52.6%	33.1%	75.7%	34.5%
TCP		81.8%	37.7%	98.2%	36.2%	99.8%	37.0%	77.9%	36.9%	50.0%	35.0%	55.5%	36.9%	77.2%	36.7%
Two_stage		65.5%	56.2%	93.0%	56.3%	98.7%	57.3%	64.9%	56.0%	34.0%	57.2%	38.9%	57.3%	65.8%	56.7%
TCP_w/o_DA	46%	75.1%	56.4%	95.8%	56.4%	99.2%	56.6%	70.8%	55.8%	34.2%	56.4%	41.5%	56.6%	69.4%	56.4%
TCP		77.4%	58.4%	96.3%	58.0%	100%	57.1%	72.0%	59.0%	36.1%	57.8%	46.3%	58.5%	71.3%	58.1%

Table 2. The performance on ImageCLEF-DA dataset (VGG16-based and ResNet50-based).

Models	FLOPs↓	I→P		P→I		I→C		C→I		C→P		P→C		Average	
		Acc	Param↓	Acc	Param↓	Acc	Param↓	Acc	Param↓	Acc	Param↓	Acc	Param↓	Acc	Param↓
VGG16-base		71.3%		80.0%		88.5%		77.0%		61.1%		87.2%		77.5%	
Two_stage		68.0%	29.0%	77.7%	29.9%	88.5%	27.5%	64.5%	29.0%	56.0%	29.3%	83.3%	26.9%	73.0%	28.6%
TCP_w/o_DA	26%	70.5%	33.6%	79.5%	34.2%	89.0%	33.1%	77.1%	34.5%	62.2%	35.0%	85.1%	33.1%	74.5%	33.9%
TCP		72.0%	39.0%	80.5%	32.2%	90.5%	35.1%	77.8%	36.3%	64.8%	36.6%	87.5%	34.5%	78.9%	35.6%
Two_stage		58.6%	65.8%	69.2%	62.9%	80.6%	64.3%	57.7%	57.0%	43.2%	67.8%	74.5%	61.5%	63.9%	63.2%
TCP_w/o_DA	70%	61.0%	55.4%	69.1%	65.7%	80.5%	66.5%	55.1%	63.3%	47.0%	66.5%	70.9%	65.8%	63.9%	63.9%
TCP		61.9%	66.7%	69.5%	66.0%	81.8%	65.7%	59.8%	68.7%	49.7%	68.8%	75.9%	67.2%	66.4%	67.1%
ResNet50-base		74.8%		82.2%		92.3%		83.3%		70.0%		89.8%		82.1%	
Two_stage		71.8%	29.4%	81.3%	31.5%	92.1%	34.4%	76.5%	32.4%	64.0%	29.2%	84.0%	30.8%	78.2%	31.2%
TCP_w/o_DA	12%	73.0%	33.5%	80.5%	34.6%	92.0%	33.8%	76.1%	31.2%	64.3%	30.1%	86.3%	36.3%	78.7%	33.2%
TCP		75.0%	37.5%	82.6%	36.5%	92.5%	35.5%	80.8%	36.7%	66.2%	36.6%	86.5%	37.6%	80.6%	36.7%
Two_stage		65.6%	53.2%	71.8%	56.5%	85.2%	54.2%	68.2%	54.0%	57.4%	51.5%	78.1%	54.0%	71.1%	53.9%
TCP_w/o_DA	46%	66.6%	55.4%	73.0%	57.4%	85.5%	55.4%	67.7%	55.5%	55.5%	53.6%	77.0%	57.1%	70.8%	55.7%
TCP		67.8%	57.2%	77.5%	58.0%	88.6%	56.2%	71.6%	58.5%	57.7%	55.7%	79.5%	58.2%	73.8%	57.3%

4.3 Results and Analysis

Firstly, we evaluate all the tasks on Office-31 dataset. The results are shown in Table 1. As can be seen, the full TCP method outperforms other methods under the same compression rate (FLOPs reduction) and can reduce more parameters. It is important and interesting that TCP achieves even better performance than the baseline model (which is not pruned). This is probably because some redundant channels in the base model are removed thus negative transfer is reduced. Especially for the results of ResNet50-based models, our ResNet50-based baseline achieves comparable or better performance compared with the discrepancy-based deep UDA approaches result of DDC [38] and DAN [19] in [21]. However, we can get better performance on half of the tasks and we even get 100% on task W→D after 46% FLOPs have been reduced.

Secondly, we evaluate our methods on ImageCLEF-DA dataset and the results are shown in Table 2. We can draw the same conclusion that TCP performs better on all tasks on ImageCLEF-DA dataset. We get higher accuracy than the baseline on all the VGG16-based experiments after 26% FLOPs have

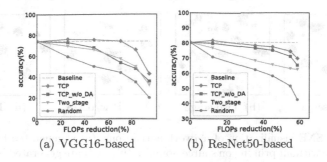

Fig. 3. The pruning result on task A→W with more compression rate.

been reduced, and we also get higher accuracy on the target dataset on half of the tasks on ResNet50-based experiments after 12% FLOPs have been reduced, compared with the baseline which is almost the same as the discrepancy-based deep UDA approaches results of DAN in [21].

Apart from Tables 1 and 2, Fig. 3 shows the comparison for all methods. And we also add a Random method which randomly removes a certain number of channels to achieve the same reduction of FLOPs. Combining these results, more conclusions can be made. (1) Compared with Two_stage, TCP is more efficient because it is a unified framework and treat the pruning as a single optimization problem, while Two_stage is a split method and it does not take the target domain into consideration while pruning. (2) Compared with TCP_w/o_DA, the full TCP uses the transfer channel evaluation to represent the discrepancy between the source and target domains. We try to remove those less important channels for both source and target domains and reduce negative transfer by reducing domain discrepancy. (3) As can be seen from Fig. 3, our TCP outperforms other methods on unlabeled target dataset under different compression rate. (4) This indicates that TCP is generic, accurate, and efficient, which can dramatically reduce the computational cost of a deep UDA model without sacrificing the performance.

4.4 Visualization Analysis

To evaluate the effectiveness of TCP in reducing negative transfer, in Fig. 4, we follow [4] to visualize the model activations of task A→W pruned by different methods using t-SNE [4]. Figure 4(a) shows the results of ResNet50-based baseline without pruning on the source domain. And Fig. 4(b), (c) and (d) denote the result of ResNet50-based models on the target domain, which have been pruned by 12% FLOPs with our three methods Two_stage, TCP_w/o_DA and TCP respectively. The colored digits represent the ground truth of the examples, so the number is from 0 to 30, which denotes target dataset has 31 categories. Here we randomly pick 10 categories to visualize. As can be seen, the target categories are discriminated much more clearly with the model pruned by our

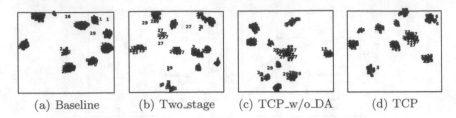

(a) Baseline   (b) Two_stage   (c) TCP_w/o_DA   (d) TCP

Fig. 4. The t-SNE visualization of network activations. (a) is generated by ResNet50-based baseline without pruning on source domain. (b)(c)(d) are generated by ResNet50-based (with 12% FLOPs pruned) with our three methods on target domain respectively. Best view in color.

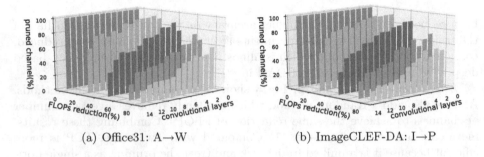

(a) Office31: A→W   (b) ImageCLEF-DA: I→P

Fig. 5. The pruned structure of all the 13 convolutional layers of VGG16-based network on different dataset for deep unsupervised domain adaptation. Best view in color.

TCP method. This suggests that our TCP method is effective in learning more transferable features by reducing the cross-domain divergence.

To explore if there is any pattern in the structure of the pruned models, we show the structure of pruned models on task A→W and I→P in Fig. 5 with TCP. As we can see, higher layers have more redundancy than lower layers in VGG16-based models, and our TCP prefer pruning the channels of higher layers. This is reasonable for unsupervised domain adaptation because the lower layers usually encode many common and important features for both source and target domains. Moreover, with more parameters in higher layers, our TCP thus can prune more parameters under the same compression rate. The same result can be observed on ResNet50-based models.

5   Conclusion and Future Work

In this paper, we propose a unified Transfer Channel Pruning (TCP) approach for accelerating deep unsupervised domain adaptation models. TCP is capable of compressing the deep UDA model by pruning less important channels while simultaneously learning transferable features by reducing the cross-domain distribution divergence. Therefore, it reduces the impact of negative transfer and maintains competitive performance on the target task. TCP is a generic, accurate, and efficient compression method that can be easily implemented by most deep learning libraries. Experiments on two public benchmark datasets demonstrate the significant superiority of our TCP method over other methods.

In the future, we plan to extend TCP in pruning adversarial networks and apply it to heterogeneous UDA problems.

References

1. Bengio, Y., Courville, A., Vincent, P.: Representation learning: a review and new perspectives. IEEE Trans. Pattern Anal. Mach. Intell. **35**(8), 1798–1828 (2013)
2. Chollet, F.: Xception: deep learning with depthwise separable convolutions. arXiv preprint arXiv:1610.02357 (2017)
3. Denton, E.L., Zaremba, W., Bruna, J., LeCun, Y., Fergus, R.: Exploiting linear structure within convolutional networks for efficient evaluation. In: Advances in Neural Information Processing Systems, pp. 1269–1277 (2014)
4. Donahue, J., et al.: Decaf: a deep convolutional activation feature for generic visual recognition. In: ICML, pp. 647–655 (2014)
5. Fernando, B., Habrard, A., Sebban, M., Tuytelaars, T.: Unsupervised visual domain adaptation using subspace alignment. In: Proceedings of the IEEE International Conference on Computer Vision, pp. 2960–2967 (2013)
6. Ganin, Y., Lempitsky, V.: Unsupervised domain adaptation by backpropagation. In: ICML (2015)
7. Ganin, Y., et al.: Domain-adversarial training of neural networks. J. Mach. Learn. Res. **17**(1), 2030–2096 (2016)
8. Gong, B., Grauman, K., Sha, F.: Connecting the dots with landmarks: discriminatively learning domain-invariant features for unsupervised domain adaptation. In: ICML, pp. 222–230 (2013)
9. Han, S., Mao, H., Dally, W.J.: Deep compression: compressing deep neural networks with pruning, trained quantization and huffman coding. arXiv preprint arXiv:1510.00149 (2015)
10. Han, S., Pool, J., Tran, J., Dally, W.: Learning both weights and connections for efficient neural network. In: Advances in Neural Information Processing Systems, pp. 1135–1143 (2015)
11. He, K., Zhang, X., Ren, S., Sun, J.: Deep residual learning for image recognition. In: Proceedings of the IEEE Conference on Computer Vision and Pattern Recognition, pp. 770–778 (2016)
12. He, Y., Zhang, X., Sun, J.: Channel pruning for accelerating very deep neural networks. In: International Conference on Computer Vision (ICCV), vol. 2 (2017)
13. Hoffman, J., et al.: CyCADA: cycle-consistent adversarial domain adaptation. In: ICML (2018)
14. Hou, C.A., Tsai, Y.H.H., Yeh, Y.R., Wang, Y.C.F.: Unsupervised domain adaptation with label and structural consistency. IEEE Trans. Image Process. **25**(12), 5552–5562 (2016)
15. Hu, Y., Sun, S., Li, J., Wang, X., Gu, Q.: A novel channel pruning method for deep neural network compression. arXiv preprint arXiv:1805.11394 (2018)
16. Ioffe, S., Szegedy, C.: Batch normalization: accelerating deep network training by reducing internal covariate shift. In: ICML (2015)

17. Krizhevsky, A., Sutskever, I., Hinton, G.E.: Imagenet classification with deep convolutional neural networks. In: Advances in Neural Information Processing Systems, pp. 1097–1105 (2012)
18. Lin, J., Rao, Y., Lu, J., Zhou, J.: Runtime neural pruning. In: Advances in Neural Information Processing Systems, pp. 2181–2191 (2017)
19. Long, M., Cao, Y., Wang, J., Jordan, M.I.: Learning transferable features with deep adaptation networks. In: ICML (2015)
20. Long, M., Wang, J., Ding, G., Sun, J., Yu, P.S.: Transfer feature learning with joint distribution adaptation. In: Proceedings of the IEEE International Conference on Computer Vision, pp. 2200–2207 (2013)
21. Long, M., Zhu, H., Wang, J., Jordan, M.I.: Deep transfer learning with joint adaptation networks. In: ICML (2017)
22. Luo, J.H., Wu, J.: Autopruner: an end-to-end trainable filter pruning method for efficient deep model inference. arXiv preprint arXiv:1805.08941 (2018)
23. Luo, J.H., Wu, J., Lin, W.: Thinet: a filter level pruning method for deep neural network compression. arXiv preprint arXiv:1707.06342 (2017)
24. Molchanov, P., Tyree, S., Karras, T., Aila, T., Kautz, J.: Pruning convolutional neural networks for resource efficient inference. In: ICLR (2017)
25. Pan, S.J., Kwok, J.T., Yang, Q.: Transfer learning via dimensionality reduction. In: AAAI, vol. 8, pp. 677–682 (2008)
26. Pan, S.J., Tsang, I.W., Kwok, J.T., Yang, Q.: Domain adaptation via transfer component analysis. IEEE Trans. Neural Netw. 22(2), 199–210 (2011)
27. Pan, S.J., Yang, Q., et al.: A survey on transfer learning. IEEE Trans. Knowl. Data Eng. 22(10), 1345–1359 (2010)
28. Paszke, A., et al.: Automatic differentiation in pyTorch (2017)
29. Patel, V.M., Gopalan, R., Li, R., Chellappa, R.: Visual domain adaptation: a survey of recent advances. IEEE Signal Process. Mag. 32(3), 53–69 (2015)
30. Rastegari, M., Ordonez, V., Redmon, J., Farhadi, A.: XNOR-Net: imagenet classification using binary convolutional neural networks. In: Leibe, B., Matas, J., Sebe, N., Welling, M. (eds.) ECCV 2016. LNCS, vol. 9908, pp. 525–542. Springer, Cham (2016). https://doi.org/10.1007/978-3-319-46493-0_32
31. Saenko, K., Kulis, B., Fritz, M., Darrell, T.: Adapting visual category models to new domains. In: Daniilidis, K., Maragos, P., Paragios, N. (eds.) ECCV 2010. LNCS, vol. 6314, pp. 213–226. Springer, Heidelberg (2010). https://doi.org/10.1007/978-3-642-15561-1_16
32. Simonyan, K., Zisserman, A.: Very deep convolutional networks for large-scale image recognition. In: ICLR (2015)
33. Sun, B., Feng, J., Saenko, K.: Return of frustratingly easy domain adaptation. In: AAAI, vol. 6, p. 8 (2016)
34. Sun, B., Saenko, K.: Subspace distribution alignment for unsupervised domain adaptation. In: BMVC, pp. 24.1–24.10 (2015)
35. Sun, B., Saenko, K.: Deep CORAL: correlation alignment for deep domain adaptation. In: Hua, G., Jégou, H. (eds.) ECCV 2016. LNCS, vol. 9915, pp. 443–450. Springer, Cham (2016). https://doi.org/10.1007/978-3-319-49409-8_35
36. Tahmoresnezhad, J., Hashemi, S.: Visual domain adaptation via transfer feature learning. Knowl. Inf. Syst. 50(2), 585–605 (2017)
37. Tzeng, E., Hoffman, J., Saenko, K., Darrell, T.: Adversarial discriminative domain adaptation. In: Computer Vision and Pattern Recognition (CVPR), vol. 1, p. 4 (2017)
38. Tzeng, E., Hoffman, J., Zhang, N., Saenko, K., Darrell, T.: Deep domain confusion: maximizing for domain invariance. arXiv preprint arXiv:1412.3474 (2014)

39. Venkateswara, H., Eusebio, J., Chakraborty, S., Panchanathan, S.: Deep hashing network for unsupervised domain adaptation. In: Proceedings of CVPR, pp. 5018–5027 (2017)
40. Wang, J., Chen, Y., Hao, S., Feng, W., Shen, Z.: Balanced distribution adaptation for transfer learning. In: 2017 IEEE International Conference on Data Mining (ICDM), pp. 1129–1134. IEEE (2017)
41. Wang, J., Feng, W., Chen, Y., Yu, H., Huang, M., Yu, P.S.: Visual domain adaptation with manifold embedded distribution alignment. In: 2018 ACM Multimedia Conference on Multimedia Conference, pp. 402–410. ACM (2018)
42. Wang, J., et al.: Everything about transfer learning and domain adapation. http://transferlearning.xyz
43. Yosinski, J., Clune, J., Bengio, Y., Lipson, H.: How transferable are features in deep neural networks? In: Advances in Neural Information Processing Systems, pp. 3320–3328 (2014)
44. Zhou, A., Yao, A., Guo, Y., Xu, L., Chen, Y.: Incremental network quantization: towards lossless CNNS with low-precision weights. arXiv preprint arXiv:1702.03044 (2017)
45. Zhuang, F., Cheng, X., Luo, P., Pan, S.J., He, Q.: Supervised representation learning: transfer learning with deep autoencoders. In: IJCAI, pp. 4119–4125 (2015)

A Heterogeneous Domain Adversarial Neural Network for Trans-Domain Behavioral Targeting

Kei Yonekawa[1(✉)], Hao Niu[1], Mori Kurokawa[1], Arei Kobayashi[1],
Daichi Amagata[2], Takuya Maekawa[2], and Takahiro Hara[2]

[1] KDDI Research, Inc., Chiyoda-ku, Tokyo, Japan
{ke-yonekawa,ha-niu,mo-kurokawa,
kobayasi}@kddi-research.jp
[2] Osaka University, Suita, Osaka, Japan
{amagata.daichi,maekawa,hara}@ist.osaka-u.ac.jp

Abstract. To realize trans-domain behavioral targeting, which targets interested potential users of a source domain (e.g. E-Commerce) based on their behaviors in a target domain (e.g. Ad-Network), heterogeneous transfer learning (HeTL) is a promising technique for modeling behavior linkage between domains. It is required for HeTL to learn three functionalities: representation alignment, distribution alignment, and classification. In our previous work, we prototyped and evaluated two typical transfer learning algorithms, but neither of them jointly learns the three desired functionalities. Recent advances in transfer learning include a domain-adversarial neural network (DANN), which jointly learns distribution alignment and classification. In this paper, we extended DANN to be able to learn representation alignment by simply replacing its shared encoder with domain-specific types, so that it jointly learns the three desired functionalities. We evaluated the effectiveness of the joint learning of the three functionalities using real-world data of two domains: E-Commerce, which is set as the source domain, and Ad Network, which is set as the target domain.

Keywords: Behavioral targeting · Heterogeneous transfer learning · Domain adversarial training

1 Introduction

Behavioral targeting (BT) is an online advertising technique that delivers advertisements (namely Ads) of particular items (namely Ad-Items) to potential users who may purchase the Ad-Items through touchpoints such as web browsers or applications on smart devices held by users. The potential users are identified based on user behaviors related to the Ad-Items using data collected on the touchpoints. However, searching the potential users in only one domain has limitations in its range of possible potential users i.e. limitations in scale [1].

To expand the range, trans-domain BT that targets potential users of one (source) domain who lie in another (target) domain is a promising method. Under the condition that user behaviors in each domain are represented independently and heterogeneously,

© Springer Nature Switzerland AG 2019
L. H. U and H. W. Lauw (Eds.): PAKDD 2019 Workshops, LNAI 11607, pp. 274–285, 2019.
https://doi.org/10.1007/978-3-030-26142-9_24

trans-domain BT is defined as BT that identifies potential users of the source domain with user behaviors in the target domain related to the Ad-Items of the source domain. In the case of (source) E-Commerce → (target) Ad Network, trans-domain BT finds potential users of E-Commerce who have accessed websites related to E-Commerce items in Ad Network.

To realize the trans-domain BT, our previous work [13] proposed a method called Virtual Touch-Point (VTP). In VTP, the trans-domain BT is enabled by employing transfer learning techniques, which model behavior linkage across domains based on the limited information of user correspondence across domains, so that one can reach interested potential users who lie in the target domain beyond the ID linkage across domains. Once the model is estimated, the model is securely transmitted to another domain and used to predict potential users to be reached, because the model has no ID information.

Transfer learning techniques utilized by VTP are desired to be heterogeneous transfer learning (HeTL) types rather than homogeneous transfer learning (HoTL) ones. While HeTL allows feature spaces in both domains to be different, HoTL assumes feature spaces in both domains are the same. Since the available data regarding the behavior of the users of the domains do not necessarily match nor even have any part in common with each other, handling data heterogeneity is necessary for retaining wider applicability of the trans-domain BT. Otherwise the applicability of the trans-domain BT is limited to the domains having homogeneous data. HeTL has two roles: representation alignment and distribution alignment. While the former aligns feature spaces between domains, the latter aligns data distributions between domains. Besides these roles, classification is required for the final prediction.

In [13], we created prototypes by implementing typical transfer learning algorithms, specifically, HFA [8] and HEGS [2]. Each algorithm has its drawbacks. While HFA jointly learns the representation alignment by feature augmentation and the classifier by SVM formulation, HFA does not explicitly learn the distribution alignment. Although HEGS learns the distribution alignment by instance selection based on clustering, HEGS cannot jointly learn the representation alignment and the distribution alignment. Since HEGS assumes that the feature spaces in both domains are the same, the representation alignment needs to be performed as a preprocessing step.

Recent advances in transfer learning techniques include the Domain-Adversarial Neural Network (DANN) [4], which jointly learns the distribution alignment by Domain-Adversarial Training and the classifier by standard backpropagation. Although the original DANN is a HoTL method sharing its encoder among domains, we can extend DANN to a HeTL method by simply replacing its shared encoder with domain-specific ones, which in turn enables joint learning of the representation alignment, the distribution alignment, and the classifier, and thus performance improvement is expected.

Another advantage of DANN is that it can be extended to a fully end-to-end model by replacing its encoders with deeper ones such as convolutional neural networks (CNNs) and recurrent neural networks (RNNs). In this paper, as an initial evaluation, we evaluate the effectiveness of joint learning of the representation alignment, the distribution alignment, and the classifier, using pre-processed features as input.

The remainder of this paper is organized as follows: Sect. 2 illustrates problem formulation, Sect. 3 describes related works, Sect. 4 describes our prototype, Sect. 5 shows our experimental results using real-world data on trans-domain BT: (source) E-Commerce → (target) Ad Network, Sect. 6 is the discussion, and Sect. 7 concludes this paper.

2   Problem Formulation

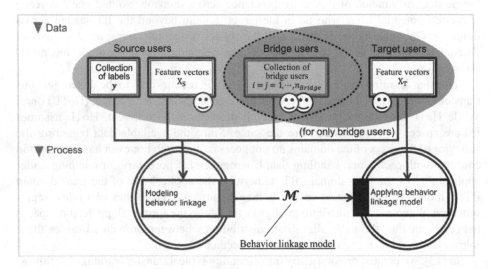

Fig. 1. The data and the processes involved in trans-domain BT in VTP

Let us denote source and target feature spaces as $\mathcal{X}_S$ and $\mathcal{X}_T$, respectively. The source feature vectors representing source behaviors are written as $X_S = \left\{ x_i^{(S)} \in \mathcal{X}_S; i = 1, \cdots, n_S \right\}$, where $x_i^{(S)}$ is a feature vector of a user i in the source domain S and n_S is the number of users in the source domain. Similarly, the target feature vectors representing target behaviors are written as $X_T = \left\{ x_j^{(T)} \in \mathcal{X}_T; j = 1, \cdots, n_T \right\}$, where $x_j^{(T)}$ is a feature vector of a user j in the target domain T and n_T is the number of users in the target domain. The label data is $y = \{ y_i; i = 1, \cdots, n_S \}$, where y_i indicates whether a user i in the source domain S has a pre-defined behavior, namely conversion (e.g. whether the user has viewed or purchased an Ad-Item in a certain period).

Here, we make two assumptions:

(1) As defined above, the label data are assumed to be available only in the source domain.
(2) ID linkage across domains is partially available. $i = j = 1, \cdots, n_{Bridge}$ indicates users with ID linkage, namely bridge users. Therefore, the target feature vectors within the bridge users are available in the source domain.

To construct VTP, we applied two processes to the data as shown in Fig. 1.

Modeling Behavior Linkage. This process abstracts the relationship between the source feature vectors X_S, the label data y, and the target feature vectors X_T within the bridge users to learn a behavioral linkage model $\mathcal{M}$ that represents the relationship. By this, private ID-behavior linkage is eliminated and only essential information to tie the two domains i.e. the behavior linkage across domains can be transmitted to the target domain.

Applying Behavior Linkage Model. This process maps information of the label data y across domains S → T via the behavioral linkage model M and finds target users in the target domain.

The objective of VTP is to minimize information loss in the above process and to maximize the probability of conversion by the target users.

3   Related Works

In this section, we describe previous works that aimed to minimize the information loss across domains. Transfer learning is a machine learning technique to leverage the knowledge in the source domain to improve learning efficiency in the target domain. Transfer learning in the case when feature spaces in both domains are the same is called homogeneous transfer learning (HoTL). HoTL has the role of aligning data distributions between domains, which is called domain adaptation. HEGS [2] is a domain adaptation method based on instance selection using a clustering method, where the clusters with a high proportion of instances from the source domain are eliminated due to the low relevance to the target domain. [3] is a general method (which means not dedicated to domain adaptation, e.g. outlier detection) based on instance weighting using the approximated probability density ratio between the target domain and source domain. DANN [4] is a state-of-the-art method of HoTL, which learns a domain invariant feature space based on domain adversarial training.

On the other hand, transfer learning in the case when feature spaces in both domains differ is called heterogeneous transfer learning (HeTL). HeTL has two roles: representation transformation and domain adaptation. The primary concern of HeTL lies in representation transformation, the role of which is to align feature spaces between domains. The representation transformation is largely categorized into two approaches. The first one maps X_T to X_S, which is referred to as asymmetric transformation, including Arc-t [5] and FSR [6]. The second one maps both X_S, X_T to a common latent space, which is referred to as symmetric transformation, including HeMap [7], HFA [8], and SHFA [9]. Each method differs in assumptions on the existence of common features and the necessity of labels. FSR requires some common features shared across domains. Arc-t, HeMap, HFA, and SHFA work without common features. Label information is required for Arc-t, FSR, HFA, and SHFA but not for HeMap. Among the methods that require label information, SHFA considers the semi-supervised setting, where the instances in the target domain are only partially labeled.

Transfer learning can be categorized into four types based on different situations between the source and target domains with respect to the availability of label information. In the case that labeled data are available in both domains, it is called inductive transfer learning setting. In the case that labeled data are available only in the source domain, it is called transductive transfer learning. In the case that labeled data are available only in the target domain, it is called self-taught learning. In the case that labeled data are available in neither of the domains, it is called unsupervised transfer learning.

In our setting, it is fundamentally the transductive transfer learning setting due to the fact that the view or purchase histories of products in the source domain, which are used as label information, are observable only in the source domain. However in this work, as a relaxation of the setting, we assume that ID linkage across domains is partially available and thus we address the inductive transfer learning setting.

4   Prototype

We created a prototype by implementing typical transfer learning algorithms:

- **Prototype method 1 (PM1)** A step-by-step method, in which HeMap, HEGS, and Xgboost [10] correspond to representation alignment, distribution alignment, and supervised learning of a classifier, respectively.
- **Prototype method 2 (PM2)** HFA, an end-to-end method, in which representation alignment and a classifier are learned jointly. Distribution alignment is not learned explicitly.
- **Prototype method 3 (PM3)** Heterogeneous DANN, which we modified from the original DANN to be extended to HeTL, is an end-to-end method, in that representation alignment, distribution alignment, and a classifier are learned jointly.

While these three methods share a commonality that they do not require common features between domains, these methods are contrastive in terms of the extent to which these methods learn the functionalities jointly, so we evaluate which method is suitable for real-world data.

PM1 and PM2 were evaluated in our previous work [13], and PM3 is newly evaluated in this paper. The details of PM1 and PM2 are described in [13] and reintroduced in the Appendix of this paper for completeness. The details of PM3 is as follows.

Modeling Behavior Linkage. DANN learns a domain invariant feature space based on domain adversarial training. While the original DANN shares its encoder among domains, we replaced its encoder with the domain-specific types as shown in Fig. 2 in order to extend DANN to a HeTL method. The feature extractors (encoders) of both domains G_{fS}, G_{fT} and the class predictor $G_y(z)$ ($z \in \mathcal{Z}$, the common latent space) are trained to minimize class prediction loss L_y, and the domain discriminator G_d is trained to minimize domain discrimination loss L_d. On the other hand, G_{fS}, G_{fT} are trained to maximize domain discrimination loss L_d. This conflicting objective is for improving the domain invariance of the common feature space $\mathcal{Z}$, and the objective can be turned into a minimization problem shown by Eq. (1), introducing a gradient reversal layer R, such that $R(x) = x$ and $dR/dx = -\alpha I$. I is an identity matrix, α is a gradient scaler, and

γ is a balancing parameter. α and γ are tuned using cross validation. We set both L_y and L_d to be the binary cross-entropy loss. d_k is a domain label of the k-th instance. The optimization can be performed using a stochastic gradient descent method such as Adam [12]. It is observed that Eq. (1) joint learns representation alignment, distribution alignment, and classification. The resulting behavior linkage model $\mathcal{M} = (F_{DANN})$.

$$\min_{G_y, G_{f*}, G_d} L = \sum_{i=1}^{n_S} L_y\left(G_y\left(G_{fS}\left(x_i^{(S)}\right)\right), y_i\right) + \sum_{j=1}^{n_T} L_y\left(G_y\left(G_{fT}\left(x_j^{(T)}\right)\right), y_j\right)$$
$$+ \gamma \sum_{k=1}^{n_S+n_T} L_d\left(G_d\left(R\left(G_{f*}\left(x_k^{(*)}\right)\right)\right), d_k\right) \quad (1)$$

Applying Behavior Linkage Model. We compute the predictive probability of conversion $P\left(y_j | x_j^{(T)}\right) = F_{DANN}\left(x_j^{(T)}\right)$ for each user j in the target domain. The resulting target users are defined by $\left\{j; P\left(y_j | x_j^{(T)}\right) > \theta\right\}$ with an arbitrary threshold $0 < \theta < 1$.

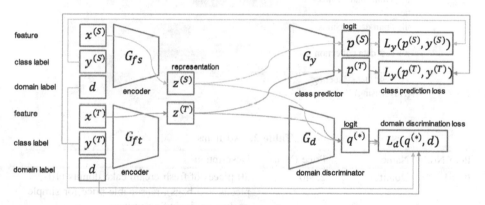

Fig. 2. Heterogeneous DANN

5 Experiments

The proposed prototype of VTP for trans-domain BT was evaluated through the real data of two domains: (source) E-Commerce $\rightarrow$ (target) Ad Network, by finding the potential users in the target domain who may purchase the Ad-Items of the source domain. Although we strictly followed the setting in our previous work, we describe the experimental setting here for completeness.

5.1 Experiment Setup

Periods. The experiment involved five distinct periods as shown in Table 1. In Table 1, Period i is the period of ID linkage of the two domains, Periods ii and iii are the periods for collecting and labeling the train data, and Periods iv and v are the periods for collecting and labeling the test data. In addition, the advertising of the Ad-Items was performed during Period v.

Ad-Items. The experiment was conducted for four Ad-Items shown in Table 2, which were selected from the items sold from 2018/03 in the source domain. The selection was performed by finding the items that were sold from 2018/03 and have similar names to the top 10 most popular items of 2017/03, from the items of the categories "delivered gourmet", "healthcare", and "beauty".

Table 1. Periods

	2017 11	2017 12	2018 01	2018 02	2018 03	2018 04
i) ID linkage	▓					
ii) Train data collection		▓	▓			
iii) Train data labeling				▓		
iv) Test data collection			▓	▓		
v) Test data labeling (Advertising)					▓	

Table 2. Ad-Items

Item No.	Name	Price (Yen)	Description
Item1	Donut	2,340	30 pieces of fresh cream cake donuts (10 pieces × 3 bags), reasonable price for simple packaging, including postage
Item2	Wine	11,800	10 sets of French & Spanish red wine, including postage
Item3	Beauty Salon	7,300	Cellulite repulsion course 120 min × 2 times, 79% OFF
Item4	Thermometer	1,980	One electronic thermometer, 53% OFF

ID Linkage of Two Domains. For the ID linkage between the source domain and target domain, the advertising tag of the target domain was inserted into the websites of the source domain. If a user of the source domain had accessed these websites, the cookie ID of the target domain was generated, and then the user ID of the source domain and the ID (cookie ID) of the target domain were linked given that the user logged in to the source domain. This type of linkage was performed during Period i.

Train Data Collection. The train data of the source domain and the target domain were collected during Period ii. Specifically, the train data of the source domain is the data of the users' customer master data, and their view and purchase logs of items. The train data of the target domain is the data of the users' access logs of the websites that have the advertising tags of the target domain. The statistical information of the train data is described as follows:

- Source domain: about 250,000 IDs, 15 million records;
- Target domain: about 2,500,000 IDs, 130 million records;
- ID linkage: about 40,000 ID pairs.

The feature representation of the train data for each domain in the embedding space is described as follows.

- **Feature representation for the source domain** Firstly, a morphological analysis of all the items' descriptions was performed, and the feature vectors were generated for these items by using doc2vec [11]. Then, the feature vector for every user was generated by averaging the feature vectors of the items that were included in the users' view and purchase logs. There were some free items besides the general items, whose purchase trends were different, so we averaged feature vectors separately for the free items and the general items. In addition, users' gender, age, average price and categories of the purchased items were also included. As denoted in Sect. 2, the feature vectors of the source domain are $X_S = \left\{ x_i^{(S)}; i = 1, \cdots, n_S \right\}$, where $x_i^{(S)}$ is the feature vector of user i, and n_S is the number of users in the source domain. The total number of dimensions of the feature vector is 285: the averaged doc2vec features (128 for the free items and 128 for the general items), gender(1), age(1), the average price(1), and the categories of the purchased items (13 for the free items and 13 for the general items).

- **Feature representation for the target domain** A morphological analysis of the websites' descriptions was performed first, and then the feature vectors were generated for the websites by using doc2vec [11]. The feature vector per user was generated by averaging the feature vectors of their accessed websites. The feature vectors of the target domain are $X_T = \left\{ x_j^{(T)}; j = 1, \cdots, n_T \right\}$, where $x_j^{(T)}$ is the feature vector of user j, and n_T is the number of users in the target domain. The number of dimensions of the feature vectors is 256.

Train Data Annotation. For the users (samples) included in the train data, the label y was determined based on whether they viewed similar items of the Ad-Item. Here, the similar items were these items with the top 10 nearest feature vectors to the feature vector of the Ad-Item.

Test Data Collection. The advertising of the Ad-Items was randomly distributed to all the users of the target domain during Period v. For the users who viewed the websites with the Ads of the Ad-Items, the access logs of these users during Period iv were collected as the test data. The feature representation X'_T of these users was generated by the same process as applied to the users of train data.

Test Data Annotation. For the above users who viewed the websites with the Ads of the Ad-Items, they were labeled y' according to whether they clicked the Ads or not.

Methods. As a baseline method (BL), we evaluated Xgboost, which is a non-transfer learning method, trained only using (X_T, y). PM1, PM2, and PM3 were trained using (X_S, X_T, y). After training, each method was applied to X'_T, then performance measures were calculated using y'.

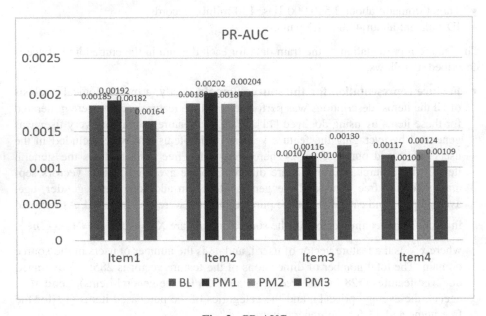

Fig. 3. PR-AUCs

5.2   Experimental Results

The performance evaluation was performed for the four Ad-Items separately, by comparing the PR-AUC of BL, PM1, PM2, and PM3. Figure 3 and 4 illustrate the results of PR-AUC and Lift curve, respectively. The PR-AUC is the value of the area under the PR-curve, and Lift curve is plotted by calculating the ratio of the precision of the model to that of random guessing as the lift and then connecting the lift-recall points with different values of the threshold θ. Here, the precision of random guessing means the appearance ratio: N_{click}/N_{view}, where N_{click} and N_{view} are the number of the users who clicked the Ads and the number of the users who viewed the websites with the Ads of the Ad-Item, respectively (i.e., the numbers of the positive samples and the number of all the samples of the test data, respectively).

Figure 3 shows that PM3 outperforms the other methods in terms of PR-AUC in some cases, namely Item2 and Item3. The possible reason why PM3 underperforms more than any of the other methods even BL for Item 1 and Item 4, is discussed in Sect. 6.

Since this is a performance evaluation of a targeting technique for advertisements, we are interested in the lift rather than recall, and therefore we plotted lift curves only in the range of recall from 0% to 20%. The two vertical green lines indicate the recall of 1% and 2% respectively, and the two horizontal green lines indicate the lift of 1.0 and 1.5, respectively. Figure 4 shows that a higher lift value than that of the other methods can be achieved by PM3 for Item2, Item3, and Item4, given a suitable threshold θ, which could be used to improve the efficiency and reduce the cost of advertising for practical commercial application.

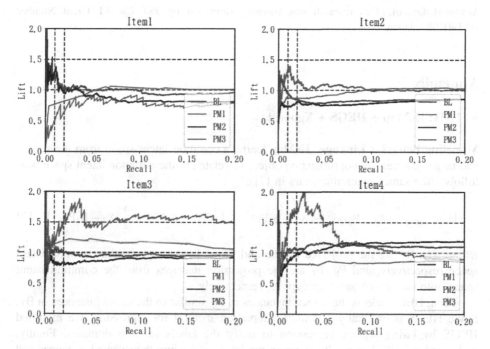

Fig. 4. Lift curves

6 Discussions

After the experiments, it was revealed that some of the items used as the training labels were deceptive. We calculated the odds ratio (OR), for each Ad-Item, of viewing the Ad-Item given the history of viewing its similar items. As a result, we found that there was an item with OR = 0.696 in the similar items of Ad-Item1 and an item with OR = 0.89 in those of Ad-Item4, respectively. This may be the reason for the underperformance of PM3 on Item1 and Item4 in terms of PR-AUC, and on Item1 in terms of lift.

7 Conclusion

In this paper, to realize trans-domain behavioral targeting, we extended the domain-adversarial neural network (DANN) to enable joint learning of three desired functionalities of heterogeneous transfer learning (HeTL): representation alignment, distribution alignment, and classification. We evaluated it using the real-world data of two domains: (source) E-Commerce → (target) Ad Network. As a result, heterogeneous DANN outperformed the conventional methods in terms of PR-AUC and lift curve in some cases. For further study, we shall research how to make reliable training labels for the cold start problem, and also extend heterogeneous DANN to a fully end-to-end model by replacing its encoders with deeper versions.

Acknowledgment. This research was partially supported by JST CREST Grant Number J181401085, Japan.

Appendix

- **PM1: HeMap + HEGS + Xgboost**

Modeling Behavior Linkage. HeMap learns a common latent space from the source and target features. The optimization objective related to the common latent space is as follows (the same equation appears in [7]):

$$\min_{B_T^T B_T=I, B_S^T B_S=I} \|X_S - B_S P_S\|^2 + \|X_T - B_T P_T\|^2 + \beta\left(\frac{1}{2}\|X_T - B_S P_T\|^2 + \frac{1}{2}\|X_S - B_T P_S\|^2\right) \quad (2)$$

where, B_S, B_T are the projected source and target instances onto the common latent space respectively, and P_S, P_T are the projection matrices from the common latent space onto the source and target space, respectively.

Then, HEGS selects the source instances in B_S similar to the target instances in B_T. Since HEGS is originally a domain adaptation method for regression, we modified HEGS by hiring logistic regression to unify the labels in both domains. Finally, Xgboost learns the binary classification model $F_{Xgb}(\cdot)$ using the selected instances and labels. Hyper-parameters of HeMap, HEGS, and Xgboost are tuned using cross validation. The resulting behavior linkage model $\mathcal{M} = (P'_T, F_{Xgb})$, where P'_T is a pseudo-inverse of the projection matrix P_T obtained by HeMap.

Applying Behavior Linkage Model. We compute the predictive probability of conversion $P\left(y_j|x_j^{(T)}\right) = F_{Xgb}\left(P'_T x_j^{(T)}\right)$ for each user j in the target domain. The resulting target users are defined by $\left\{j; P\left(y_j|x_j^{(T)}\right) > \theta\right\}$ with an arbitrary threshold $0 < \theta < 1$.

- **PM2: HFA**

Modeling Behavior Linkage. HFA learns a multiple kernel classifier $F_{HFA}(\cdot)$ on a common latent space using the source and target features and the labels. The projection matrices P, Q from the source and target space onto the common latent space are coupled as $H = [P, Q]'[P, Q]$, and then kernel matrices are computed and optimized based on H. Hyper-parameters of HFA are tuned using cross validation. The resulting behavior linkage model $\mathcal{M} = (F_{HFA})$.

Applying Behavior Linkage Model. Similar to PM1, we compute the predictive probability of conversion $P\left(y_j|\boldsymbol{x}_j^{(T)}\right) = F_{HFA}\left(\boldsymbol{x}_j^{(T)}\right)$ for each user j in the target domain. The resulting target users are defined by $\left\{j; P\left(y_j|\boldsymbol{x}_j^{(T)}\right) > \theta\right\}$ with an arbitrary threshold $0 < \theta < 1$.

References

1. The promise of first-party data. Econsultancy. https://cdn2.hubspot.net/hubfs/370829/Campaigns_and_Emails/Archived_Emails/2015_Emails/2015_Econsultancy_Promise_of_First_Party_Data_June/The_Promise_of_First_Party_Data_Signal_Econsultancy_Report.pdf. Accessed 15 Oct 2018
2. Shi, X., Liu, Q., Fan, W., Yang, Q., Yu, P.S.: Predictive modeling with heterogeneous sources. In: Proceedings of the 2010 SIAM International Conference on Data Mining, pp. 814–825 (2010)
3. Yamada, M., Suzuki, T., Kanamori, T., Hachiya, H., Sugiyama, M.: Relative density-ratio estimation for robust distribution comparison. In: Advances in Neural Information Processing Systems, vol. 24, pp. 594–602 (2011)
4. Ganin, Y., et al.: Domain-adversarial training of neural networks. J. Mach. Learn. Res. **17**, 1–35 (2016)
5. Kulis, B., Saenko, K., Darrell, T.: What you saw is not what you get: domain adaptation using asymmetric kernel transforms. In: Proceedings of the 30th IEEE Conference on Computer Vision and Pattern Recognition, pp. 1785–1792 (2011)
6. Feuz, K.D., Cook, D.J.: Transfer learning across feature-rich heterogeneous feature spaces via feature-space remapping (FSR). ACM Trans. Intell. Syst. Technol. **6**, 1–27 (2015)
7. Shi, X., Liu, Q., Fan, W., Yu, P.S., Zhu, R.: Transfer learning on heterogenous feature spaces via spectral transformation. In: Proceedings of the 10th IEEE International Conference on Data Mining, pp. 1049–1054 (2010)
8. Duan, L., Tsang, I.W.: Learning with augmented features for heterogeneous domain. In: Proceedings of the 29th International Conference on Machine Learning (2012)
9. Li, W., Duan, L., Xu, D., Tsang, I.W.: Learning with augmented features for supervised and semi-supervised heterogeneous domain adaptation. IEEE Trans. Pattern Anal. Mach. Intell. **36**, 1134–1148 (2014)
10. Chen, T., Guestrin, C.: Xgboost: a scalable tree boosting system. In: Proceedings of the 22nd ACM SIGKDD International Conference on Knowledge Discovery and Data mining, pp. 785–794 (2016)
11. Rehurek, R., Sojka, P.: Software framework for topic modelling with large corpora. In: LREC 2010 Workshop on New Challenges for NLP Frameworks (2010)
12. Kingma, D.P., Ba, J.: Adam: a method for stochastic optimization. In: Proceedings of the International Conference on Learning Representations (2014)
13. Kurokawa, M., et al.: Virtual touch-point: trans-domain behavioral targeting via transfer learning. In: Workshop on Big Data Transfer Learning in conjunction with IEEE International Conference on Big Data (2018)

Natural Language Business Intelligence Question Answering Through SeqtoSeq Transfer Learning

Amit Sangroya[✉], Pratik Saini[✉], Mrinal Rawat[✉], Gautam Shroff[✉], and C. Anantaram[✉]

TCS Innovation Labs, Chennai, India
{amit.sangroya,pratik.saini,rawat.mrinal,gautam.shroff,
c.anantaram}@tcs.com

Abstract. Enterprise data is usually stored in the form of relational databases. Question Answering systems provides an easier way so that business analysts can get data insights without struggling with the syntax of SQL. However, building a supervised machine learning based question answering system is a challenging task involving large manual annotations for a specific domain. In this paper we explore the problem of transfer learning for neural sequence taggers, where a source task with plentiful annotations (e.g., Training samples (NL questions) on IT enetrprize domain) is used to improve performance on a target task with fewer available annotations (e.g., Training samples (NL questions) on pharmaceutical domain). We examine the effects of transfer learning for deep recurrent networks across domains and show that significant improvement can often be obtained. Our question answering framework is based on a set of machine learning models that create an intermediate sketch from a natural language query. Using the intermediate sketch, we generate a final database query over a large knowledge graph. Our framework supports multiple queries such as aggregation, self joins, factoid and transnational.

1 Introduction

Various enterprise applications such as finance, retail, pharmacy etc. store a vast amount of data in the form of relational databases. However, accessing relational databases requires an understanding of query languages such as SQL, which, while powerful, is not easy for business analysts to master. Building a data science assistant that captures context and semantic understanding is an important and challenging problem. The problem is multi-dimensional as it involves complexities at multiple levels e.g. relational database level, natural language interface level and semantic understanding level. At relational database level, the system must be able to handle the complexities related to data representation. At natural language interface level, the system should support NL related issues such as handling ambiguity, variations and semantics. Most importantly,

L. H. U and H. W. Lauw (Eds.): PAKDD 2019 Workshops, LNAI 11607, pp. 286–297, 2019.
https://doi.org/10.1007/978-3-030-26142-9_25

the system must be able to understand user's context and should respond back in a meaningful way.

An important challenge for sequence tagging is how to transfer knowledge from one task to another, which is often referred to as transfer learning. Transfer learning can be used in multiple scenarios such as Multi-task where labels are used to perform a different task or using the labels in a new domain. Transfer learning can improve performance by taking advantage of more plentiful labels from related tasks/domains. Even on datasets with relatively abundant labels, multi-task transfer can sometimes achieve improvement over state-of-the-art results [1].

There are a number of recent works which are trying to build Conversational AI systems that support human-like cognitive capabilities such as context-awareness, personalization, and ability to handle complex NL inputs. Apple Siri, Microsoft Cortana, Amazon Alexa are some of the widely used personal assistants that are already in market. However, there are still a variety of open research issues while adapting conversation AI based digital assistant tools to new domains [2]. To address the limitations of existing techniques towards designing conversational AI based personal digital data assistants, we make following key contributions in this paper:

- We present framework that exploits different levels of representation sharing and provides a unified framework to handle cross-application and cross-domain transfer.
- We present Curie, a data science assistant that supports NL conversational interface for effective data assistance across a range of tasks i.e. NL question answering and dialogue, data visualization and data transformation. The Curie platform is designed to support context, intent identification, multiple ways of interaction through dialog, question answering and a variety of NL query processing capabilities.
- We also introduce a novel approach for parsing natural language sentences and generating an intermediate representation (query sketch) using a combination of machine learning models. This intermediate representation is further used to execute low level database queries.
- We demonstrate our results on two enterprise datasets and one public dataset i.e. WikiSQL. Our results show state of art performance on these datasets while supporting transfer learning that ensures that our approach can be directly used across a variety of domains with minimal effort.

2   Related Work

There are a lot of conversational AI based personal assistant tools already in market for the end users. Some of these tools are designed for open-domain conversations e.g. Cortana, Siri, Alexa and Google Now [3,4]. On the other hand there are also tools where focus is on supporting interactive data assistance e.g.

Microsoft Power BI [5]. However, there are still limitations of the existing solutions in terms of natural language based conversational capabilities while supporting data assistance. There are still various open issues in terms of semantic understanding and context-awareness user's queries. Most important task for building a NL based data assistant is to generate structural query language (SQL) queries from natural language.

The study of translating natural language into SQL queries has a long history. Popescu et al. introduce *Precise NLI*, a semantic parsing based theoretical framework for building reliable natural language interfaces [6]. This approach rely on high quality grammar and is not suitable for tasks that require generalization to new schema. Recent works consider deep learning as the main technique. There are many recent works that tackle the problem of building natural language interfaces to relational databases using deep learning [7–14]. Zhong et al. [9] propose *Seq2SQL* approach that uses reinforcement learning to break down NL semantic parsing task to several sub-modules or sub-SQL incorporating execution rewards. Yavuz et al. introduce *DialSQL*, a dialogue based structured query generation framework that leverages human intelligence to boost the performance of existing algorithms via user interaction [15]. The flexibility of our approach enables us to easily apply sketches to a new domain. Our framework also does not require large corpus of NL sentences as training input.

Transfer learning for natural language processing (NLP) tasks can be broadly divided into two categories: *resource-based transfer* and *model-based transfer*. Resource-based transfer utilizes additional linguistic annotations as weak supervision for transfer learning, such as cross-lingual corpora. Resource-based transfer is mostly limited to cross-lingual transfer. Model-based transfer, on the other hand, does not require additional resources. Model-based transfer exploits the similarity and relatedness between the source task and the target task by adaptively modifying the model architectures, training algorithms, or feature representation [16]. Our approach falls into the category of model-based transfer.

3   Approach for Transfer Learning Across Domains

In this section, we introduce our transfer learning approach. We first introduce an abstract framework for neural sequence tagging and then discuss the proposed transfer learning architecture. Since different domains have domain-specific regularities, sequence taggers trained on one domain might not have optimal performance on another domain. The goal of cross-domain transfer is to learn a sequence tagger that transfers knowledge from a source domain to a target domain. We assume that few labels are available in the target domain. We exploit the condition where two domains have label sets that can be mapped to each other. For example, some predicates in the pharmaceutical domain can be mapped to IT industry such as employees, company etc. while some predicates can be very much domain specific like drugs, pharmacy, store etc. If the two domains have mappable label sets, we share all the model parameters and feature representation in the neural networks, including the word and character embedding.

We build sequence to sequence Bidirectional LSTM (Long Short Term Memory) model to generate a query sketch. The intent is to identify the concepts, entities, predicates, values etc. in a given sentence. We build separate machine learning models for discovering the predicates, values, entities, operations etc. In order to build these models, we rely on separate annotations for each model. For example, A natural language query is annotated as follows. {*What is the employee id of John*} is annotated as {0 0 0 A A 0 B} (See Fig. 1 for example sequence).

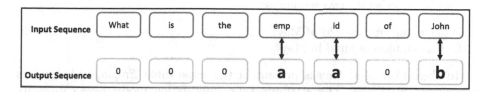

Fig. 1. Example of input output sequence

This example shows that there are two concepts *employee id* and *John* marked as A and B respectively. The same token is used for the concepts which consist of more than one word e.g. *employee id* consists of two words, so we mark them with token 'A'. We follow this representation in our models.

Similarly, we have other models such as **aggregations and operators** model to discover aggregations and operators for predicates and entities respectively. Our framework currently supports following set of aggregation functions: count, groupby, min, max, sum, and sort. {*How many <predicate> work in <predicate> <value>*} is annotated as: {0 0 count 0 0 0 equal}. Note that, Model-level transfer learning is achieved through exploiting the semantics shared by the two domains. For example, *"Employee"* in IT domain and *"Agent"* in pharmaceutical domain refer to the same named entity, and the semantic similarities can be leveraged for NER.

3.1   SeqtoSeq Model for Meta-Types Identification

This model identifies the **type of concepts** (predicates and values) at the node or table level. For example, *employee id* belongs to *Employee* node and *Employee* is a *PERSON*, so the type of *employee id* is *PERSON*.

If a concept is present in more than one table, **type** information helps in the process of disambiguation. For example, consider following sentences:

- What is the employee id of Washington?
- List all the stores in Washington.

Here, in first example *Washington* refers to the name of a person, whereas in second example it is the name of a location. In such cases, this model is useful to disambiguate that in first example the node level type is PERSON and in second example it is LOCATION. This specific feature helps in making the overall framework domain agnostic. In this model, all the entities in input are marked with tag *<value>*. For example, natural language sentence, {*How many employees work in project <value>*} is annotated as {0 0 person 0 0 project project}.

3.2 SeqtoSeq Model for Attribute Level Type Identification

This model identifies the **attribute** type of concepts (predicates and values). For example, let's take two sentences:

- What is the employee id of May?
- List all employees hired in May?

In these examples, *May* is present in the same table *Employee*, but refers to different attributes. The attribute type information model here can easily distinguish based on the nature of query that in first example, *May* is the name of a person and in second example it is *date*. The attribute type information in combination with node type information helps in increasing the accuracy. In this model, all the entities in input are marked with tag *<value>*. For example, {*How many employees work in project <value>*} is annotated as: {0 0 name 0 0 name name}.

4 Curie Framework for Question Answering

To this end, we develop *Curie* as a data assistance framework to help business analysts in getting insights about data in a user friendly way. Instead of meandering through the database for a small detail, Curie provides an interface where business analysts just need to type their query in a natural language through a conversational interface. Curie is designed to handle variety of NL queries such as aggregations, factoid, transnational and ad-hoc queries (See Table 5). Depending on the nature of query, Curie can also respond with appropriate visualized interpretation of the data in the form of Pie Charts, Graphs etc. Curie architecture is primarily divided into two parts. (a) Mechanism to generate an intermediate form (**Query Sketch**) given a NL sentence and (b) Approach of transforming a sketch to database query (Fig. 2 and Table 1).

4.1 Generating Database Query from Sketch

The process of generating the query is independent of underlying database i.e. the same approach can be used for generating queries across databases. We demonstrate this concept using two popular relational database query languages: SQL (structured query language) and CQL (cipher query language). The general idea to generate a query is as follows. Sketch S is a set {P,C}, where P is a set {$p_1,p_2, p_3,... p_n$}   and each p_i is {n_i, a_i, h_i, g_i}, where

Table 1. Variety of natural language queries

Transactional	How many employees are in Mumbai?
FAQ	How can I apply sick leave?
Factoid	What is the population of Mars?
Aggregations	How many employees in each project?
Visualization	Tell me the monthly profit of company in last decade?
Sorting	List all employees based on their age

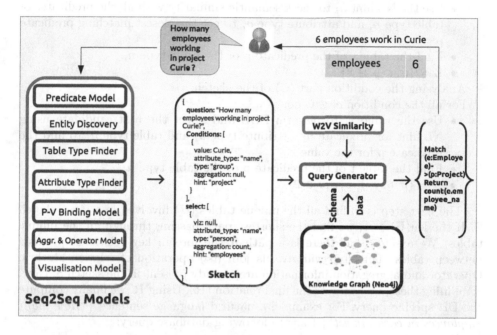

Fig. 2. Models for NL to query generation

n_i = Node/table level type of predicate p_i
a_i = Attribute/predicate level type of predicate p_i
h_i = NL Hint for the predicate p_i
g_i = Aggregation information for the predicate p_i

Similarly, **C** is a set $\{c_1, c_2, c_3, \ldots c_n\}$ where each c_i is $\{n_i, a_i, h_i, v_i, o_i\}$

n_i = Node/table level type for the predicate involved in condition c_i
a_i = Attribute/attribute level type for the predicate involved in condition c_i
h_i = NL hint for the predicate involved in condition c_i
g_i = Value involved in condition c_i
o_i = Operator involved in condition c_i

Provided with a sketch **S**, we generate a database query **Q**. As explained earlier, sketch contains information about all the required predicates **L**, i.e. the

list of attributes/columns that need to be extracted from Table **T**. This component will give a new relation that only contains the columns in **L**. This operation
is projection (Π) in relation algebra. The sketch has also information about all
conditions φ which gives us a new relation that contains only those rows satisfying φ in **T**. This is the selection (σ) operation of relation algebra. To generate
the query following set of operations are executed over the sketch:

- Analysing the predicate part (P) of the sketch:
  For all predicates $p_i \in P$ do:
    - Use the NL hint h_i to check semantic similarity with all the predicates of
      table type n_i and attribute type a_i to get the closest matching predicate
      p
    - Find the table t of the predicate p of the table type n_i
    - **L** = {p | p $\in$ P}
- Analysing the condition part (C) of the sketch:
  For all the condition $c_i \in C$ do:
    - Use the value v_i to perform EDL and select the best candidate using
      NL hint for predicate h_i, attribute type a_i and table type n_i to find the
      predicate p for the value v_i
    - Find the table t of the predicate p of the table type n_i
    - φ = {p o v | p,o,v $\in$ P}

The next step is to find all the unique tables (U) involved in a given sketch
S. If the len(U) > 1, we find the shortest path passing through all the unique
tables. We assume that there is a path if the foreign key relationship exists
between tables. This step will give us join ($\bowtie$) operation of relation algebra.
Operator and aggregation information are already present in the sketch S. Now,
combining these, we compute the final relation (**R**). Using **R**, we finally compute
the DB specific query. For example, a natural language sentence { *"How many
employees in each project"*} leads to following database query:

$$\Pi_{count(employee\_id),g(project\_name)}$$
$$(Employee_{pid\_fk} \bowtie_{pid} Project)$$

a. Database Query SQL:
SELECT COUNT(emp_id), proj_name
FROM **Employee** INNER JOIN **Project**
ON (pid_fk = pid) GROUP BY proj_name

b. Database Query CQL:
MATCH
  (e:**Employee**)-[:WORKS]->(p:**Project**)
RETURN
  COUNT (e.employee_id), p.project_name

5 Results and Discussion

We conducted our experiments on an Intel Xeon(R) computer with E5-2697
v2 CPU and 64 GB memory, running Ubuntu 14.04. We evaluated `Curie` on

three real-world datasets, out of which two are our internal enterprise datasets (See Table 2). First dataset is related to employees in a large software services company and their allocations. Our second internal dataset is about a large pharmaceutical company's stores and employee's information. Our third dataset for experimental evaluation is widely used WikiSQL dataset. Ent. Dataset-1 & 2, composed of following type of queries: factoid, self joins, Group By, sorting, and aggregations. WikiSQL dataset is composed of factoid, aggregation. Queries in WikiSQL were only over single columns.

Table 2. Datasets and their statistics

Dataset	# Tables	# Columns	Train. set	Test set
Ent. Dataset-1	5	42	1700	300
Ent. Dataset-2	3	70	1700	300
WikiSQL	13000	1000	31000	8000

Note that, in order to generate the intermediate form(sketch), all models are not necessary. The predicate, value, aggregation & operator models are sufficient to generate the sketch. The type, attribute and PV binding models are just used to improve the accuracy. As seen in Table 3 and Fig. 3, Curie has shown a very good performance for Ent. Dataset-1 & 2. However, in case of WikiSQL dataset, the value model did not give good accuracy. The reason of this is the fact that WikiSQL attribute names are not proper English words and contains some non ascii characters. Because of this, we were not able to find a good semantic understanding. However, this problem can be solved by training the model on a corpus specialized for a particular domain. Also, our entity discovery model was based on the **Lucene** search. It could be further improved by combining **Lucene** search with deep learning based entity extraction. Interestingly, for evaluating Ent. Dataset-2, we just used the same model that was trained on Ent. Dataset-1. We can see that almost similar results were achieved on the latter dataset demonstrating the transfer learning capabilities of our ML models.

In Table 4, we demonstrate sample queries from IT and pharmaceutical domain. Note that both domains are semantically overlapping to some extent and this knowledge is exploited by the deep neural networks. In order to test our transfer learning approach, we perform various experiments by taking both these domains as source domain and target domain. In first experiment IT domain is taken as source domain i.e. training data is not available for pharmaceutical domain. We compute the test accuracy which is close to close to 40%. Thereafter, we use training data for pharmaceutical domain and test on same domain, the accuracy in this case is improved to 45%. Next, we train the model on both IT and pharmaceutical domain and we get an accuracy improvement and it reaches to 52% (See Fig. 4 for details). Similar experiment is performed by interchanging the domains and we get similar behavior in terms of transfer learning (See Fig. 5).

Table 3. Accuracy across models for all datasets

	Ent. Dataset-1	Ent. Dataset-2	WikiSQL
Predicate Acc	97.6	96.9	88.4
Value Acc	97.56	95	71.2
Type Acc	87	85.86	NA
Attribute Acc	82	80.2	NA
PV Binding Acc	96.57	96.34	NA
Aggr. & Operator Acc	97.1	96.4	87
Overall Sketch Acc.	89.64	88.74	50.1
Overall Exec. Acc.	94.78	92.3	54.24

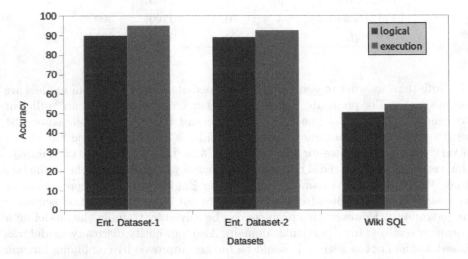

Fig. 3. Logical form and Execution Accuracy across datasets

Table 4. Sample Queries from IT domain and Pharmaceutical domain

Domains	Example Queries
IT domain	Who is the supervisor of pratik
IT domain	Tell me all the projects under machine learning group
IT domain	How many people in each research group and research area
IT domain	List all the female married terminated employees in project P_1001
Pharma domain	List the employees name who are terminated
Pharma domain	How many female employees are there in each store
Pharma domain	What all locations are active
Pharma domain	What is the expiry date of paracetamol

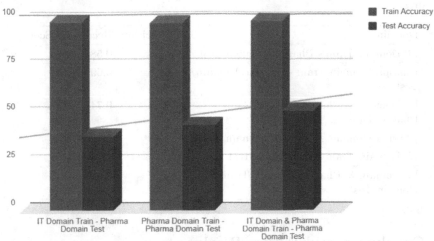

Fig. 4. Training and Test Accuracy of Transfer Learning on two domains - Source domain is IT domain and target domain is Pharmaceutical domain

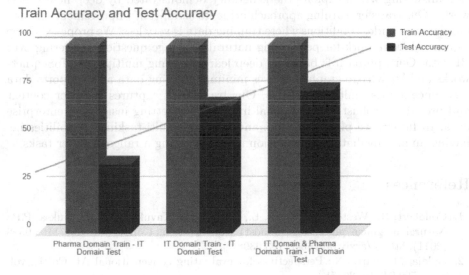

Fig. 5. Training and Test Accuracy of Transfer Learning on two domains - Source domain is Pharmaceutical domain and target domain is IT domain

Table 5. Precision, Recall and F1 Score for two Domains: With and Without Transfer Learning

Domains	Precision	Recall	F1 Score
IT Domain Train - Pharma Domain Test	0.64	0.58	0.61
Pharma Domain Train - Pharma Domain Test	0.68	0.69	0.69
IT Domain & Pharma Domain Train - Pharma Domain Test	0.74	0.74	0.74
Pharma Domain Train - IT Domain Test	0.58	0.57	0.58
IT Domain Train - IT Domain Test	0.77	0.78	0.77
IT Domain & Pharma Domain Train - IT Domain Test	0.77	0.81	0.79

6   Conclusion and Future Work

In this paper we develop a transfer learning approach for cross-domain question answering which exploits the generality demonstrated by deep neural networks. Our transfer learning approach achieves significant improvement on various domains under conditions where training data is very less. We proposed Curie as a novel framework for performing natural language question answering over BI data. Our approach is based on deep learning using multiple SeqToSeq networks and knowledge graph that uses minimal training data and supports data assistance across multiple domains. Our framework captures the user context and provides a robust conversational interface for getting insights in enterprise data. In future, we plan to explore and evaluate multi-tasking capabilities i.e. having an intermediate representation and supporting a range of other tasks.

References

1. Collobert, R., Weston, J., Bottou, L., Karlen, M., Kavukcuoglu, K., Kuksa, P.P.: Natural language processing (almost) from scratch. CoRR, vol. abs/1103.0398 (2011). http://arxiv.org/abs/1103.0398
2. Jadeja, M., Varia, N.: Perspectives for evaluating conversational AI. CoRR, vol. abs/1709.04734 (2017)
3. Sadun, E., Sande, S.: Talking to Siri: Learning the Language of Apple's Intelligent Assistant, 2nd edn. Que Publishing Company, Indianapolis (2013)
4. Ehrenbrink, P., Osman, S., Möller, S.: Google now is for the extraverted, Cortana for the introverted: investigating the influence of personality on IPA preference. In: Proceedings of the 29th Australian Conference on Computer-Human Interaction, ser. OZCHI 2017, pp. 257–265. ACM, New York (2017). https://doi.org/10.1145/3152771.3152799
5. Parks, M.: Microsoft Business Intelligence. POWER BI. CreateSpace Independent Publishing Platform, USA (2014)

6. Popescu, A.-M., Etzioni, O., Kautz, H.: Towards a theory of natural language interfaces to databases. In: Proceedings of the 8th International Conference on Intelligent User Interfaces, ser. IUI 2003, pp. 149–157. ACM, New York (2003). https://doi.org/10.1145/604045.604070

7. Xu, X., Liu, C., Song, D.: SQLNet: generating structured queries from natural language without reinforcement learning (2017)

8. Yaghmazadeh, N., Wang, Y., Dillig, I., Dillig, T.: SQLizer: query synthesis from natural language. Proceedings of ACM Programming Languages, vol. 1, no. OOP-SLA, pp. 63:1–63:26, October 2017. https://doi.org/10.1145/3133887

9. Zhong, V., Xiong, C., Socher, R.: Seq2SQL: generating structured queries from natural language using reinforcement learning (2017)

10. Yin, P., Lu, Z., Li, H., Kao, B.: Neural enquirer: Learning to query tables with natural language (2015)

11. Pasupat, P., Liang, P.: Compositional semantic parsing on semi-structured tables (2015)

12. Li, F., Jagadish, H.V.: Constructing an interactive natural language interface for relational databases. In: Proceedings of VLDB Endowment, vol. 8, no. 1, pp. 73–84, September 2014. https://doi.org/10.14778/2735461.2735468

13. Yu, T., Li, Z., Zhang, Z., Zhang, R., Radev, D.: TypeSQL: knowledge-based type-aware neural text-to-SQL generation. In: Proceedings of the 2018 Conference of the North American Chapter of the Association for Computational Linguistics: Human Language Technologies, Volume 2 (Short Papers), pp. 588–594. Association for Computational Linguistics (2018). http://aclweb.org/anthology/N18-2093

14. Dong, L., Lapata, M.: Coarse-to-fine decoding for neural semantic parsing. CoRR, vol. abs/1805.04793 (2018)

15. Yavuz, S., Gur, I., Su, Y., Yan, X.: DialSQL: dialogue based structured query generation. In: Proceedings of the 56th Annual Meeting of the Association for Computational Linguistics, ACL 2018, Melbourne, Australia, 15–20 July 2018, Volume 1: Long Papers, pp. 1339–1349 (2018). https://aclanthology.info/papers/P18-1124/p18-1124

16. Yang, Z., Salakhutdinov, R., Cohen, W.W.: Transfer learning for sequence tagging with hierarchical recurrent networks. CoRR, vol. abs/1703.06345 (2017). http://arxiv.org/abs/1703.06345

Robust Faster R-CNN: Increasing Robustness to Occlusions and Multi-scale Objects

Tao Zhou, Zhixin Li$^{(\boxtimes)}$, and Canlong Zhang

Guangxi Key Lab of Multi-source Information Mining and Security,
Guangxi Normal University, Guilin 541004, China
lizx@gxnu.edu.cn

Abstract. Recognizing objects at vastly different scales and objects with occlusion is a fundamental challenge in computer vision. In this paper, we propose a novel method called Robust Faster R-CNN for detecting objects in multi-label images. The framework is based on Faster R-CNN architecture. We improve the Faster R-CNN by replacing ROIpoolings with ROIAligns to remove the harsh quantization of RoIPool and we design multi-ROIAligns by adding different sizes' pooling(Aligns operation) in order to adapt to different sizes of objects. Furthermore, we adopt multi-feature fusion to enhance the ability to recognize small objects. In model training, we train an adversarial network to generate examples with occlusions and combine it with our model to make our model invariant to occlusions. Experimental results on Pascal VOC 2012 and 2007 datasets demonstrate the superiority of the proposed approach over many state-of-the-arts approaches.

1 Introduction

Object detection is a fundamental task in computer vision area which aims to locate and recognize every object instance with a bounding box. One kind of object detection method rely on region proposal, such as SPPnet [4], Fast R-CNN [2]. The region-based methods divide the object detection task into two sub-problems: At the first stage, a dedicated region proposal generation network is grafted on deep convolutional neural networks (CNNs) which could generate high quality candidate boxes. Then at the second stage, a region-wise subnetwork is designed to classify and refine these candidate boxes. Another family of object detectors such as SSD [9] and RON [8], they do not rely on region proposal and directly estimate object candidates. After the original R(region)-CNN [3], Fast R-CNN [2] proposes ROIPooling (Spatial-Pyramid-Pooling) allowing the classificaiotn layers to reuse features computed over CNN feature maps. Faster R-CNN [11], which incorporates region proposal generation in the framework has achieved good results currently. The performance is 69.9 on Pascal VOC 2007 dataset. Under Fast R-CNN pipeline, several works try to improve the detection speed and accuracy, with more effective region proposals, multi-layer

© Springer Nature Switzerland AG 2019
L. H. U and H. W. Lauw (Eds.): PAKDD 2019 Workshops, LNAI 11607, pp. 298–310, 2019.
https://doi.org/10.1007/978-3-030-26142-9_26

fusion, multi-scale pooling and more effective training strategy. However, it is still a problem that how to learn an object detector that is invariant to occlusion and is suitable for objects with different scales. To address these issues and take full advantage of CNN for multi-label image detection, we propose an improved model named Robust Faster R-CNN, we design four different sizes of Poolings as well as multi-feature fusion to extract the features of different scales' objects in multi-label data. In addition, we train the adversarial network to solve the occlusion problem. Furthermore, recent complementary advances such as ROI (Region of interest) Align [5] and multi-ROI pooling [7] could be naturally employed. The experimental results on Pascal VOC 2012 and Pascal VOC 2007, show that our approach performs more effectively and more accurately.

2   Improved Model

The Robust Faster R-CNN model proposed in this paper is based on the Faster RCNN [11] network structure. As shown in Fig. 1, in our improved model, we transfer the parameters pre-trained on Faster R-CNN to our model and improve the structure of the model by replacing the last pooling layer with a multi-scale ROIAligns layer. And we also consider multi-feature fusion. During the training phase, we train an adversarial network to generate occlusion examples.

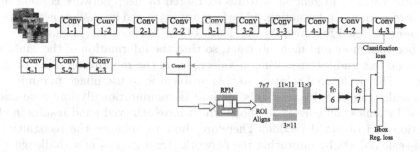

Fig. 1. The improved Faster R-CNN Model with multi-scale ROIAligns

2.1   Model Parameter Transfer

As shown in Fig. 2, we use parameter transferring to reuse the parameters of Faster R-CNN, the shared model contains 13 convolution layers and contains tens of millions of parameters. Therefore, sufficient training time and iteration are needed to obtain an effective detection model. By using parameters transferring [10], the parameters pre-trained on Faster RCNN are directly transferred to our model except for the last fc6 layer as the parameters of this layer between these two network are different. Based on the pre-trained parameters by Faster RCNN, fine-tuning is enforced to adjust the parameters. That is, the parameters of the Faster RCNN are used to continue training our model, which can reduce the training time.

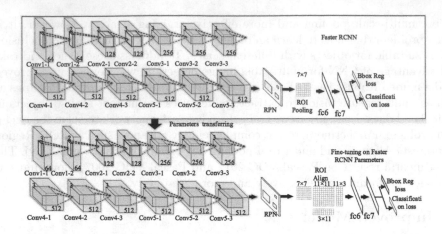

Fig. 2. The improved RCNN Model with Parameter transferring.

2.2 Multi-feature Fusion

Recognizing objects of different sizes is a fundamental challenge in classification problems. For CNN, different depth features correspond to different levels of semantic features. In general, features extracted by deep network contain more high-level semantic information, while features extracted by shallow network contain more detailed features. As the depth of the network increases, the feature map becomes more and more abstract, so that the information of the contained features is less and less, resulting in a decrease in the recognition effect of small objects. The traditional solution to this problem is to use image pyramids, i.e. multi-scale training. However, this method is computationally intensive and it is used by almost all current methods which have achieved good results in classification and object detection. Therefore, how to enhance the recognition of multi-scale objects by improving the network structure is a new challenge.

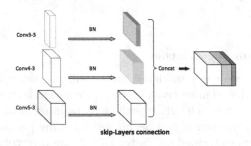

Fig. 3. Multi-feature fusion structure.

The traditional VGG16 model has a single route for feature extraction. It does not make good use of the feature layer before Conv5-3, so the recognition ability for small objects is insufficient. The reason for these is that the pooling layer filter filters most of the information. In order to make full use of the multi-layer convolution features and improve the recognition ability of small objects, we integrate the pooled features from Conv4-3 and Conv3-3. The improved network structure is shown in Fig. 3. The concat layer splices the feature map and keep the feature maps' sizes unchanged, so we have more feature representations. The batch normalization (BN) and scale are added after each convolution layer. This operation can improve the speed of training and increase the classification effect which has been proved in [6]. Through simple network connection changes, the performance of recognizing small object is improved without substantially increasing the amount of calculation of the original model.

2.3 Multi-scale ROIAligns

As is known to all, RoIPool [11] is a standard operation for extracting a small feature map (e.g., 7×7) from each RoI. RoIPool first quantizes a floating-number RoI to the discrete granularity of the feature map, this quantized RoI is then subdivided into spatial bins which are themselves quantized, and finally feature values covered by each bin are aggregated (usually by max pooling). Quantization is performed, e.g., on a continuous coordinate x by computing $[x/16]$, where 16 is a feature map stride and $[\cdot]$ is rounding; likewise, quantization is performed when dividing into bins (e.g., 7×7). These quantizations introduce misalignments between the RoI and the extracted features. While this may not impact classification, which is robust to big objection, it has a large negative effect on predicting pixel-accurate objection box(for small objects). so we use ROIAlign, which is proposed in mask rcnn [5] to remove the harsh quantization of RoIPool, properly aligning the extracted features with the input. As shown in Fig. 4, ROIAlign avoids any quantization of the RoI boundaries or bins (i.e., it uses $x/16$ instead of $[x/16]$). It use bilinear interpolation to compute the exact values of the input features at four regularly sampled locations in each RoI bin, and aggregate the result (using max).

The framework after Fast R-CNN [10] has a common problem in the detection of small objects. That is, the information of the object will be lost badly. For example, the original 32×32 object has only 2×2 left to the last layer of the feature map. The methods proposed to deal with this problem is enlarging the feature map and utilizing a smaller anchor scale in RPN. The Faster R-CNN framework does ROIPooling on the feature map with pooled size 7×7 for each RPN proposal. While it is difficult for a single size ROIPooling to catch the features of different objects scales. Inspired by R2CNN [7], we add two pooled sizes: 11×3 and 3×11. The pooled size 3×11 is supposed to catch more horizontal features and help the detection of the horizontal objects whose width is much larger than its height. The pooled size 11×3 is supposed to catch more vertical features and be useful for vertical objects detection that the height is much larger than the width. Furthermore, to enhance our model's robustness of recognizing objects at small scales, we add a pooled sizes: 11×11. And we use the strategy: adding a new anchor scale and utilizing the anchor scale of

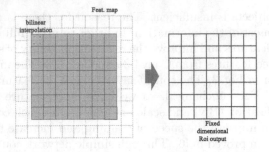

Fig. 4. The dashed grid represents a feature map, the solid lines a RoI (with 7×7 bins in this example), and the dots the 4 sampling points in each bin. RoIAlign computes the value of each sampling point by bilinear interpolation from the nearby grid points on the feature map. No quantization is performed on any coordinates involved in the RoI, its bins, or the sampling points.

$(4,8,16,32)$. R2CNN has confirmed that the adoption of the smaller anchor is helpful for small objects detection.

With multi-scale ROIAligns, we can pool features extracted at variable scales, which can improve the accuracy of object detection.

3   How to Improve Accuracy by Adversarial Network

As our detector's goals is to make our model has robustness to objects with occlusion, it is a problem how to cover all potential occlusions due to the reason that some occlusions are rare. Fortunately, an Adversarial Spatial Dropout Network (ASDN) has been proposed by the authors of A Fast RCNN [13]. It can generate examples of occlusion to train network instead of relying the dataset or sifting through data to find hard examples, which has demonstrated a good performance rencently [12].

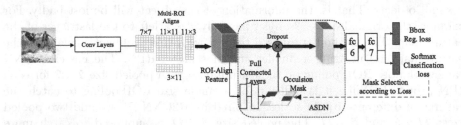

Fig. 5. Adversarial network architecture combining our network with ASDN. Occlusion masks are created to generate hard examples for training.

We use $F(x)$ to represent the original object detector network and X is a object proposals. The F_c and F_l represent output of class and predicted bounding box location respectively. C and L are the ground-truth class for X and its' spatial location. The original detector loss can be written down as,

$$E_F = E_{soft\max}(F_c(X), C) + E_{bbox}(F_l(X), L) \qquad (1)$$

where the first term is the SoftMax loss and the second term is the loss based on predicted bounding box location and ground truth box location. The adversarial network can be represented as $A(x)$, when given a feature X computed on an image, it generates new adversarial examples which are added to the training samples. The adversarial network's purpose is to learn how to predict the feature on which the detector would fail. The adversarial network is trained via the following loss function,

$$E_A = -E_{soft\max}(F_c(A(X)), C) \qquad (2)$$

Therefore, when the detector classify the feature generated by the adversarial network easily, we get a high loss for the adversarial network. On the other hand, if after adversarial feature generation it is difficult for the detector to classify, we get a high loss for the detector and a low loss for the adversarial network.

3.1   Training Detail of Adversarial Spatial Dropout Networkg

Before using ASDN to improve our network, we pre-train it to create occlusions. During training, we apply stage-wise training [13] as same as A Fast RCNN. After being pre-trained on multi-label image set, our detector has a sense of the objects in the dataset, then we train the ASDN by fixing all the layers in our network.

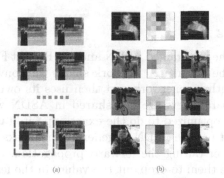

(a)   (b)

Fig. 6. (a) Examples of occlusions that are sifted to select the hard occlusions and used as ground-truth to train the ASDN network (b) Examples of occlusion masks generated by ASDN network. The black regions are occluded when passed on to our network.

As is shown in Fig. 5, the adversarial network has the same structure in convolutional layers, ROIPoolings layer as well as fully connected layers with our network. The convolutional features for each feature map after the ROIPoolings layer are obtained as the inputs for the adversarial network. Given a feature map, the ASDN will try to generate a mask indicating which parts of the feature to dropout (assigning zeros) so that the classification of the network will be harder. The specific process is as follows, given a feature map with size of $d \times d$ and a sliding window with size of $d/3 \times d/3$ is applied. The sliding window slides, overriding the position of the space and deletes the values in all the channels of the corresponding window, so a new feature vector generates. Based on all the missing $d/3 \times d/3$ windows, it passed all the new feature vectors obtained above to the Softmax loss layer to calculate the loss and selected the highest one. Then the window create a single $d \times d$ mask (with 1 for the window location and 0 for the other pixels) for it. The sliding window process is represented by mapping the window back to the image as Fig. 6(a). In this way, it generates these spatial masks for n feature maps and gets n pairs of training samples so that ASDN can generate masks that have high losses. The binary cross entropy loss is used to train ASDN, the formula is as follows,

$$E = -\frac{1}{n} \sum_p^n \sum_{i,j}^d [\tilde{M}_{ij}^p A_{ij}(X^p) + (1 - \tilde{M}_{ij}^p)(1 - A_{ij}(X^p))] \tag{3}$$

where $A_{ij}(X^p)$ represents the outputs of the ASDN in location (i, j) given input feature map X^p. The output generated by ASDN is not a binary mask but rather a continuous heatmap. The ASDN uses importance sampling to select the top $1/3$ pixels to mask out. More specifically, given a heatmap, $1/3$ pixels out of them are selected to assign the value 1 and the rest of $2/3$ pixels are set to 0. As is showed in Fig. 6(b), the network starts to recongnize which parts of the objects are significant for classification. In this case, we use the masks to occlude these parts to make the classification harder.

3.2 Joint Training

We jointly optimize the pre-trained ASDN and our Robust Faster R-CNN model. In the joint model, the adversarial network shares the convolutional layers and ROIPoolings layer with our network and then uses its own separate fully connected layers. The parameters are not shared in ASDN with our network as the two networks are optimized to do the exact opposite tasks. For training the RCNN model, we first use the ASDN to generate the masks on the features after the ROI Poolings layer during the forward propagation, the ASDN generates binary masks and use them to drop out the values in the features, then forward the modified features to calculate losses and train our model. Although our features are modified, the labels remain the same. So the "harder" and more diverse examples are created for training our model. For training the ASDN, since the sampling strategy is applied to convert the heatmap into a binary mask, which is not differentiable, it cannot directly back-prop the gradients from the classification loss. Same as A Fast rcnn, only those hard example masks are used as ground-truth to train the adversarial network by using the same loss as described in Eq. 3 to compute which binary masks lead to significant drops in Robust Faster R-CNN classification scores.

4   Experiment

4.1   Datasets and Evaluation Measures

We evaluate the proposed Robust Faster R-CNN on the PASCAL Visual Object Detection Challenge (VOC) datasets [1], which are widely used as the benchmark for object detection. In this paper, PASCAL VOC 2007 and VOC 2012 are employed for experiments. These two datasets, which contain 9,963 and 22,531 images respectively, are divided into train, val and test subsets. We conduct our experiments on the trainval/test splits (5,011/4,952 for VOC 2007 and 11,540/10,991 for VOC 2012). The evaluation metric is Average Precision (AP) and mean of AP (mAP) complying with the PASCAL challenge protocols.

4.2   Image-fine-tuning

We initializing our network with the parameters from Faster R-CNN trained on VOC 07+12 trainval. Since we used multi-scale ROIPooling, the dimensions inputting to the fc6 is changed. So the fully connected layers fc6 is initialized from zero-mean Gaussian distributions with standard deviations 0.01, and the learning rate is set to 0.01, and we decrease the learning rates to one tenth of the current ones after every 20 epoches (60 epoches in all). In the training of our model, the number of iterations is set to 60 epochs, each epoch has 2000 iterations. Figure 7 shows that the mAP scores on VOC 07+12 begin to convergence after almost 30 epochs' iteration.

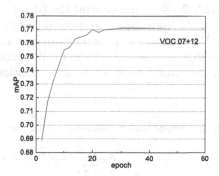

Fig. 7. The changing trend of mAP scores on and VOC 07+12 during I-FT.

4.3   Joint Model Training

After trained on datasets, our network has a sense of the objects, we train ASDN for 12K iterations. Given the pre-trained ASDN and our improved model, we train the joint model for 120K iterations. Similar to image-fine-tuning, we adopt a discriminating learning rate, which starts with 0.001 and decreases to 0.0001 after 60K iterations. The momentum and the weight decay are set as 0.9 and 0.0005, the same as in the image-fine-tuning step. In the joint model training, After the ROIPoolings-layer during the forward propagation we first use the ASDN to generate the masks on the features. The ASDN generates binary masks and use them to drop out the values in the features, then forward the modified features to calculate losses and train the Robust Faster R-CNN model. Although our features are modified, the labels remain the same. So the "harder" and more diverse examples are created for training the Robust Faster R-CNN model.

4.4   Ablation Experiments

The ablation experiments is designed to evaluate the influence of different anchor scales and different ROIPooling sizes on object detection. We add a smaller anchor scale to the original scales (8,16,32) so that the anchor scales become (4,8,16,32), which would generate 12 anchors in RPN. It is clearly that, the model Faster RCNN+multi ROIPooling with four pooled sizes (3×11, 11×3, 7×7, 11×11) is better than Faster RCNN with one pooled size (7×7). Table 1 reports results of several models on VOC 2007 test set. Compared with the model with ROIAligns and feature fusion layer, we can find that the model with ASDN obtains a significant improvements of 2.3%, reflecting the effectiveness of ASDN. Besides, we consider the impact of pooled size (11×11). We give the results of the model with three pooled sizes (3×11, 11×3, 7×7) layer and four pooled sizes (3×11, 11×3, 7×7, 11×11). It shows that the improved network structure through the pooled size (11×11) is superior to the three pooled size structure with 0.3% increase. And the (3×11, 11×3, 7×7) have a significant increase of 0.6% than the single pooled size (7×7). This is mainly because with additional pooled sizes (11×11) can partially catch the objects with smaller regions in VOC dataset. And the results present the effectiveness of multi-feature fusion and ROIAligns, with an promotion of 0.4% and 0.3% respectively.

Table 1. Comparing the results under different settings on VOC 2007 test set.

Approach	Anchor scales	Pooled sizes	mAP
Faster RCNN (FRCN)	(8,16,32)	7×7	73.2
Faster RCNN	(4,8,16,32)	7×7	73.3
FRCN+multi ROIPooling	(4,8,16,32)	3×11, 11×3, 7×7	73.9
FRCN+ASDN	(8,16,32)	7×7	75.6
FRCN+multi ROIPooling	(4,8,16,32)	3×11, 11×3, 7×7, 11×11	74.2
FRCN+multi ROIAligns	(4,8,16,32)	3×11, 11×3, 7×7, 11×11	74.6
FRCN+multi ROIAligns+ASDN	(4,8,16,32)	3×11, 11×3, 7×7, 11×11	76.8
FRCN+feature fusion+multi ROIAligns (I-FT)	(4,8,16,32)	3×11, 11×3, 7×7, 11×11	74.8
FRCN+feature fusion+multi ROIAlign+ASDN	(4,8,16,32)	3×11, 11×3, 7×7, 11×11	77.1

4.5   Results

Tables 2 and 3 reports the experimental results compared with the state-of-the-art approaches on VOC 2007 test set and VOC 2012 test set. It is noticeable that our model is significantly better than the benchmark, Faster R-CNN, and the detection accuracy for small objects is significantly improved, such as bottle and plant. Although the map is lower than the state-of-the-art approach RON in VOC 2007, it is also 3.9% better than Faster R-CNN, which proves the effectiveness of our innovation.

Table 2. Detection results on Pascal VOC 2007 test set, comparing with other approaches.

	Faster R-CNN	A-Fast-RCNN	SSD	RON	I-FT	(I-FT+ASDN) Ours
aero	76.5	75.7	79.8	86.0	79.0	79.8
bike	79.0	83.6	79.5	82.5	82.1	84.3
bird	70.9	68.4	74.5	76.9	74.3	76.0
boat	65.5	58.0	63.4	69.1	65.4	68.0
blt	52.1	44.7	51.9	59.2	56.3	57.2
bus	83.1	81.9	84.9	86.2	85.6	87.2
car	84.7	80.4	85.6	85.5	85.6	88.0
cat	86.4	86.3	87.2	87.2	86.7	88.7
chair	52.0	53.7	56.6	59.9	52.8	58.9
cow	81.9	76.1	80.1	81.4	81.8	84.4
tabel	65.7	72.5	70.0	73.3	67.9	71.3
dog	84.8	82.6	85.4	85.9	85.7	86.9
hrs	84.6	83.9	84.9	86.8	87.0	90.8
mbk	77.5	77.1	80.9	82.2	81.6	83.2
per	76.7	73.1	78.2	79.6	82.9	81.6
plant	38.8	38.1	49.0	52.4	39.8	43.4
shp	73.6	70.0	78.4	78.2	75.1	77.2
sofa	73.9	69.7	72.4	76.0	74.8	77.0
train	83.0	78.8	84.6	86.2	84.4	85.2
tv	72.6	73.1	75.5	78.0	75.4	77.5
mAP	73.2	71.4	75.1	77.6	74.8	77.1

From the two tables we can see that our model trained during Image-fine-tuning (I-FT) with the pooled sizes of *(3 × 11, 11 × 3, 7 × 7, 11 × 11)* has a better result than Faster R-CNN model, which confirms feasibility of our multi-scale ROIAligns. It is also clearly that the model combing with ASDN has a significant improvement than the model during the Image-fine-tuning (I-FT) on

Table 3. Detection results on Pascal VOC 2012 test set, comparing with other approaches.

	Faster R-CNN	A-Fast-RCNN	SSD	RON	I-FT	(I-FT+ASDN) Ours
aero	84.9	82.2	84.9	86.5	86.8	87.0
bike	79.8	75.6	82.6	82.9	81.9	83.5
bird	74.3	69.2	74.4	76.6	77.6	78.9
boat	53.9	52.0	55.8	60.9	57.2	60.1
blt	49.8	47.2	50.0	55.8	55.1	57.2
bus	77.5	76.3	80.3	81.7	80.4	83.2
car	75.9	71.2	78.9	80.2	76.4	80.5
cat	88.5	88.5	88.8	91.1	89.5	90.2
chair	45.6	46.8	53.7	57.3	47.6	50.9
cow	77.1	74.0	76.8	81.1	80.6	82.4
tabel	55.3	58.1	59.4	60.4	60.7	61.3
dog	86.9	85.6	87.6	87.2	88.5	89.9
hrs	81.7	80.3	83.7	84.8	84.8	89.8
mbk	80.9	80.5	82.6	84.9	82.4	82.8
per	79.6	74.7	81.4	81.7	84.7	86.6
plant	40.1	41.5	47.2	51.9	40.6	47.4
shp	72.6	70.4	75.5	79.1	73.9	74.2
sofa	60.9	62.2	65.6	68.6	61.6	70.0
train	81.2	77.4	84.3	84.1	84.7	86.2
tv	61.5	67.0	68.1	70.3	62.2	69.9
mAP	70.4	69.0	73.1	75.4	72.8	75.6

image set which proved the effectiveness of ASDN. Figure 8 shows some examples of results on Pascal VOC 2007 and Pascal VOC 2012. It is evident that Robust Faster R-CNN can recognize the objects with different scales and can predict their locations well, especially like plane, bird and people. This is because these objects usually have different sizes and aspect ratios. We can also see the robustness of our approach to occlusions, such as car, plant and people which often with occlusions.

Detection speed is an important index for evaluating the performance of an object detection model. We compare the Robust Faster R-CNN with Faster R-CNN on PASCAL VOC 2007. We collected each detection time of the model, and averaged all detection times. The detection speed of Robust Faster R-CNN is about 252 ms per image, while the figure for Faster R-CNN is 200 ms. Although the running time is a bit longer, it is an expected result, as Robust Faster R-CNN consumes more time than Faster R-CNN in feature extraction in Multi

Fig. 8. Selected examples of object detection results on the PASCAL VOC 2007 and VOC 2012.

ROIAligns. The difference between 200 ms and 252 ms is little, so it still meets the requirements of real-time detection of target.

5   Conclusion

In this paper, we present a Robust Faster R-CNN, an effective object detection framework for detecting objects with occlusion and objects with different scales. The multi-ROIAligns as well as multi-feature fusion are used to learn semantic multi-scale feature representation, which makes our model invariant to objects with different sizes and width-height aspect ratios like people, cars and planes. And the ASDN is combined with our network to form the adversarial network to generate occlusion training samples, which makes the model robust to occlusions. In comparison with many state-of-the-art approaches, experimental results show that our approach has a significant increase in accuracy on PASCAL VOC 2012 and PASCAL VOC 2007, and the detection speed is not significantly reduced.

Acknowledgments. This work is supported by the National Natural Science Foundation of China (Nos. 61663004, 61762078, 61866004), the Guangxi Natural Science Foundation (Nos. 2016GXNSFAA380146, 2017GXNSFAA198365, 2018GXNSFDA281009), the Research Fund of Guangxi Key Lab of Multi-source Information Mining and Security (16-A-03-02, MIMS18-08), the Guangxi Special Project of Science and Technology Base and Talents (AD16380008), the Guangxi Bagui Scholar Teams for Innovation and Research Project, and Innovation Project of Guangxi Graduate Education under grant (XYCSZ2018077).

References

1. Everingham, M., Williams, C.: The PASCAL visual object classes challenge 2010 (VOC2010). In: International Conference on Machine Learning, pp. 117–176 (2010)
2. Girshick, R.: Fast R-CNN. In: Advances in Neural Information Processing Systems, pp. 91–99 (2015)

3. Girshick, R., Donahue, J., Darrell, T., Malik, J.: Rich feature hierarchies for accurate object detection and semantic segmentation. In: Proceedings of IEEE International Conference on Computer Vision and Pattern Recognition, pp. 580–587 (2014)

4. He, K., Zhang, X., Ren, S., Sun, J.: Spatial pyramid pooling in deep convolutional networks for visual recognition. IEEE Trans. Pattern Anal. Mach. Intell. **37**(9), 1904 (2015)

5. He, K., Gkioxari, G., Dollar, P., Girshick, R.: Mask R-CNN. IEEE Trans. Pattern Anal. Mach. Intell. **PP**(99), 1 (2017)

6. Huang, G., Liu, Z., Laurens, V.D.M., Weinberger, K.Q.: Densely connected convolutional networks. In: Proceedings of IEEE International Conference on Computer Vision and Pattern Recognition, pp. 2261–2269 (2016)

7. Jiang, Y., et al.: R2CNN: rotational region CNN for orientation robust scene text detection. In: Proceedings of IEEE International Conference on Computer Vision and Pattern Recognition, pp. 2261–2269 (2017)

8. Kong, T., Sun, F., Yao, A., Liu, H., Lu, M., Chen, Y.: RON: reverse connection with objectness prior networks for object detection. In: Proceedings of IEEE International Conference on Computer Vision and Pattern Recognition, vol. 1 (2017)

9. Liu, W., et al.: SSD: single shot multibox detector. In: Leibe, B., Matas, J., Sebe, N., Welling, M. (eds.) ECCV 2016. LNCS, vol. 9905, pp. 21–37. Springer, Cham (2016). https://doi.org/10.1007/978-3-319-46448-0_2

10. Oquab, M., Bottou, L., Laptev, I., Sivic, J.: Learning and transferring mid-level image representations using convolutional neural networks. In: Proceedings of IEEE International Conference on Computer Vision and Pattern Recognition, pp. 1717–1724 (2014)

11. Ren, S., He, K., Girshick, R., Sun, J.: Faster R-CNN: towards real-time object detection with region proposal networks. In: Advances in Neural Information Processing Systems, pp. 91–99 (2015)

12. Zhou, T., Li, Z., Zhang, C., Lin, L.: An improved convolutional neural network model with adversarial net for multi-label image classification. In: Geng, X., Kang, B.-H. (eds.) PRICAI 2018. LNCS (LNAI), vol. 11013, pp. 38–46. Springer, Cham (2018). https://doi.org/10.1007/978-3-319-97310-4_5

13. Wang, X., Shrivastava, A., Gupta, A.: A-fast-RCNN: hard positive generation via adversary for object detection. In: Proceedings of IEEE International Conference on Computer Vision and Pattern Recognition, pp. 21–26 (2017)

Effectively Representing Short Text via the Improved Semantic Feature Space Mapping

Ting Tuo[1(✉)], Huifang Ma[1,2(✉)], Haijiao Liu[1(✉)], and Jiahui Wei[1(✉)]

[1] College of Computer Science and Engineering, Northwest Normal University,
Lanzhou, China
{nwnutuot,mahuifang,nwnulhj,nwnuweijh}@yeah.net
[2] Guangxi Key Lab of Multi-source Information Mining and Security,
Guangxi Normal University, Guilin, China

Abstract. Short text representation (STR) has attracted increasing interests recently with the rapid growth of Web and social media data existing in short text form. In this paper, we present a new method using an improved semantic feature space mapping to effectively represent short texts. Firstly, semantic clustering of terms is performed based on statistical analysis and word2vec, and the semantic feature space can then be represented via the cluster center. Then, the context information of terms is integrated with the semantic feature space, based on which three improved similarity calculation methods are established. Thereafter the text mapping matrix is constructed for short text representation learning. Experiments on both Chinese and English test collections show that the proposed method can well reflect the semantic information of short texts and represent the short texts reasonably and effectively.

Keywords: Semantic feature space · Similarity computation ·
Text mapping matrix · Short text representation

1 Introduction

With the rapid development of technology, an increasing number of short texts have been generated including search result snippets, forum titles, image or video titles, tweets, microblogs and so on. However, unlike normal texts, short texts do not provide enough contextual information, the data sparsity problem is easily encountered [1]. Directly applying the traditional methods based on bag-of-words (BOW) model is not satisfactory, as BOW model ignores the order and semantic relations between terms. Therefore, how to acquire effective representations of short texts has been an active research issue.

Various methods have been proposed to deal with the sparseness issue. They can be divided into two broad categories: knowledge-based and corpus-based. The former is to import external information such as WordNet, Probase, or other user constructed knowledge bases peripheral information sources to enrich the representations of short texts. Some noticeable works include: Piao et al. [2] employ synsets from WordNet and concepts from DBpedia for representing user interests. Moreover, existing research

© Springer Nature Switzerland AG 2019
L. H. U and H. W. Lauw (Eds.): PAKDD 2019 Workshops, LNAI 11607, pp. 311–321, 2019.
https://doi.org/10.1007/978-3-030-26142-9_27

have shown that Probase consists of millions of concepts, which explicitly model the context of semantic relationships. Li et al. [3] propose an efficient and effective approach for semantic similarity using a large scale probabilistic semantic network, known as Probase. In [4], the authors propose a Wikitop system to automatically generate topic trees from the input text by performing hierarchical classification using the Wikipedia Category Network. In addition, short texts can be mapped to a semantic space learned from a set of common repositories, with the new representation in this semantic space being a combination of source information and external information.

However, these external knowledge-enhanced methods are mostly domain-based, which limits short text representation. Though better performance can be obtained, the performance heavily relies on the amount and quality of outer or additional information. In the opposite direction of knowledge-based approaches, a large number of corpus-based methods have been proposed. Adopting corpus-based methods have several advantages: carefully selected external data are not required, meanwhile, when the amount of data is large, the process of finding relevant information is more efficient. For example, Shen et al. [5] point out that SetExpan method can deal with noisy context features derived from free-text corpora, which may lead to entity intrusion and semantic drifting. In [6], Wang designs a new system for knowledge bases containing incomplete and uncertain information due to limitations of information sources and human knowledge. Besides, Mikolov et al. [7] presented word2vec representation model to encode word knowledge and achieved significant results for text representation.

In this work, we propose an effectively representing short text method using the improved semantic feature space mapping (i.e., SFSM). Firstly, semantic clustering of terms is performed based on statistical analysis and word2vec, and the semantic feature space can then be represented via the cluster center. Secondly, the context information is integrated via the semantic feature space, then three kinds of similarity calculation methods are established. Finally, the text mapping matrix is constructed to project the raw short text vectors into a common low-dimensional semantic feature vector space.

The rest of this paper is organized as follows: Sect. 2 shows the construction of semantic feature space. Section 3 proposes the short text representation method based on the improved semantic feature space mapping. Section 4 demonstrates the experimental results on both Chinese and English test collections. Finally, conclusion and future work are described in Sect. 5.

2 Semantic Feature Space Construction

In this section, first of all, we develop a semantic similarity calculation method between terms, which is taken as the initial similarity for further processing, and then illustrate the process of the construction of the semantic feature space, which can be considered for converting the raw short text snippets into modeled vectors.

2.1 Calculation of Initial Similarity

The initial similarity between terms determines the performance of the clustering results, we combine statistical analysis and deep learning approach to calculate the initial similarity between two terms. From the perspective of statistical analysis, we mainly consider mutual information and co-occurrence; and in terms of deep learning aspect, words embedding are adopted. Finally, the initial similarity between the terms is obtained by synthesizing the above two aspects.

Statistical Analysis Aspects. It is believed that the larger the MI value between terms, the more information they share and the greater the similarity. Thus, mutual information can be used to measure the similarity between two terms.

Let $D = \{d_1, d_2, \ldots, d_m\}$ be the short text corpus, where m is the number of texts in D. $T = \{t_1, t_2, \ldots, t_n\}$ denotes the vocabulary of D, where n is the number of unique terms in D. We apply the normalized mutual information (NMI) [8] $NMI(t_1, t_2)$ to measure the similarity between term t_1 and t_2 as follows.

Furthermore, we also introduce the co-occurrence distribution [9] of terms to calculate the similarity between terms. In detail, a term t_i is presented by a term co-occurrence vector $[cor(t_1, t_2), \ldots, cor(t_1, t_n)]$. Formally, the co-occurrence distribution of the term t1 is as follows:

$$co(t_1) = [cor(t_1, t_1), cor(t_1, t_2), \ldots, cor(t_1, t_n)]$$

Where $cor(t_1, t_n)$ is decided by the co-occurrence of terms t_1 and t_n. In this work, we apply the method in [10] to calculate $cor(t_1, t_n)$.

Then, the mutual information and the co-occurrence are in targeted to calculate the statistical similarity $sim(t_1, t_2)$, which is formalized as:

$$sim(t_1, t_2) = \frac{1}{2} NMI(t_1, t_2) + \frac{1}{2} co(t_1, t_2) \tag{1}$$

where $co(t_1, t_2)$ is the co-occurrence similarity between the vector t_1 and t_2, i.e. cosine similarity between $co(t_1)$ and $co(t_2)$. And the mutual information between terms is considered as equally important as co-occurrence in our method.

Deep Learning Aspects. In recent years, several deep learning based approaches have been found to be promising for compute terms similarity. Among them, Word2vec models have been proven to be highly efficient in finding words embedding templates and uncover various semantic and syntactic relationships. Mikolov et al. [7] showed that such words embedding have the capability to capture linguistic regularities and patterns. In this paper, we use CBOW model to acquire the vector representation of the term ti, i.e., $V(t_i) = Word2vec(t_i)$. Ultimately, the initial similarity can be obtained from these two aspects as follows:

$$S_0(t_1, t_2) = \lambda sim(t_1, t_2) + (1 - \lambda)V(t_1, t_2) \tag{2}$$

Where $\lambda \in [0,1]$ determines the relative importance of statistical and deep learning aspects. It is worth noting that the cosine similarity computation method is adopted to calculate the similarity between words embedding vectors.

2.2   The Construction of Semantic Feature Space

Ideally, using feature selection would make it possible to choose a feature subset from the original feature set, which best represents the semantic concepts. In our work, we develop an effective semantic feature space consisting of terms with small semantic gap and high distinguishing ability. The key idea is to aggregate terms with higher similarity as a cluster, then the most centrally located term in each cluster is chosen as the representative to form the semantic feature space, which can be used to convert the original sparse and noisy short text to the low dimension and semantically richer feature space.

In this section, the modified k-Medoids [11] algorithm is performed to cluster terms. Good initial centers are essential for the success of partitioning clustering algorithms. Instead of using random initial centers, we identify good initial centers incrementally by a refined method from Moore [12]. The first medoid is randomly selected among all terms. Then we select the point that has the maximum of the minimum of the distances from each of the existing medoids to be the next medoid, i.e.,

$$M^0 = \{t_j|\max_{t_j}\{\min_i\{d_{sem}(m_i, t_j) > \alpha\}\}, t_j \in T, m_i \in M^0\} \tag{3}$$

Where, $d_{sem}(m_i, t_j)$ is the distance between m_i and t_j, which equals to $1 - S_0(m_i, t_j)$, t_j is the j-th candidate term, m_i is the i-th medoid of existing medoids, and α is the threshold in the limit of initial medoid count. This process continues until we do not find any medoids satisfying (3). In the end, we get the initial $M^0 = \{m_1^0, m_2^0, \ldots, m_k^0\}$. It is worth noting that the larger the threshold of α, the small the value of k.

With k medoids in the t-th iteration, each term $t_i \in T$ is assigned to its closest medoid $m^* \in M^t = \{m_1^t, \ldots, m_k^t\}$. The convergence condition is that the difference between the two adjacent clustering results less than threshold β or reaches the predetermined iteration number.

Where $w_{ij} \in \{0,1\}$, $\sum_{i=1}^k w_{ij} = 1, 0 < \sum_{i=1}^k w_{ij} < n$, $k(<n)$ is a known number of class centers, n is the count of objects to cluster. $W = [w_{ij}]$ is a $k \times n$ binary matrix, $M = [m_1, m_2, \ldots, m_k]$ is a set of cluster medoids and m_i is the i-th cluster medoid.

According to the above processing of k-Medoids, we can get k clusters for all given T, i.e. $K = \{K_1, K_2, \ldots, K_k\}$, and the semantic feature space can then be represented via the cluster center, formalized as $FS(m_1, m_2, \ldots, m_k)$.

3   Short Text Representation Based on Semantic Feature Space Mapping

In this section, we define the context of terms based on k clusters and develop three similarity calculation methods, namely max, average, and weighted similarity. Thereafter the text mapping matrix is constructed for short text representation learning.

3.1   Context-Based Similarity Calculation Methods

In general, co-occurred terms provide strong and consistent clues to the sense of a target term. The context of a term is thus obtained according to its co-occurrence relation with other terms in the whole short text D. The context of t_i in the short text D is defined as:

$$context(t_i) = \{t_j | t_i \in d_s, t_j \in d_s, s \in 1, 2 \ldots, m, 1 \leq j \leq n\} \tag{4}$$

The intersection of t_1 and t_2 with each cluster in the feature space, namely K_{t1} and K_{t2}, is formally defined as follows:

$$K_{t_1} = \{x | x = K_i \cap context(t_1), \forall K_i \in K \cap x \neq \emptyset\} \tag{5}$$

$$K_{t_2} = \{y | y = K_i \cap context(t_2), \forall K_i \in K \cap y \neq \emptyset\} \tag{6}$$

It is worth noting that the three similarity calculation methods are the similarity between the sets. To be specific, the similarity between two sets is the average similarity between its corresponding elements in the set. Three improved similarity calculation methods are defined as follows:

Definition 1 (Max similarity). Given terms t_1 and t_2, the **max similarity** between t_1 and t_2 is as:

$$S_1(t_1, t_2) = Max_{x \in K_{t_1}, y \in K_{t_2}} \{F(x, y)\} \tag{7}$$

Where $F(.)$ is the relation strength function, and cosine similarity is introduced to quantify the similarity between x and y. The max similarity function tends to select smaller clusters because it is easier for small clusters to look similar by the cosine similarity and hence dominate the *Max similarity* score.

Definition 2 (Average similarity). Given terms t_1 and t_2, the **average similarity** between t_1 and t_2 is as follows:

$$S_2(t_1, t_2) = \frac{\sum_{x \in k_{t_1}, y \in K_{t_2}} F(x, y)}{|K_{t_1}| \times |K_{t_2}|} \tag{8}$$

Definition 3 (Weighted similarity). Given terms t_1 and t_2, the **weighted similarity** of the set is calculated on K_{t1} and K_{t2} as the weighted similarity as follows:

$$S_3(t_1, t_2) = \sum_{x \in K_{t_1}} W_x \sum_{y \in K_{t_2}} W_y F(x, y) \tag{9}$$

Where $W_x = \frac{w_x}{\sum_{z \in K_{t_1}} w_z}$, $W_y = \frac{w_y}{\sum_{z \in K_{t_2}} w_z}$. The corresponding value vector of x (or y) indicates the average value of the set similarity for t_1 (or t_2) and each context in K_{t1} (or K_{t2}). We can utilize Eq. (2) to obtain the value of w_x, then normalize the weights of elements x and y. W_x and W_y reflect the importance of element x and y on t_1 and t_2 in K_{t1} and K_{t2} respectively.

3.2 Short Text Representation

The vector space model is adopted to represent short text, and each short text d is considered to be a vector, in which each term is associated with a weight indicating its importance. However, the main weakness of this term-vector representation is that different but semantically related terms are not matched and the dimension is always high.

Our goal is to construct a projection matrix that maps the corresponding term-vectors into a low-dimensional semantic feature space such that similar short texts are close when projected into this space. The text mapping matrix is constructed based on three improved similarity calculation methods. In the end, the text mapping matrix is used to map the short text vectors into the new feature space.

In particular, for an original short text which consists of l terms $d_{org}= (t_1,t_2,\ldots,t_l)$, we employ the TF-IDF term weighting model, in which the short text snippet d_{org} is represented as $\mathbf{V}_{org}$:

$$\mathbf{V}_{org} = [w_{t_1}, w_{t_2}, \ldots, w_{t_i}, \ldots, w_{t_l}] \tag{10}$$

In the refined approaches, all the features terms in the semantic feature space FS and all the terms of original text are considered to construct a text mapping matrix M. Each entry S_{ij} reflects the semantic proximity between terms t_i and t_j. Then, the refined short text semantic mapping vector $\mathbf{V}_{ref}$ can be obtained via the multiplication of $\mathbf{V}_{org}$ and the mapping matrix M:

$$\mathbf{V}_{ref} = \mathbf{V}_{org} \times M = \begin{bmatrix} w_{t_1} \\ w_{t_2} \\ \vdots \\ w_{t_l} \end{bmatrix}^T \times \begin{bmatrix} S_{1,m_1} & S_{1,m_2} & \cdots & S_{1,m_k} \\ S_{2,m_1} & S_{2,m_2} & \cdots & S_{2,m_k} \\ \vdots & \vdots & \ddots & \vdots \\ S_{l,m_1} & S_{l,m_2} & \cdots & S_{l,m_k} \end{bmatrix} \tag{11}$$

Where S_{i,m_i} can be calculated according to the Eq. (7), (8) or (9). Thence, our method has three variants of SFSM-MS, SFSM-AS, and SFSM-WS, which correspond to max, average, and weighted similarity methods, respectively.

Therefore, after mapping, each short text is represented by a less sparse and low dimensional vector that has non-zero entries for all terms that are semantically similar to those that appear in the semantic feature space.

4 Experiments and Results Analysis

In this section, we first give the experimental setup and some parameter analyses in Sects. 4.1 and 4.2, and then compare our approaches with baselines methods on both Chinese and English data sets in Sect. 4.3.

4.1   Experimental Setup

For the purpose of evaluating the performance of our algorithm, we carry out experiments on both Chinese and English test collections, respectively. We adopt 10 classes obtained from DBLP [10], with 1000 paper titles obtained from CCF recommended list in Rank A and B as English data sets. Besides, we select 10 categories from the Sogou corpus as Chinese data sets. For each category, 1000 news abstracts are selected as experimental data. A series of preprocessing works on both data sets including data denoising, stop words removal, and for Chinese data set, we utilized ICTCLAS [13], a Chinese text segmentation tool, as the tokenizer for the task of text segmentation to acquire the vocabulary.

To evaluate the performance of our algorithm and compare it with other short text presentation algorithms in a fair and reasonable way, we set up our experimental study as follows.

(1) The proposed algorithm is compared with two different types of representative short text presentation algorithms. They are (i) the traditional TF-IDF [5] short text representation method, (ii) leveraging term co-occurrence distance and strong classification features for short representation method, i.e., CDCFS [10], respectively.
   The traditional TF-IDF representation method has been widely used because it is simple, but it completely ignores the order of terms. In contrast, the CDCFS representation method which leveraging term co-occurrence distance and strong classification features, so the effect is improved, but this method fails to consider the context information of the terms.

(2) Two different types of classification algorithms KNN and SVM are employed to classify data sets. Existing short text classification work suggests that most of the term-weighting schemes shows its best performance in the range of 20–45. Therefore, we parameterize k-NN by choosing different value k in this range and demonstrated the best performance using optimal k. As for the SVM algorithm, we use the linear kernel functions and implement it on the libsvm tool, the other parameters of SVM are set to their default values. We then take 5-fold cross-validation to show the effectiveness of our method.

(3) When evaluating the performance of the short text classification algorithms, we use the MacroF1 and MacroP [14] as the evaluation of metrics.

4.2   Parameters Analysis

In this subsection, we describe some experiments on three important parameters (i.e., λ, α, β) involved in our method. From previous analysis, we know that the parameter λ controls the importance of mutual information and co-occurrence, the parameter α is related to the number of the initial class centers, and the parameter β is a clustering termination threshold. Because of the limitation of the paper, we only show the results of SFSM-WS method on SVM.

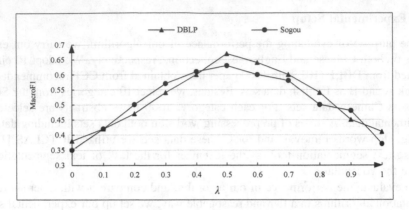

Fig. 1. MacroF1 varying with the parameter λ

Figure 1 reports the values of MacroF1 on two data sets varying with the parameter λ from 0 to 1 with a 0.1 step. We observe that as the value of λ increase, the MacroF1 rapidly increases up to the peaks (in the case of $\lambda = 0.5$), and then decreasing continuously. Besides, from Eq. (2) we know that when $\lambda = 0$, it means that only the mutual information between the terms is considered; likewise, when $\lambda = 1$, only the word co-occurrence is considered. Thus, we set $\lambda = 0.5$ as the optimal experimental parameter.

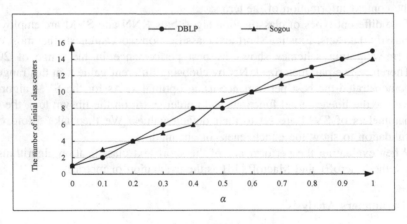

Fig. 2. The impact of α on the number of initial class centers

As is shown in Fig. 2, with the increase of α, the number of initial class centers increases on both data sets. We note that when $\alpha = 0.6$, the initial class center number is exactly the same as the predefined class number. Therefore, we set $\alpha = 0.6$ as the best experimental parameter.

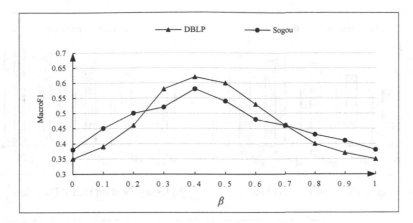

Fig. 3. MacroF1 varying with the parameter β

To find the best threshold value β, we can observe the relationship between parameter β and MacroF1 in classification algorithm, that is to say, when MacroF1 achieves its maximum value (similarity effect is the best), the value of β will be the best. In Fig. 3, we set the parameter β varying from 0 to 1 in steps of 0.1 on two data sets. With the increase of the parameter β, the value of MacroF1 will increase and then arrive its peak (in the case of $\beta = 0.4$), so $\beta = 0.4$ is selected as the optimal experimental parameter.

4.3 Experimental Results and Analysis

In this section, we aim to observe the effectiveness of our approaches comparing with the two baseline methods on SVM and KNN.

To evaluate the differences between our method and baseline method performance, we perform 5-fold cross-validation on both DBLP and Sogou data sets. The results of the MacroP and the MacroF1 are as follows: Firstly, in all cases, we can see that our methods performed significantly better than the baseline methods. Furthermore, we find that the experimental results on KNN and SVM have slight difference, which also demonstrates that the proposed representation method is insensitive to the classification algorithm and has good robustness.

Secondly, from the experimental results, we can clearly see that the values of MacroP and MacroF1 in SFSM-MS, SFSM-AS, and SFSM-WS, have reached more than 50%. At the same time, the values of MacroF1 in SFSM-WS are better than SFSM-MS and SFSM-AS on both data sets, the reason can be derived from Eq. (9).

Again, it can be seen from the Fig. 4 (a) to (d), our most advanced approaches, SFSM, leads the competition against the peers by large margins in two data sets, especially in DBLP data sets. The classification results of the five different methods can be roughly divided into three levels. The worst performance is the traditional TF-IDF method, which is actually due to the TF-IDF model completely ignores the order of terms. Then we can see that the CDCFS method is significantly improved compared to

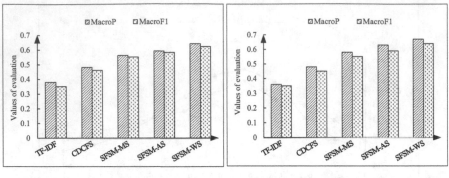

(a) Using KNN on DBLP Data Sets   (b) Using SVM on DBLP Data Sets

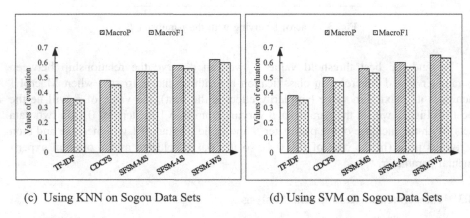

(c) Using KNN on Sogou Data Sets   (d) Using SVM on Sogou Data Sets

Fig. 4. The performance comparison between our approaches and baselines

TF-IDF, and the MacroP and the MacroF1 values can reach to 40%. Even in this case, our approaches is still significantly better than the CDCFS approach.

In conclusion, the results of these experiments demonstrate that introducing the term contexts is beneficial to improve semantic feature space, thereby improving the short text representation ability.

5 Conclusion

In this paper, we present a new method using an improved semantic feature space mapping to effectively represent short texts. We achieve this in terms of a three-step procedure: firstly, semantic clustering of terms is performed based on the mutual information and co-occurrence between terms, and the semantic feature space can then be represented via the cluster center. Then, the context information is integrated via the semantic feature space. On this basis, three improved similarity calculation methods are established. Thereafter the text mapping matrix is constructed for short text representation learning. Experiments on both Chinese and English data sets show that our method is efficient to be applied on large scale data sets.

In the future, we will further improve our work, mainly considering parameter optimization and combining deep learning techniques, to provide more contextual semantic information for terms to effectively represent short texts.

Acknowledgements. This work is supported by the National Natural Science Foundation of China (No. 61762078, No. 61363058, No. 61663004, No. kx201705) and Guangxi Key Lab of Multi-source Information Mining and Security (No. MIMS18-08).

References

1. Lu, H.Y., Xie, L.Y., Kang, N, et al.: Don't forget the quantifiable relationship between words: using recurrent neural network for short text topic discovery. In: AAAI 2017, pp. 1192–1198 (2017)
2. Piao, G.Y, Breslin, J.G.: User modeling on Twitter with WordNet Synsets and DBpedia concepts for personalized recommendations. In: CIKM 2016, pp. 2057–2060 (2016)
3. Li, P., Wang, H., Zhu, K.Q., et al.: A large probabilistic semantic network based approach to compute term similarity. IEEE Trans. Knowl. Data Eng. **27**(10), 2604–2617 (2015)
4. Kumar, S., Rengarajan, P., Annie, A.X.: Using Wikipedia category network to generate topic trees. In: AAAI 2017, pp. 4951–4952 (2017)
5. Shen, J., Wu, Z., Lei, D., Shang, J., Ren, X., Han, J.: SetExpan: corpus-based set expansion via context feature selection and rank ensemble. In: Ceci, M., Hollmén, J., Todorovski, L., Vens, C., Džeroski, S. (eds.) ECML PKDD 2017. LNCS (LNAI), vol. 10534, pp. 288–304. Springer, Cham (2017). https://doi.org/10.1007/978-3-319-71249-9_18
6. Wang, D.Z.: Archimedes: efficient query processing over probabilistic knowledge bases. ACM SIGMOD **46**(2), 30–35 (2017)
7. Mikolov, T., Sutskever, I., Chen, K., et al.: Distributed representations of words and phrases and their compositionality. In: NIPS 2013, pp. 3111–3119 (2013)
8. Jiang, H.D., Turki, T., Wang J.T.L.: Reverse engineering regulatory networks in cells using a dynamic bayesian network and mutual information scoring function. In: ICMLA 2017, pp. 761–764 (2017)
9. Amagata, D., Hara, T.: Mining top-k co-occurrence patterns across multiple streams. IEEE Trans. Knowl. Data Eng. **29**(10), 2249–2262 (2017)
10. Ma, H.F., Xing, Y., Wang, S., et al.: Leveraging term co-occurrence distance and strong classification features for short text feature selection. In: KSEM 2017, pp. 67–75(2017)
11. Song, H., Lee, J.G., Han, W.S.: PAMAE: parallel k-medoids clustering with high accuracy and efficiency. In: KDD 2017, pp. 1087–1096 (2017)
12. DBLP Dataset [EB/OL], 20 Apr 2016. http://dblp.uni-trier.de/xml/
13. ICTCLAS, ICTCLAS2012-SDK-0101, rar[EB/OL] (2016). http://www.nlpir.org/download/
14. Ali, C.M., Khalid, S., Aslam, M.H.: Pattern based comprehensive urdu stemmer and short text classification. IEEE Access **6**, 7374–7389 (2018)

Probabilistic Graphical Model Based Highly Scalable Directed Community Detection Algorithm

XiaoLong Deng[✉] [ID], ZiXiang Nie, and JiaYu Zhai

Beijing University of Posts and Telecommunications, Beijing, China
{shannondeng,niezixiang,zhaijiayu}@bupt.edu.cn

Abstract. Community detection algorithms have essential applications for character statistics in complex network which could contribute to the study of the real network, such as the online social network and the logistics distribution network. But traditional community detection algorithms could not handle the significant characteristic of directionality in real network for only concentrating on undirected network. Based on Information Transfer Probability method of classic Probabilistic Graphical Model (PGM) theory from Turing Award Owner Pearl, we propose an efficient local directed community detection method named Information Transfer Gain (ITG) from basic information transfer triangles which composed the core structure of community. Then, aiming at processing the large scale directed social network with high efficiency, we propose the scalable and distributed algorithm of Distributed Information Transfer Gain (DITG) based on GraphX model in Spark. Finally, with extensive experiment on directed artificial network dataset and real social network dataset, we prove that our algorithm have good precision and efficiency in distributed environment compared with some classical directed detection algorithms such as FastGN, OSLOM and Infomap.

Keywords: Distributed computing · Directed community detection · Information transfer gain · Probabilistic graphical mode · Scalable algorithm

1 Introduction

Social network is one of the most important complex networks on the internet and in the last decade it has become into a very huge network connecting people around the world. The form of social networks is diverse, like the Twitter network which is a directed following relationship network and the Facebook network which is an undirected friend network. But the traditional community detection algorithms always focus on analysing undirected social network while the community detection on directed networks is becoming a gigantic challenge at present. In this paper, in order to promote the algorithm running efficiency and accuracy, starting from the basic triple structure of directed community basing on Probabilistic Graphical Model (PGM) theory [1], we propose a new directed community detection algorithm ITG modeling on the directed information transfer process to detect communities precisely in directed networks. Furthermore, in order to handle large scale network data, we develop the parallelized

© Springer Nature Switzerland AG 2019
L. H. U and H. W. Lauw (Eds.): PAKDD 2019 Workshops, LNAI 11607, pp. 322–340, 2019.
https://doi.org/10.1007/978-3-030-26142-9_28

version of ITG and name it Distributed Information Transfer Gain (DITG) algorithm which can be deployed in distributed system with clustering depended time-consuming and acceptable precision. The implementation of DITG is constructed on Spark architecture and GraphX graph architecture. The visualization of the partition result of ITG algorithm implementation is designed as support software to demonstrate the vertex cluster and its densely connected community. Some open-source libraries are used to get a clearer interface. Finally, from the supported visualization result and comparison result of experiment, we achieve the precision and efficiency of ITG algorithm in distributed environment.

Section 2 demonstrates related work of directed community detection algorithm. Section 3 gives the details of Distributed Architecture of designing and implementation of our DITG algorithm. Section 4 shows the result of visualization and comparison result of DITG algorithm in distributed environment with other classic algorithms mentioned in Sect. 2. Section 5 gives the conclusion and further work.

2 Related Work

Directed Community Detection (DCD) is the algorithm to solve the problem in complex network that to find significant vertex groups as the communities which has same attributes or strong connection with each other. The connection shows the information transfer between the focus vertex and its neighbors. The classification of related directed community detection has two main types, which are density-based and pattern-based [2, 20–24]. In this paper, our main focus is on density-based type in order to fit the real truth that the social network considers the density as the community attributes mainly.

In 2004, Newman [3] put forward an algorithm FastGN (FN) which is based on module degree optimization. This algorithm makes use of the Q value gain of each edge exchanging among different communities to find the direction of optimal module degree. And then Newman and Clauset proposed the CNM algorithm [4] based on heap structure to improve FastGN. Complexity of CNM is nearly linear to network scale in large-scale networks.

In 2006, Pons proposed the random walk community detection algorithm Walktrap [5], which was based on the similarity of nodes in large-scale networks. By using the definition of Euclidean distance for the distance among different communities, it has good time complexity.

In 2007, Raghvan improved LPA by providing the RAK algorithm [6], which was based on community detection operation with an approximate linear direct ratio when network scale increased. Through the predefined target function, it simplifies the complexity of LPA (Label Propagation Algorithm) and uses network structure as a guide to detect community structure. However, RAK algorithm has some special drawbacks in benchmark networks experiments, and it needs improvement.

In 2008, Rosvall [8] summarized the introduction of random walk based community detection algorithm in details and set up a model for the probability of information owing in different nodes by using information entropy function in Information Theory and put forward the Infomap algorithm. On the LFR [9] standard dataset, the Infomap algorithm has been proven to better perform than some overlapping community detection algorithms [7, 10, 11, 16, 18].

In 2011, Radicchi [12] proposed a measuring function which used Q value based on Significance function as detection conformity and put forward Order Statistics Local Optimization Method (OSLOM). OSLOM algorithm is the best algorithm for community detection in directed weighting edges networks. In 2014, Prat-Pérez [13] proposed a Scalable Community Detection (SCD) algorithm which is aimed for constructing the target function of undirected community detection algorithm to optimize the target function in iterations. The calculation of the target function Weighted Community Clustering (WCC) is consists of some parameters that can be paralleled. It used paralleling thread to do the implementation of SCD algorithm.

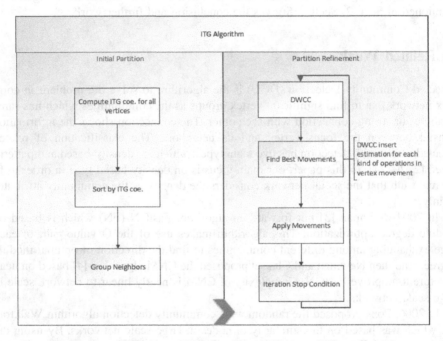

Fig. 1. Local structure of ITG algorithm (ITG coe = ITG coefficient)

However, all the algorithms above do not focus on directed large-scale networks for directed community detection and have some drawbacks, it is necessary to construct a more efficient and parallel algorithm to achieve promotion.

In 2017, we proposed a model of directed vertex clustering coefficient named Information Transfer Gain (ITG) to construct new directed community detection algorithm [26, 27]. Our algorithm has two main steps which are initial partition and partition refinement. Figure 1 shows the algorithm structure of our ITG algorithm. One of the sub-steps in the algorithm has the time-consuming calculation. By researching on the algorithm calculation details, we found that it can be parallelized into distributed system. In this paper, we proposed the distributed framework of ITG algorithm to promotion algorithm running time and efficiency.

3 Distributed Architecture

3.1 Distributed Framework Selection

(1) Spark
Spark is a new cluster computing framework which supports allocations with working sets while providing similar scalability and fault tolerance properties to MapReduce [15]. The key technology to handle abstracted computing resource of Spark is Resilient Distributed Dataset (RDD). RDD is designed to manage the actual resource located in different clusters to reduce time in reading disk file data. It has made great promotion to MapReduce in distributed memory management mechanism.

(2) GraphX
GraphX is a distributed graph computation framework which unifies graph-parallel and data-parallel computation in a single system [17]. GraphX has plenty of advancing characteristics such as GraphX Data Model, operators of graph, graph-cut for storage in Spark environment and so on. In this paper, we focus on the point of the operators of vertex parallel-computing and edge parallel-computing. The details of the two parallel-computing modes can be found in Sect. 3.2.

3.2 Distributed ITG Framework

Our graph structure in GraphX framework is composed of vertex RDD and edge RDD. These two RDDs have the pre-defined operations provided by GraphX for application design. To implement the directed community detection algorithm, we study deep into the source code of GraphX.

GraphX has different data organization methods in vertex and edge data organization from local graph implementation. To fit the Spark data organization method, graph of GraphX will not be stored in a single cluster, but stored in multiple clusters in sub graphs which is the cut version of the whole graph both the vertex table and edge table. Due to the possible information loss in single cluster, the routing table is also stored by each cluster. The corresponding triple in our proposed DITG algorithm is a distributed structure stored in different clusters obviously. So we chose the vertex parallel-computing and edge parallel-computing [17] to make efficiency promotion which can be found in Figs. 2 and 3.

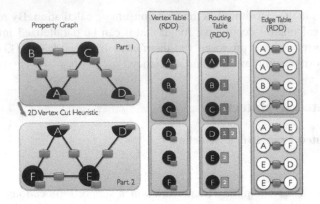

Fig. 2. Vertex routing edge tables

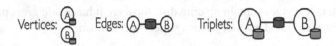

Fig. 3. Vertex, edge and triple model

The vertex parallel-computing is a straightforward approach of traditional graph operations. It regards vertex set as the RDD object and does graph operations like RDD operations to get the result. While the edge parallel-computing is using triple as the basic computing target, it uses the triple to get the idea of map operation of edges, reduce operation of vertex. To do a graph operation with edge parallel computing, the data will be the parameters in map operation of triples. The triple will be mapped to data with edge scope and send the result to the source or destination vertex to do following reduce operation.

3.3 Parallelized Computing Function Units

In distributed designing of ITG algorithm, the following parts require parallelized computing to promote algorithm running speed. The subsections will show the details of those parts with their parallel-computing mode.

(1) Initial Community Partition by ITG

In this part, the main task is to find the ITG coefficient of each vertex. The straight forward consideration is to do vertex parallel-computing and count the corresponding triple and triangle structure of the vertex and aggregate the weighted number with the ITG coefficient calculation formula. The information transfer gain clustering coefficient (ITGC) of node in a directed network can be found in Formula (1) as follows:

$$ITGC_i = \frac{\sum\limits_{t=1}^{15} ITG_{i\_triangle}(t) \times Number(t)}{\sum\limits_{t'=1}^{6} ITG_{i\_triple}(t') \times Number(t')} \tag{1}$$

$ITGC_i$ is the ITG value of node i in a directed network. $\sum\limits_{t=1}^{15} ITG_{i\_triangle}(t) \times$ $Number(t)$ is the weighted number of triangles which use node i as the top vertex (i.e., the information transfer source node), and its weight is the ITG (information transfer gain) contribution $ITG_{i\_triangle}(t)$ from the 15 different types of weighted triangles multiplied by its counted number $Number(t)$. $\sum\limits_{t'=1}^{6} ITG_{i\_triple}(t') \times Number(t')$ is the weighted number of the triples using node i as the top vertex; its weight is the weighted sum of the six $ITG_{i\_triple}(t')$ values of different types of triples multiplied by its counted number $Number(t')$.

But in a practical manner, this approach of ITG coefficient calculation is not available in GraphX framework because of its data store method. The count of the two structures (Part. 1 and Part. 2) in Fig. 2 is so difficult to do the simple operation in independent calculation. So we process this task by using edge parallel-computing with two different types of edge function in Fig. 4 to promote efficiency.

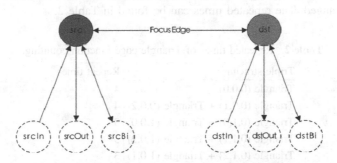

Fig. 4. Edge function operating model

Triple structure edge function has the edge context which is the developed version of triple as the input. The context has the information of the focus edge and the two end-points of the vertices. It has the neighbor sets of the source and destination vertex classified by the edge direction from the neighbor to these two vertices which are SrcIn set, SrcOut set, DstIn set and DstOut set. These sets form the basic relationship of the focus edge and the two end-point vertices. To construct the ITG coefficient of source vertex and destination vertex, there is also a generated set Src/DstBi set to represent the set of neighbors that has bidirectional edge with Src/Dst vertex. Moreover, we count all different arrangements with different Src/Dst set and the focus edge type as the triple result in one triple for one vertex. The next calculation is the reduce part focus on vertices.

Each vertex will get the map result from the message sent by the triple as source vertex or destination vertex. And if the focus edge is bidirectional, it will be calculated twice for the two directions. Table 1 shows the repeated times of all triple structure in Fig. 10.

Table 1. Triple structure edge function counting repeat times.

Triple structure	Repeat times
Triple (0,0)	4
Triple (0,1) + Triple (1,0)	3
Triple (0,2) + Triple (2,0)	3
Triple (1,1)	2
Triple (1,2) + Triple (2,1)	2

The triangle count is similar to triple count but still has two differences. One is the different arrangements which become the different Src/Dst set joint result and the focus edge type. The joint result of those sets can be regard as triangle structure. When the joint result set size is larger than zero, it shows that such arrangement of triangle exists and the set size is the triangle number. The other difference is that the reduce calculation has changed. The repeated times can be found in Table 2.

Table 2. Repeated times of Triangle edge function counting.

Triple structure	Repeat times
Triangle (0,0,0)	4
Triangle (0,0,1) + Triangle (0,0,2)	4
Triangle (0,1,0) + Triangle (1,0,0)	3
Triangle (0,1,1) + Triangle (1,0,2)	3
Triangle (0,1,2) + Triangle (1,0,1)	3
Triangle (0,2,0) + Triangle (2,0,0)	3
Triangle (0,2,1) + Triangle (2,0,2)	3
Triangle (0,2,2) + Triangle (2,0,1)	3
Triangle (1,1,0)	2
Triangle (1,1,1) + Triangle (1,1,2)	2
Triangle (1,2,0) + Triangle (2,1,0)	2
Triangle (1,2,1) + Triangle (2,1,2)	2
Triangle (1,2,2) + Triangle (2,1,1)	2
Triangle (2,2,0)	2
Triangle (2,2,1) + Triangle (2,2,2)	2

Finally, using the ITG calculation formula (1), we can combine all structure of triples and triangles into the ITG coefficient.

(2) DWCC Calculation

DWCC is the Directed Weighted Community Clustering coefficient used in ITG algorithm as the target function. Our algorithm optimizes DWCC as the iteration procedure to do partition refinement. Two functions are used in the calculation of DWCC. One is $t(v, C)$ which is used to calculate the triangle number that vertex v formed within the community C. It also uses the edge function defined in Sect. 3.1. But the joint result of set calculation has been changed into community version which joint the Community C as well to control the result generated within the Community C. The other is $vt(v, C)$ which is used to calculate the number of vertex in the Community C that can form a triangle with vertex v. Although the edge function is also available, the counting object is the vertex number. The calculation of $vt(v, C)$ can be implemented by sending out connection value to neighbor of Vertex v in Spark very quickly. While considering vertex v is the source of the focused edge, when it can form triangle with the destination within the community, it will be added by 0.5. When Vertex v is the destination vertex and the situation is similar, but the value also considers the Focus edge type. For the edge is bidirectional, it will be added 0.5; for the edge is the In type edge only, it can be added only by 0.25.

When constructing the target function of vector influence clustering coefficient model, we referred the definition of Arnau Prat-Prezs [13] in Weighted Community Clustering (WCC). We focus on property of directed graph and have completed the directed improvement of the model to define the new target function as Directed Weighted Community Clustering (DWCC) coefficient. When defining DWCC in the relationship of vertex to community, we defined $wt(x, C)$ as the formed weighted triangle number by vertex x within community C. And we defined $wvt(x, C)$ as the weighted neighbor number which can formed triangles by vertex x within community C.

The weighted triangle number $wt(x, C)$ means that based on the focus vertex x, the formed triangles with the definition of ITG in the triangle structure which can be recorded as a weighted triangle. The weighted node means that the edge of the node and focus vertex x with definition of ITG can be recorded as a weighted node.

In the optimized iteration of target function by Arnau Prat-Prezs [13], our partition refinement step is related to three functions of possible increase of WCC value and the oneness of them which means that three functions can be convert to the calculation of WCC_I. In the process of directional improvement, we also need the estimation of $DWCC_I$ to reduce the time complexity of our algorithm and which can be found in the following formula (2):

$$
\begin{aligned}
DWCC_I'(v, C) &= DWCC(P') - DWCC(P) \\
&= \frac{1}{V} \cdot (d_{in} \cdot \Theta_1 + (r - d_{in}) \cdot \Theta_2 + \Theta_3)
\end{aligned}
\tag{2}
$$

Table 3. Statistical value meaning in $DWCC_I$ estimation.

Parameter	Original statistic meaning	Directed statistic meaning
r	Node Num of Com	Node Num of Com
δ	Edge density of community	Weighted edge density of community
d_{in}	Neighbour Num of node in the community	Weighted neighbour Num of node in the community
d_{out}	Neighbour Num of node out of the community	Weighted neighbour Num of node out of the community
b	Boundary edge Num of community	Weighted boundary edges Num of community
ω	Avg clustering coefficient of whole graph	Avg ITG of whole graph

(Num: Number, Com: Community, Avg: Average)

We promoted the statistic parameters in Table 3 for directional improvement from Formula (1). With the promoted statistics of parameters; Formula (1) has constant time complexity. The update of statistics only occurs when the structure of community has been changed. After promotions and experiments, the whole time complexity of our algorithm of the entire graph is $O(nm)$, which m is the number of times a community structures changes.

(3) Parameters in DWCC **Insert Estimation Calculation**

Parameter d_{in} and d_{out} are two values used in DWCC insert estimation calculation which describes the neighbor of vertex v inside the community C and the neighbor out of the community C. These two parameters also use edge functions to calculate in a simple way. It only finds the weighted neighbor numbers by using value adding like Sect. 3.2.

Another two parameters are also needed to be calculated in parallel-computing, but edge function is not the best implementation. The parameter b in Table 3 describes the weighted boundary edges of community C. The other parameter δ describes the density of community. For these two parameters, we the map function on edge and reduce function on edge as well. For the situation that the source vertex is in the community and the destination vertex is out of the community, b will be added by 0.5. For another situation that the source vertex is out of the community and the destination vertex is in the community, b will be added by 0.25 only. Parameter δ counts the edge number straightforwardly with that the source vertex and the destination vertex are both in the community.

3.4   Parallel Implementation of ITG Algorithm

Figure 5 demonstrates the paralleled ITG algorithm structure in distributed environment with the process operation as distributed design and edge functions mentioned above. The algorithm maintains the two steps of ITG algorithm, the initial partition and partition refinement, but the detail in the two steps has been changed into a distributed version. Figure 5 gives the paralleled structure which improved from the local version mentioned in Fig. 1.

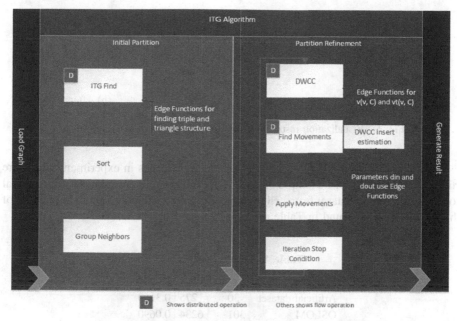

Fig. 5. Distributed implementation architecture of ITG

4   Experiment Result and Comparison

4.1   Experiment Environment

CPU frequency of the Master node is Intel(R) Xeon(R) CPU E5-2440 v2 @ 1.90 GHz. Memory is 16 GB with 4 TB hard disk. JDK version is 1.8.0_131. Spark version is 2.1.0. Hadoop version is 2.7.3. CPU of Slave nodes were Intel64 Family 6 Model 44 Genuine Intel 1584. Their memories are 16 GB each. Eight slave nodes are used in our experiment deployment.

4.2   Experiment Result

The two sub-figures in Fig. 6(a) and (b) show the visualized result of the artificial network dataset in Table 4 before and after ITG algorithm running.

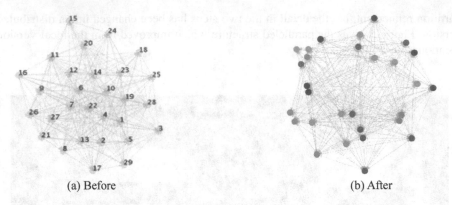

(a) Before   (b) After

Fig. 6. Visualization result of community detection of ITG algorithm

For the comparison result, we adopt four classic datasets in experiment which are artificial dataset, OSLOM example dataset [12], subject reference dataset [8] and real world mobile calling dataset from cell phone calling records in one month in one city of China which can be found in Table 4.

Table 4. Dataset attributes.

Dataset	Vertex	Edge	Density
Artificial dataset	30	275	0.3160
OSLOM	301	6234	0.0690
Subject reference	40	306	0.1962
Calling record	284	3934	0.0489

With the comparison of ITG algorithm with OSLOM, Infomap and FastGN algorithm, the following table gives the result in directed modularity as the index.

Table 5. Directed modularity.

Dataset	FastGN	OSLOM	Infomap	ITG
Artificial dataset	0.3322	Low quality	1.0000	0.3588
OSLOM	0.8651	0.8794	0.9963	0.6910
Subject reference	0.8007	0.8072	0.8007	0.8235
Calling record	0.5486	0.4822	0.3915	0.4080

In addition, Table 5 shows that the precision of ITG algorithm is acceptable and sometimes it is even better than traditional community detection algorithms. And Fig. 7 is the box figure of Table 5 by the statistic indexes that could be found our ITG algorithm is more stable while having the less value scope of Directed Modularity.

While Table 6 shows the time-consuming improvement of Distributed ITG (DITG) algorithm with local ITG algorithm, Table 7 shows the prediction of time-consuming improvement when using eight slave nodes instead of four nodes. The improvement efficiency prediction is based on Amdahl law [25] which could be found in Formula (3). And the distributed task promotion is about 0.25 from the four nodes experiment result in Table 6. While the node number increasing to eight, the efficiency improvement is around 0.28 in Table 7.

$$S_{lanntency}(s) = \frac{1}{(1-p)+p/s} \tag{3}$$

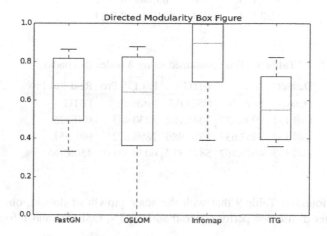

Fig. 7. Directed modularity box of four algorithms

Table 6. Time consumed (Unit: second).

Dataset	ITG time	DITG time	Efficiency improvement
Artificial dataset	1161.507	994.315	23%
OSLOM	3946.397	3234.751	22%
Subject reference	1532.677	1287.964	19%
Calling record	3019.623	2516.353	20%

Table 7. Time consumed of using 8 nodes (Unit: second).

Dataset	ITG	DITG	Est Effi Pro	Real Effi Pro
Artificial dataset	1161.507	907.427	28%	25%
OSLOM	3946.397	3083.123	28%	29%
Subject reference	1532.677	1197.404	28%	26%
Calling record	3019.623	2359.080	28%	29%

(Est: Estimated, Effi: Efficiency, Pro: promotion)

Besides, aim to test the parallel performance of our Distributed ITG algorithm (DITG), we use some real directed large calling record network from calling graph in one southern city of China, and it can be found in Table 8. Then we could find the compared experimental result of DITG with other classical directed community detection algorithm in Table 9.

Table 8. Large directed network datasets.

Dataset	Vertex	Edges	Edges density
Call L-1	13,310	34,591	0.3847
Call L-2	29,624	55,423	0.5345
Call L-3	61,510	65,202	0.9433
Call L-4	512,024	1,021,861	0.5011

Table 9. Time consumed using 8 nodes (Seconds).

Dataset	ITG	DITG	Est Effi Pro	Real Effi Pro
Dataset	FastGN	OSLOM	Infomap	DITG
Call L-1	2906.124	3002.762	3230.453	950.125
Call L-2	2867.634	2994.986	2898.872	901.651
Call L-3	4676.767	5877.877	6030.331	1500.765

It can be found in Table 9 that with the scale growth of dataset, our DITG algorithm has better distributed performance than FastGN, OSLOM and Informap in time consumed.

Finally, we discuss the best calculation conditions for ITG algorithm itself. We create many types of artificial datasets from this experiment. The first topic is the relation of ITG algorithm precision with graph density. The artificial datasets are created with the same vertices number, same probability to have an edge among communities, same community number but different probability to have an edge within a community. We carry out ITG algorithm on these datasets and use Directed Modularity [19], Jaccard [1, 14] and F-1 [1, 14] as the comparison indexes. And we have got the comparison result in Fig. 8.

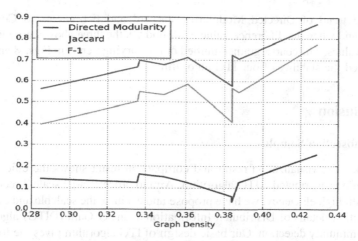

Fig. 8. Relationship of graph density and algorithm precision

Another topic is that the balance of community size has effect on the precision of ITG algorithm or not. We create datasets with the same vertices number, same probability to have an edge within a community, same probability to have an edge among communities and same community number, but different community size. The difference of the largest community size to the smallest community size is varied and experiment result can be found in Fig. 9.

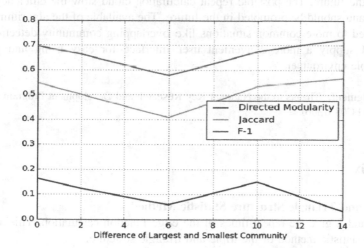

Fig. 9. Community size difference and algorithm precision

It can be found from Figs. 8 and 9 that with the continuing growth of Graph Density, Directed Modularity, Jaccard and F-1 are also increasing while F-1 index has the fastest growth. And when Graph Density is close to value of 0.38, Directed Modularity, Jaccard and F-1 curves all step into a sudden fall which may be attributed to some special dense structure of community composition. Furthermore, we can found the varying trend of Directed Modularity, Jaccard and F-1 curves to Community Difference and when the difference value is around 10, the three above curve will reach their peak values. The coarse-grained unified curve varying trends in Figs. 8 and 9 have demonstrated the stability of our ITG algorithm.

5   Conclusion

5.1   A Subsection Sample

In this paper, the visualization result and comparison result shows the efficiency and precision of our distributed DITG algorithm. Aim to process large scale directed social network with high efficiency, we have propose and evaluate the scalable and distributed implementation details of Distributed Information Transfer Gain (DITG) algorithm in directed community detection. Our basic design of ITG algorithm gives the foundation of implementation and the distributed design shows the engineering part of the implementation which is easy to deploy. The edge function design in DITG implementation constructed on Spark and GraphX for parallel-computing architecture gives more space for other graph related algorithms. With extensive experiment on directed artificial network dataset and real social network dataset, it proves that our algorithm have good precision and efficiency in non-distributed and distributed environment compared with some classical directed detection algorithms such as FastGN, OSLOM and Infomap.

There are still some problems that may influent this architecture which should be solved in the future. The possible repeat calculation cloud slow the efficiency of the algorithm and should be promoted in the future. The available of the algorithm should be expanded to more common situations like overlapping community detection. And we should supply a more convenient user interface for easy using and possible incompatible environment.

Acknowledgment. Thanks to the National Key Research and Development Program of China (No. 2018YFC0831306).

Appendix

A. Triple and Triangle Structure Statistic Method
Tables 1 and 2 give the repeat times of the edge function calculation. This appendix shows the statistic method of the triple and triangle structure.

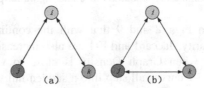

Fig. 10. Two triple forms based vertex i in directed graph

Figure 10 shows the structure basis with three vertices. For the statistic of triple, two numbers are used to represent the two edges connect vertex and its two neighbours vertex and vertex. The number has three versions which are 0, 1 and 2 exactly. 0 represents the bidirectional edge while 1 and 2 represent the out direction edge and in direction edge respectively. The in and out attribute is observed by the focus vertex. When it comes to triangle structure, three numbers are used. The first two numbers remains the meaning. While the third number represents the edge attribute of the opposite edge of vertex. 0 is for bidirectional edge as well. 1 and 2 represent the directions from vertex to vertex and from vertex to vertex respectively. So the statistic of the triple and triangle structure is obvious for the counting of edge function repeat times. We set up the model of the situation in Fig. 10(a), and we can get the all nine ITG figures respectively.

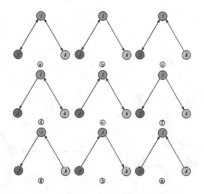

Fig. 11. Vertex based directed triples (all sub graphs of Fig. 10(a))

It can be found in Fig. 11 the basic nine sub graphs of Fig. 10(a) and the other eighteen sub graphs of Fig. 10(b) can be found in Fig. 12. All the twenty seven sub graphs are classified to two types of weighted triangles which is the computation fundamental of Formula (1).

Table 11. All ITG computation in sub graphs of Figure A.10.

Number	Figure number	ITG_i	$ITG_{i \leftrightarrow j}$	$ITG_{i \leftrightarrow k}$
1	Figure 11-a	2	1	1
2	Figure 11-b	1.5	1	0.5
3	Figure 11-c	1.25	1	0.25
4	Figure 11-d	1.5	0.5	1
5	Figure 11-e	1.25	0.25	1
6	Figure 11-f	1	0.5	0.5
7	Figure 11-g	0.75	0.5	0.25
8	Figure 11-h	0.75	0.25	0.5
9	Figure 11-m	0.5	0.25	0.25

B. Parameters Details of Formula (2)

$$\Theta_1 = \frac{((r-1)\delta + 1 + q)(d_{in}-1)\delta}{(r+q)((r-1)(r-2)\delta^3 + (d_{in}-1)\delta + q(q-1)\delta\omega + q(q+1)\omega + d_{out}\omega)} \tag{2-1}$$

$$\Theta_2 = -\frac{(r-1)(r-2)\delta^3}{(r-1)(r-2)\delta^3 + q(q-1)\omega + q(r-1)\delta\omega} \cdot \frac{(r+1)\delta + q}{(r+q)(r-1+q)} \tag{2-2}$$

$$\Theta_3 = \frac{d_{in}(d_{in}-1)\delta}{d_{in}(d_{in}-1)\delta + d_{out}(d_{out}-1)\omega + d_{out}d_{in}\omega} \cdot \frac{d_{in} + d_{out}}{r + d_{out}} \tag{23}$$

$$q = (b - d_{in})/r \tag{2-4}$$

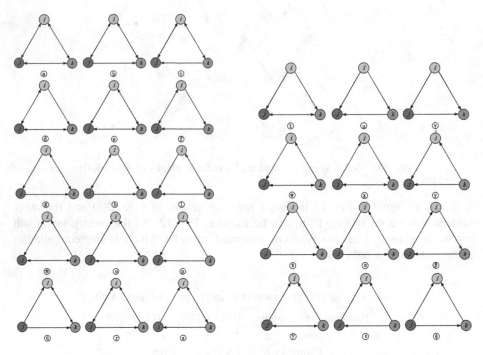

Fig. 12. Vertex based directed triples (all sub graphs of Fig. 10(b))

References

1. Koller, D., Friedman, N.: Probabilistic Graphical Models: Principles and Techniques. MIT Press, Cambridge (2009)
2. Malliaros, F.D., Vazirgiannis, M.: Clustering and community detection in directed networks: a survey. Phys. Rep. **533**(4), 95–142 (2013)

3. Newman, M.E.: Fast algorithm for detecting community structure in networks. Phys. Rev. E **69**(6), 066133 (2004)
4. Clauset, A., Newman, M.E.J., Moore, C.: Finding community structure in very large networks. Phys. Rev. E **70**(6), 066111 (2004). https://doi.org/10.1103/PhysRevE.70.066111
5. Pons, P., Latapy, M.: Computing communities in large networks using random walks. In: Yolum, P., Güngör, T., Gürgen, F., Özturan, C. (eds.) ISCIS 2005. LNCS, vol. 3733, pp. 284–293. Springer, Heidelberg (2005). https://doi.org/10.1007/11569596_31
6. Raghavan, U.N., Albert, R., Kumara, S.: Near linear time algorithm to detect community structures in large-scale networks. Phys. Rev. E **76**(3), 036106 (2007). https://doi.org/10.1103/PhysRevE.76.036106
7. Blondel, V.D., Guillaume, J.L., Lambiotte, R., et al.: Fast unfolding of communities in large networks. J. Stat. Mech: Theory Exp. **10**, P10008 (2008). https://doi.org/10.1088/1742-5468/2008/10/P10008
8. Rosvall, M., Bergstrom, C.T.: Maps of random walks on complex networks reveal community structure. Proc. Natl. Acad. Sci. **105**(4), 1118–1123 (2008). https://doi.org/10.1073/pnas.0706851105
9. Lancichinetti, A., Fortunato, S.: Community detection algorithms: a comparative analysis. Phys. Rev. E **80**(5), 056117 (2009). https://doi.org/10.1103/PhysRevE.80.056117
10. Gregory, S.: Finding overlapping communities in networks by label propagation. New J. Phys. **12**(10), 103018 (2010)
11. Ahn, Y.-Y., Bagrow, J.P., Lehmann, S.: Link communities reveal multiscale complexity in networks. Nature **466**(7307), 761–764 (2010)
12. Lancichinetti, A., Radicchi, F., Ramasco, J.J., Fortunato, S.: Finding statistically significant communities in networks. PLoS One **6**(4), e18961 (2011)
13. Prat-Pérez, A., Dominguez-Sal, D., Larriba-Pey, J.L.: High quality, scalable and parallel community detection for large real graphs. In: The 23rd International Conference on World Wide Web, pp. 225–236. ACM, Seoul (2014). https://doi.org/10.1145/2566486.2568010
14. Levorato, V., Petermann, C.: Detection of communities in directed networks based on strongly-connected components. In: 2011 International Conference on Computational Aspects of Social Networks (CASoN), pp. 211–216. IEEE, Salamanca (2011). https://doi.org/10.1109/cason.2011.6085946
15. Zaharia, M., Chowdhury, M., Franklin, M.J., Shenker, S., Stoica, I.: Spark: cluster computing with working sets. In: Proceedings of the 2nd USENIX Conference on Hot Topics in Cloud Computing, pp. 1–10, Berkeley, CA, USA (2010)
16. Yang, J., Leskovec, J.: Overlapping community detection at scale: a nonnegative matrix factorization approach. In: Proceedings of the Sixth ACM International Conference on Web Search and Data Mining, pp. 587–596. ACM (2013)
17. Xin, R.S., Crankshaw, D., Dave, A., Gonzalez, J.E., Franklin, M.J., Stoica, I.: GraphX: unifying data-parallel and graph-parallel analytics. CoRR abs/1402.2394 (2014)
18. Sun, P.G., Gao, L.: A framework of mapping undirected to directed graphs for community detection. Inf. Sci. **298**, 330–343 (2015)
19. Zhang, X., Martin, T., Newman, M.E.: Identification of core-periphery structure in networks. Phys. Rev. E **91**(3), 032803 (2015)
20. Liu, J., Aggarwal, C., Han, J.: On integrating network and community discovery. In: Proceedings of the Eighth ACM International Conference on Web Search and Data Mining, pp. 117–126. ACM (2015)
21. Newman, M.E.J.: Community detection in networks: modularity optimization and maximum likelihood are equivalent. CoRR abs/1606.02319 (2016)

22. Grover, A., Leskovec, J.: node2vec: scalable feature learning for networks. In: Proceedings of the 22nd ACM SIGKDD International Conference on Knowledge Discovery and Data Mining, pp. 855–864. ACM (2016)
23. Leskovec, J., Sosic, R.: SNAP: a general-purpose network analysis and graph-mining library. ACM Trans. Intell. Syst. Technol. (TIST) 8(1), 1 (2016)
24. Newman, M.E., Clauset, A.: Structure and inference in annotated networks. Nat. Commun 7, 11863 (2016)
25. Amdahl, G.M.: Validity of the single processor approach to achieving large-scale computing capabilities. In: AFIPS Conference Proceedings, no. (30), pp. 483–485 (1967). https://doi.org/10.1145/1465482.1465560
26. Deng, X., Zhai, J.: Efficient vector influence clustering coefficient based directed community detection algorithm. IEEE Access 5, 17106–17116 (2017). https://doi.org/10.1109/access.2017.2740962
27. Deng, X., Dou, Y., Lv, T., Nguyen, Q.V.H.: A novel centrality cascading based edge parameter evaluation method for robust influence maximization. IEEE Access 5, 22119–22131 (2017). https://doi.org/10.1109/access.2017.2764750

Hilltop Based Recommendation in Co-author Networks

Qiong Wu[1], Xuan Ou[2], Jianjun Yu[2(✉)], and Heliang Yuan[3]

[1] Beijing Innovation Center for Mobility Intelligence (BICMI),
Beijing, China
[2] Computer Network Information Center,
Chinese Academy of Science, Beijing, China
yujj@cnic.ac.in
[3] School of Electronic Information and Engineering,
Beihang University, Beijing, China

Abstract. The scale of projects and literatures have been continuously expanded and become more complex with the development of scientific research. Scientific cooperation has become an important trend in the scientific research. Analysis of the co-author network is a big data problem. Without enough data mining, the research cooperation will be limited to some same group, named as "small group" in the co-author networks. This situation has led to the researchers' lack of openness and limited scientific research results. It is important to recommend some potential collaboration from huge amount of literature. We propose a method based on Hilltop algorithm, an algorithm in search engine, to recommend co-authors by link analysis. The candidate set is screening and scored for recommendation. By setting certain rules, the expert set formation of the Hilltop algorithm is added to the screening. And the score is calculated by the durations and times of the collaborations. The co-authors can be extracted and recommended from the big data of the scientific research literatures through the experiments.

Keywords: Co-author network · Community discovery · Modularity · Link analysis · Hilltop algorithm

1 Introduction

The analysis of scientific research papers is a big data processing process. Social Network Analysis (SNA) and Complex Network methods are all derived from graph theory in mathematics. According to the graph theory, the definition of the co-author networks is as follows. The joint network [4] G is composed of a node set N and a link set L, where N represents the collection of the authors and the nodes represent the authors of the scientific papers. If two authors co-author

---

Q. Wu—This research is supported by NSFC Grant No. 61836013 and CAS 135 Informatization Project XXH13504.

L. H. U and H. W. Lauw (Eds.): PAKDD 2019 Workshops, LNAI 11607, pp. 341–351, 2019.
https://doi.org/10.1007/978-3-030-26142-9_29

a paper, there is a link between the two nodes and L represents the collection of all these links. A community discovery and co-author recommendations based on the co-author network can be investigated. With the strengthening of scientific and technological cooperation, the scientific research cooperation has become a major trend in the academic field [2]. In another word, the researchers need to find more partners.

Zhao Juan analyzed the fund project information from computer science subject published by the NSFC website from 1988 to 2010 [10], and concluded that the average cooperation of a project is 2.4477 [5], while Zhao Yandong's survey result was 2.9 [11]. These results were lower than 4.7, the results of Welch and Melker's survey of six subject areas from the United States [9]. Therefore, the Chinese researchers have more demand for the research partners. In [3], it is found that some authors, although collaborating with others for many times, are often limited to the same group. This forms the "small group" of the co-author networks. The "small group" problem will limit the depth of scientific research. Thus, we intend to expand the co-authors' recommendation method to break through the limitations of the "small group" phenomenon and to make effective co-author recommendations for the researchers.

Typical recommendation technologies mainly include content-based recommendation and collaborative screening recommendation method [1]. The co-author network lacks the scholar evaluation information for collaborative screening recommendation. Thus, the content-based method is commonly used. But, the traditional content-based method does not involve the use of network structures. To solve this issue, we introduce the link prediction technology in the data mining area [8]. This technology predicts whether there have or may not have links between two nodes in the network through the known network structure [6]. The methods include the prospect of undiscovered links and the prediction of future possibilities [7]. There are three main types of link prediction algorithms: link prediction algorithm based on graph topology, link prediction algorithm based on the data mining classification algorithm, and link prediction algorithm based on network modeling probability model. The topology-based method is easier to calculate, but the accuracy is lower because it does not combine the self-property of the network topology. The method based on the classification method is difficult to adapt to the dynamic changes of the network in the intermediate stage between training and verification. The probabilistic model-based method combines global information and has high accuracy. It can fully discover potential relationships in the network, but the modeling phase of the algorithm takes a lot of time.

We will introduce the Hilltop algorithm from the search engine to establish a co-author recommendation model. In the subset generation stage of Hilltop, the screening of candidate communities and candidates is added. Then, the scoring sequence is combined with the number and length of cooperation. Finally, effectiveness of the recommendation model is verified on the DBLP dataset (a computer science bibliography website).

2 Co-author Recommendation Model

2.1 Introduction and Improvement of the Hilltop Algorithm

The model should break the "small group" restrictions and expand the network. And, the recommended co-authors should have the possibility to work with the recommended scholars. The co-author scores are recommended outside of the "small group", which is similar to the target page selection and target page rating in the Hilltop algorithm. In Hilltop algorithm, the number and quality of the sources referring to a page are a good measure of the page's quality. The key point of the Hilltop algorithm is it uses the "expert" sources, which have been created with the specific purpose of directing people towards resources. Thus, the Hilltop algorithm in the search engine is introduced into the co-author's recommendation model. Compared with the traditional Hilltop algorithm, the following improvements are made.

First, the generation of expert page subsets is changed. The expert page for the Hilltop algorithm is determined by the relevance of the page content to the topic. While, the close connection between the authors of the co-author network comes from the cooperation. The generation of the subset is changed to the link between the nodes as a measure.

Second, the generation of candidate subsets is changed. The development of contacts is to promote the cooperation. So, the combinations of expert collection and candidate collection are not consider, and the method of recommending friends to generate candidate collection is simply used.

Third, the candidate group and candidate screening links are added. The possibility that the two scholars can cooperate with each other is that their small groups are not compete. So the screening of such "small groups" is necessary. The recommending people who have not cooperated will be more feasible. Therefore, the screening will be carried out in the candidate sets and the "small groups". At the same time, the subsequent calculation amount can be further simplified by screening to improve the efficiency of recommendation.

Fourth, the score is performed by combining the attributes of the joint network. The co-author network is formed with a special factors and nature. The relationship between scholars can be reflected by the number and the recent cooperation time of the cooperation. Thus, the factors such as the number and time of cooperation will be consider in the scoring sorting.

2.2 Screening of Candidate Small Group

The screening of "small groups" for the not competing purpose is shown in Fig. 1. If there is not a connection or a very weak connection between two "small groups", it can be considered to have a competitive relationship between them. Such as, it lack links between the first and the fourth communities. Even if there is no competition, the cost of communication between researcher groups is high, and the effectiveness of recommendation is small. Therefore, we can define a distance between the "small groups". For example, the distance between "small group" Y and "small group" X is defined as the number of paths from the nodes of the "small group" X to the nodes of the "small group" Y within a finite step. There are two ways between the community one and the community two or three. But the number of nodes in the community three is less than that in the community two, these ways have a greater impact on the community three.

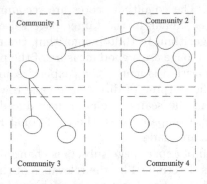

Fig. 1. Candidate "small group" screening map.

Thus, the community three is more closely connected than the community one. The formula for the "small group" screening is shown in Eq. 1.

$$DisG(X,Y) = N_Y / \sum_{x_i \in X, y_i \in Y} path(x_i, y_i) \qquad (1)$$

where N_Y is the number of nodes of the "small group" Y. The "small group" with $DisG(X,Y)$ less than a certain threshold is selected as a candidate.

In scientific research, the recommending people who have never cooperated before seems to be more valuable. So, we use the recommendation method of social network map-based friends in the recommendation system. The example of recommend researchers to the co-authors is shown in Fig. 2.

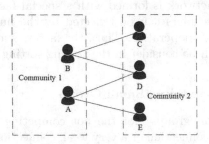

Fig. 2. Candidate collection screening map.

In Fig. 2, recommending E to A does not help the expansion of A's network, when there has been cooperation between A and E. And it has little meaning for generating new ideas. Thus, E is not a good candidate. On the other hand, B and A know each other. Then, B can act as a bridge to recommend people who have worked outside B in this community to A. Therefore, C and D can be recommended to A. But since A and D have cooperated directly, recommending D is a not good choice. However, C does not cooperate with A before, and A can contact C with lower communication cost through B. Therefore, we can recommend C to A in this example.

2.3 Scoring and Sorting the Candidates

To recommend the score outside the "small group", the Hilltop algorithm in the search engine is applied to the co-authors recommendation. It is used to complete the link analysis and realize the scoring of the recommended candidates. The scoring and sorting algorithm are shown in Figs. 3 and 4.

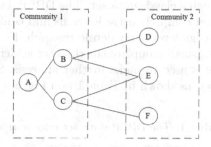

Fig. 3. Score propagation map.

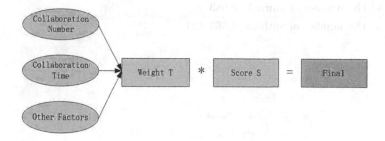

Fig. 4. Score calculation diagram.

In Fig. 3, the community one and the community two are not competing. B and C are the persons who have cooperated with A in the community. Then, B and C are scored by the number and time of cooperation with A. After that, the scores of B and C are generated. On the other hand, D, E and F in Community two have cooperated with B and C, but have not cooperated with A. So, D, E and F are the recommended candidates for A. We combine the scores of B and C by their cooperation time, etc. Then, the transfer of scores from B and C will be performed for E.

Assuming the score of B is S, a weight T can be generated for the link between B and D in Fig. 4. It can be calculated by the number and time of cooperation between B and D. The final score of D is S*T. For the candidate E, the score is carried out by accumulation of the scores from B and C. All recommended candidates are scored and ranked finally in the above manner.

3   Experiment and Verification

3.1   Data Acquisition and Cleanup

The effectiveness of the Hilltop algorithm based co-author recommendation model is validated through a classic data set. The DBLP dataset that we crawled includes high-quality and comprehensive journals and conference papers, which can reflect the cutting-edge level of academic research. First, the DBLP data is processed, and then the "small group" of the data set is identified. The co-author recommendation model is used to make further analysis. Some data is selected from the DBLP data set, as shown in Table 1 and Fig. 5.

Table 1. The DBLP data set information

	Data set parameters	Information
1	Article tag	Those papers that be tagged as #Article
2	Time duration	1993.1.1–2018.12.31
3	The number of article	1,390,758
4	The number of journal	2,083
5	The number of authors	4,563,826

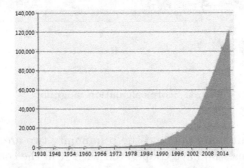

Fig. 5. Distribution of DBLP data sets.

Considering the academic research career and the distribution of papers, the data from year 2000 to 2017 was selected from the data set for experimental verification. The data of the year outside the required time range is filtered, and papers with the length of author less than 2 is also deleted. After cleanup processing, the data is constructed into a joint network. The data from year 2000 to 2012 is set as a training set, while the data from 2013 to 2017 is selected as a verification set. After cleanup and screening, the training set has 172 nodes and 341 edges, and the verification set has 590 nodes and 5695 edges.

3.2   Data Analysis and Detection

After the DBLP data sets are extracted and cleanup, the data can be used to construct a co-author network. The community detection is carried out on the DBLP training set. The result is shown by Force-Directed Graph in Fig. 6.

Fig. 6. Community detection results based on DBLP data sets.

The analysis of the community partitioning results of the DBLP training set is as follows. The modularity is used as the evaluation criteria for community partitioning in terms of partitioning effects. By hierarchical clustering, the number of communities is constantly changing, while the modularity is also changing. As shown in Fig. 7, the community detection based on the DBLP data set takes the highest modularity. The module degree of the result is between 0.5 and 0.6, and the final module degree of this training set is 0.567. Newman has described the standard of optimal community is between 0.4 and 0.7. The community detection of our results is consistent with the ideal result. In terms of the efficiency, the run time of the experiment was 73 ms. The number of iterations of the experiment was 3, which is much smaller than the number of nodes. The final number of communities is 6, which is much smaller than the number of nodes. The number of iterations and the number of communities meet the characteristics of time complexity $t \ll n$ and $k \ll n$. So, the algorithm can complete the community detection with high efficiency.

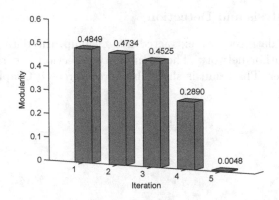

Fig. 7. The change of modularity during the process.

3.3   Verification of the Co-author Recommendation Model

We use the author H. Vincent Poor as an example of the co-author recommendation. In the community from year 2000 to 2012, the author's co-authors in the same community are shown in Table 2.

Table 2. Comparison of the improvement in distance measurement

ID	Author name	ID	Author name
1	Ping Wang	6	Ying-Chang Liang
2	Zhu Han	7	Xiaodong Wang
3	Xuemin Shen	8	Mrouane Debbah
4	Shlomo Shamai	9	Lang Tong
5	Giuseppe Caire	10	Andrea J. Goldsmith

The verifications are carried out in two aspects. First, H. Vincent Poor collaborated with 10 scholars in the same community from year 2000 to 2012. Then, we found that the cooperation between H. Vincent Poor and other 5 scholars continued in year 2013 to 2017. Combining with other factors (such as the academic career of a scholar and the time required to complete the project), it can be seen that the scholars have a strong tendency to continue to cooperate with scholars who have previously cooperated. Therefore, we can verify that there exists the "small group" phenomenon in the co-author network. We also adopts the idea of recommending friends' friends and their subsets.

The detail steps are as following. First, we find a subset of co-authors in the community. Second, we find a group of co-authors who have never cooperated outside the community. Third, the candidates are ranked and sorted based on the number and time of cooperation. Taking H. Vincent Poor as an example, the co-author recommendation results in the final scoring order are shown in Table 3.

Table 3. Co-authors recommended results for H. Vincent Poor

ID	Author name	ID	Author name
1	K. J. Ray Liu	6	Muriel M
2	Eitan Altman	7	Holger Boche
3	Ekram Hossain	8	Fumiyuki Adachi
4	Yonina C. Eldar	9	Mihaela van der Schaar
5	Mohamed-Slim Alouini	10	Moe Z. Win

In the "small group", the scholars have a tendency to continue to cooperate with people who have previously cooperated with. The goal of our research is to solve the limitations of the "small group" and to expand the network. Therefore, the evaluation results are not used for the recommendation, but the novelty is used as the evaluation criteria.

The validity of the recommendation is whether the recommended cooperation of the scholar still exists in future. It was verified that H. Vincent Poor collaborated with authors No. 4 and No. 8 in year 2013 to 2017. It can be seen that most of the referees have reached the goal of expanding their network. Therefore, the recommendation results of the experiment have a good novelty and can provide more choices to the author. Thus, the result is feasible and can provide more communication chances between the scholars.

In summary, the co-author recommendation model can break the "small group" in the co-author network by creating more communication between the scholars. A visualization display of the recommended results is shown in Fig. 8 by using Polar Heatmap from Echarts. Different colors in the figure represent the different ratings and distance from the center and the size of the area. The visualization display makes it easier for the scholars to observe and select their partners.

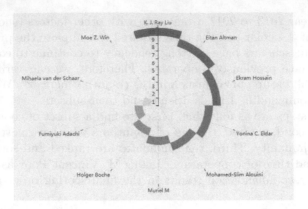

Fig. 8. Visual display of the recommended results.

4   Conclusion

In the paper, we focus on the co-authors recommendation method in the co-author network. The Hilltop algorithm in search engine to link analysis is introduced and modified. Then, the method is applied in co-authors recommendation from three aspects: the "small group" screening, the candidate collection screening, and the candidate scoring. The small group problem of co-author networks was verified through the DBLP dataset. The proposed method is verified by the possibility of cooperation between the recommender and the recommendee. It can be further extend to provide more collaboration suggests to researchers. It will has great significance with development of scientific research in future.

References

1. Cardoso, B., Sedrakyan, G., Gutiérrez, F., Parra, D., Brusilovsky, P., Verbert, K.: IntersectionExplorer, a multi-perspective approach for exploring recommendations. Int. J. Hum. Comput. Stud. **121**, 73–92 (2019)
2. Guimera, R., Amaral, L.A.N.: Functional cartography of complex metabolic networks. Nature **433**(7028), 895 (2005)
3. Lopes, G.R., Moro, M.M., Wives, L.K., de Oliveira, J.P.M.: Collaboration recommendation on academic social networks. In: Trujillo, J., et al. (eds.) ER 2010. LNCS, vol. 6413, pp. 190–199. Springer, Heidelberg (2010). https://doi.org/10.1007/978-3-642-16385-2_24
4. Newman, M.E.: Coauthorship networks and patterns of scientific collaboration. Proc. Nat. Acad. Sci. **101**(Suppl 1), 5200–5205 (2004)
5. Parthasarathy, S., Ruan, Y., Satuluri, V.: Community discovery in social networks: applications, methods and emerging trends. In: Aggarwal, C. (ed.) Social Network Data Analytics, pp. 79–113. Springer, Boston (2011). https://doi.org/10.1007/978-1-4419-8462-3_4
6. Pecli, A., Cavalcanti, M.C., Goldschmidt, R.: Automatic feature selection for supervised learning in link prediction applications: a comparative study. Knowl. Inf. Syst. **56**(1), 85–121 (2018)

7. Ren, Z.-M., Zeng, A., Zhang, Y.-C.: Structure-oriented prediction in complex networks. Phys. Rep. **750**, 1–51 (2018)
8. Wang, J., Yue, F., Wang, G., Xu, Y., Yang, C.: Expert recommendation in scientific social network based on link prediction. J. Intell **34**(6), 151–156 (2015)
9. Welch, E., Melkers, J.: Effects of network size and gender on PI grant awards to scientists and engineers: an analysis from a national survey of five fields. In: Annual Meeting of the Association for Public Policy and Management (APPAM) (2006)
10. Zhao, J., Dong, K., Yu, J., Kai, N.: Social network analysis technologies in e-science. E-Sci. Technol. Appl. (2013)
11. Zhao, Y.-D., Zhou, C.: The cooperation network of Chinese researchers: a perspective of ego-centered social network analysis. Stud. Sci. Sci. **7**, 999–1006 (2011)

Neural Variational Collaborative Filtering for Top-K Recommendation

Xiaoyi Deng[1]([✉]) [iD] and Fuzhen Zhuang[2,3] [iD]

[1] Business School, Huaqiao University, Quanzhou 362021, China
londonbell.deng@gmail.com
[2] Key Lab of Intelligent Information Processing of Chinese Academy of Sciences
(CAS), Institute of Computing Technology, CAS, Beijing 100190, China
zhuangfuzhen@ict.ac.cn
[3] University of Chinese Academy of Sciences, Beijing 100049, China

Abstract. Collaborative Filtering (CF) is one of the most widely
applied models for recommender systems. However, CF-based methods
suffer from data sparsity and cold-start, more attention has been drawn
to hybrid methods by using both the rating and content information.
Variational Autoencoder (VAE) has been confirmed to be highly effective
in CF task, due to its Bayesian nature and non-linearity. Nevertheless,
most VAE models suffer from data sparsity, which leads to poor latent
representations of users and items. Besides, most existing VAE-based
methods model either user latent factors or item latent factors, which
makes them unable to recommend items to a new user or recommend
a new item to existing users. To address these problems, we propose a
novel deep hybrid framework for top-K recommendation, named Neural
Variational Collaborative Filtering (NVCF), where user and item side
information is incorporated into the generative processes of user and
item, to alleviate data sparsity and learn better latent representations of
users and items. For inference purpose, we derived a Stochastic Gradient
Variational Bayes (SGVB) algorithm to approximate the intractable distributions of latent factors of users and items. Experiments performed on
two public datasets have showed our method significantly outperforms
the state-of-the-art CF-based and VAE-based methods.

Keywords: Neural collaborative filtering · Variational autoencoder ·
Top-K recommendation · Implicit feedback

1 Introduction

Recommender systems can help users to discover their potentially preferences
from varieties of items on the basis of their tastes. Collaborative filtering (CF) is

Supported by the National Natural Science Foundation of China (Nos. 71401058,
61773361) and the Program for New Century Excellent Talents in Fujian Province
University (NCETFJ) (No. Z1625110).

L. H. U and H. W. Lauw (Eds.): PAKDD 2019 Workshops, LNAI 11607, pp. 352–364, 2019.
https://doi.org/10.1007/978-3-030-26142-9_30

one of the key techniques to build a personalized recommender systems, due to its accuracy and scalability [1]. The essence of CF is to infer users' preferences from the behavior data of themselves and other users. Most traditional CF methods are based on matrix factorization (MF) [2,3], which projects users and items into a shared latent space and uses a latent feature vector to represent either a user or an item. However, these MF models suffer from data sparsity, so that the accuracy of user/item latent representations is limited. To address the problem of data sparsity, many researches incorporate users' and items' side information into traditional MF. To extract more latent factors of side information, previous works employ latent Dirichlet allocation (LDA) [4,5], denoising autoencoder [6, 7], marginalized denoising autoencoder [8] and stacked denoising autoencoder [9] to model side information of users or items. However, these methods use inner product to model interactions between users and items, which limits their power of capturing non-linearity [10]. To model nonlinear interaction, varieties of approaches apply deep neural networks to model these interactions and achieve promising performance, such as Neural Collaborative Filtering (NCF) [10], Deep Matrix Factorization (DMF) [11], Neural Factorization Machine (NFM) [12], DeepFM [13], JRL [14], GCMC [15], Irgan [16] and ConvNCF [17]. Nevertheless, these deep neural networks cannot capture the uncertainty of the users' and items' latent representations.

Recently, some works have take advantage of deep generative models, such as Variational Autoencoder (VAE), to perform CF task. The VAE is nonlinear probabilistic model that has the capability of capturing uncertainty, and its non-linearity enable it to explore nonlinear probabilistic latent-variable models on large-scale recommendation datasets, such as Collaborative Variational Autoencoder (CVAE) [19], CLVAE [20] and VAECF [21]. Despite the effectiveness of these VAE-based CF methods, there are still several drawbacks. CVAE directly uses inner product to model interaction hinders itself to learn nonlinear interactions between users and items. Both CLVAE and VAECF work through modeling users' behavior, which makes them cannot recommend an item to a new user. VAECF chooses the same Gaussian prior for all users and only exploit user-item feedback matrix, which leads to poor latent representations of users and items, and poor performance as the matrix sparsity is very high [22].

Consequently, to solve the problems mentioned above, we propose a deep hybrid model, Neural Variational Collaborative Filtering (NVCF), for top-K recommendation. Unlike the generative processes of user or item in most existing VAE-based methods, we model the generative process from both users and items through a unified neural variational model, which can effectively learn nonlinear latent representations of users and items for CF. The side information of users and items is incorporated into their latent factors through a neural network, which means NVCF can mitigate data sparsity and model better latent representations of users and items. The parameters of prior neural network are learned from data, leading to the fact that it is able to embed users' better preferences and items' features into latent factors of users and items, respectively. For inferring the posterior of latent factors of users and items, we derived a Stochastic Gradient Variational Bayes (SGVB) algorithm to infer these posterior, which makes the parameters of our model can be effectively learned by back-propagation. The rest of this paper is arranged as follows: Sect. 2 provides an overview of related works. Section 3 presents our model, and the parameters learning process is discussed. Section 4 shows experimental results and discussions, followed by conclusions and future work in Sect. 5.

2 Related Work

In recent years, the deep learning methods have attained tremendous achievements in various fields [23,24]. Due to the abilities of neural networks to discover nonlinear, subtle relationships in user-item feedbacks, many works utilize neural networks to address the task of CF. To incorporate item content information into item latent factors, collaborative deep learning (CDL) [6] was proposed to integrate SDAE into probabilistic matrix factorization (PMF). Deep Collaborative Filtering Framework [8] was proposed for unifying deep learning approaches with CF, which embeds the content information of items and users while CDL only considers the effects of item features. Recently, the additional stacked denoising autoencoder (aSDAE) [9] was presented to incorporate side information into MF, which jointly performs deep users and item latent factors learning from side information and collaborative filtering from the user rating. GCMC [15] considers the recommendation problem as a link prediction task with graph CNNs, which can easily integrate user/item side information (such as social networks and item relationships) into the recommendation model.

Since the above methods use inner product to model the interaction of users and items, they are not able to capture the complex structure of the interaction data between users and items. NCF framework [10] was proposed to make use of both linearity of MF and non-linearity of MLP to capture the nonlinear relationship between users and items. NFM [12] was demonstrated using Bi-Interaction layer to incorporate both feedback information and content information. Based on factorization machines, an end-to-end model DeepFM [13] was presented, which seamlessly integrates factorization machine and MLP and can model the high-order feature interactions via deep neural network and low-order interactions via factorization machine. For joint representations of user and items, JRL [14] places a MLP above the element-wise product of user embedding and item embedding, where side information is adopted to learn the corresponding user and item representations based on deep representation learning architectures. More elaborate reviews of deep learning based recommender systems can be seen in [25]. Due to the power of capturing uncertainty and non-linearity of deep generative model [18], some works utilize deep generative model to address the task of CF. CVAE [19] applies VAE to incorporate item content information into MF. CLVAE [20] incorporates auxiliary information to improve performance. VAECF [21] directly utilizes VAE for CF task. Unlike previous VAE-based recommendation methods, we model the generative process of users and items through a unified neural variational framework, which makes our model to be able to capture nonlinear user/items latent representations.

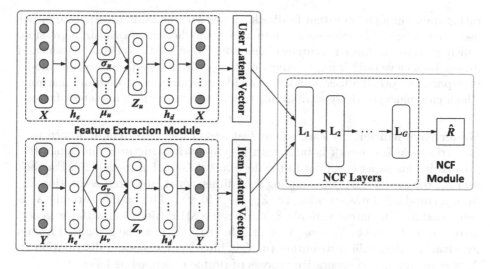

Fig. 1. The graphical model of NVCF

3 Neural Variational Collaborative Filtering

In this section, we present the neural variational collaborative filtering model (NVCF), as shown in Fig. 1. The NVCF contains two main components: the feature extraction module and the NCF module. In feature extraction process, NVCF learns and extracts user and item features through a unified deep generative framework. Then, the user and item latent vectors are fed into the NCF module to learn the user-item relation, and finally generate the rating prediction.

3.1 Notations

Given M users and N items, the user and item latent factors are denoted by $U = \{u_i | i = 1, \ldots, M\} \in \mathbb{R}^{K \times M}$ and $V = \{v_j | j = 1, \ldots, N\} \in \mathbb{R}^{K \times N}$ respectively, where K denotes the dimensions of latent factors. For implicit feedback, the user-item rating matrix is denoted by $R \in \mathbb{R}^{M \times N}$, where $R_{ij} = 1$ indicates that the i-th user has interacted with the j-th item, otherwise $R_{ij} = 0$. The user's and item's side information are denoted by two "bag-of-items" vectors over users and items, $X = \{X_i | i = 1, \ldots, M\} \in \mathbb{R}^{P \times M}$ and $Y = \{Y_j | j = 1, \ldots, N\} \in \mathbb{R}^{Q \times N}$ respectively, where P and Q are the dimensions of user side information and item side information respectively. Here, we call X and Y latent profile representation and latent content representation, respectively. Given R, X and Y, the problem is to infer user latent factor u_i and item latent factor v_j, to predict the missing ratings $\hat{R}$.

3.2 Feature Extraction

As mentioned in [21], most MF-based models assume that the prior distributions of user and item latent factors are standard Gaussian distributions, and predict rating only through user-item feedback. Some MF methods incorporate either user's or item's side information into rating prediction via linear regression, which leads to the limited accuracy of inferring latent relations between users and items. To achieve further improvement on the prediction performance, our model incorporates both the user's and item's side information into feature learning, which can make positive contributions to inferring user and item latent factors.

Generative Model. To learn robust features of user and item, a unified neural variational framework is built with a symmetric structure. The generative process in this paper is similar to the deep latent Gaussian model [26]. For each user u_i, the model starts by sampling a K-dimensional latent representation Z_{u_i} from a standard Gaussian prior, i.e. $Z_{u_i} \sim N(0, \mathbb{I}^K)$. The sample variable X_i is generated from its latent variable Z_{u_i} through a MLP (decoder) with the generative parameter θ, i.e. $X_i \sim p_\theta(X_i|Z_{u_i})$. The $p_\theta(X|Z_u)$ can be generated from a multivariate Bernoulli distribution (if X is binary) or Gaussian distribution (if X is real-value). The generative process of profile is defined as follows:

(1) For each layer $l \in [1, L]$ of the generative network,
   (a) For each column n of weight matrix W_l^d, draw $W_{l,n}^d \sim N(0, \lambda_w^{-1}\mathbb{I}_K)$
   (b) Draw bias vector $b_l^d \sim N(0, \lambda_w^{-1}\mathbb{I}_K)$
   (c) For each row i of h_l^d, draw $h_{l,j}^d \sim N(\sigma(h_{l-1,j}^d W_l^d + b_l^d), \lambda_s^{-1}\mathbb{I}_K)$

(2) For each X_i,
   (a) If X_i is binary, draw $X_i \sim B(\sigma(h_L^d W_l^d + b_{L+1}^d))$
   (b) If X_i is real-value, draw $X_i \sim N(h_L^d W_l^d + b_{L+1}^d, \lambda_X^{-1}\mathbb{I}_K)$ where λ_w, λ_s and λ_X are hyperparameters, $h(h_l^d)$ represents hidden layers of decoder.

Similar to SDAE, λ_s is taken to infinity for computational efficiency.

The latent representation Z_{u_i} can be drawn by a Gaussian prior distribution with zero mean and identity matrix: $Z_{u_i} \sim N(0, \mathbb{I}_K)$. The user latent representation u_i consists of latent user offset and the latent profile vector: $u_i = \epsilon_i + Z_{u_i}$. The generative process of content is similar to that of profile, and the item latent representation v_j is composed of latent item offset and the latent content vector: $v_i = \epsilon_j + Z_{v_j}$.

Inference Model. The inference model is also MLP (encoder) corresponding to the one in the generative model. For user, the inference process is to approximate the intractable posterior distribution $p_\theta(Z_{u_i}|X_i)$ which is determined by the generative network. Using the Stochastic Gradient Variational Bayes (SGVB) estimator, the posterior of latent profile variable Z_u can be approximated by a tractable variational distribution $q_\phi(Z_{u_i}|X_i)$.

$$q_\phi(Z_u|X_i) = N(\mu_\phi(X_i), diag(\sigma_\phi^2(X_i))) \tag{1}$$

where $\mu_\phi \in \mathbb{R}^K$ and $\sigma_\phi^2 \in \mathbb{R}^K$ are the mean and standard deviation of the approximate posterior respectively, which are nonlinear functions of X_i and the variational parameter ϕ. They are outputs of the inference mode.

Similar to [18,21], the inference process of Z_u is defined as follows:

(1) For each layer l of the inference model,
    (a) For each column n of weight matrix W_l^e, draw $W_{l,n}^e \sim N(0, \lambda_w^{-1}\mathbf{I}_K)$
    (b) Draw bias vector $b_l^e \sim N(0, \lambda_w^{-1}\mathbf{I}_K)$
    (c) For each row j of h_l^e, draw $h_{l,j}^e \sim N(\sigma(h_{l-1,j}^e W_l^e + b_l^e), \lambda_s^{-1}\mathbf{I}_K)$

(2) For each user u_i,
    (a) Draw latent mean vector $\mu_j \sim N(h_L^e W_\mu^e + b_\mu^e, \lambda_s^{-1}\mathbf{I}_K)$
    (b) Draw latent covariance vector $\log \sigma_j^2 \sim N(h_L^e W_\sigma^e + b_\sigma^e, \lambda_s^{-1}\mathbf{I}_K)$
    (c) Draw latent content vector $Z_u \sim N(\mu_j, diag(\sigma_j^2))$

As explained in [21], the evidence lower bound (ELBO) for X_i can be estimated using SGVB estimator:

$$\mathcal{L}(\theta, \phi; X_i) = \mathbb{E}_{q_\phi(Z_u|X_i)}[\log p(u_i|Z_u) + \log p_\theta(X_i|Z_u)]$$
$$- \eta \cdot \mathbb{KL}(q_\phi(Z_u|X_i)\|p(Z_u))$$
$$\simeq \log p(u_i|Z_{u_i,l}) + \frac{1}{L}\sum_{l=1}^{L} \log p_\theta(X_i|Z_{u_i,l})$$
$$- \eta \cdot \mathbb{KL}(q_\phi(Z_u|X_i)\|p(Z_u)) \tag{2}$$
$$\mathbb{KL}(q_\phi(Z_u|X_i)\|p(Z_u)) = \frac{1}{2}\sum_{i=1}^{M}(\mu_i^2 + \sigma_i^2 - \log \sigma_i^2 - 1)$$
$$Z_{u_i,l} = \mu_i + \sigma_i \odot \varepsilon_l$$

where $\mathbb{KL}$ denotes the Kullback-Leibler divergence, $\eta \in [0,1]$ is a parameter to control the regularization strength for addressing the posterior collapse problem [27], $\varepsilon_l \sim N(0, \mathbb{I})$, and $\odot$ represents the element-wise product.

The inference process of content is similar to profile inference process, and the ELBO for item network can be derived similarly:

$$\mathcal{L}(\theta, \phi; Y_j) = \mathbb{E}_{q_\phi(Z_v|Y_j)}[\log p(v_j|Z_v) + \log p_\theta(Y_j|Z_v)]$$
$$- \eta \cdot \mathbb{KL}(q_\phi(Z_v|Y_j)\|p(Z_v))$$
$$\simeq \log p(v_j|Z_{v_j,l'}) + \frac{1}{L}\sum_{l=1}^{L} \log p_\theta(Y_j|Z_{v_j,l'})$$
$$- \eta \cdot \mathbb{KL}(q_\phi(Z_v|Y_j)\|p(Z_v)) \tag{3}$$
$$\mathbb{KL}(q_\phi(Z_v|Y_j)\|p(Z_v)) = \frac{1}{2}\sum_{j=1}^{N}(\mu_j^2 + \sigma_j^2 - \log \sigma_j^2 - 1)$$
$$Z_{v_j,l'} = \mu_j + \sigma_j \odot \varepsilon_l$$

3.3   Generalized Matrix Factorization with Side Information

Inspired by the NCF, we propose a new generalized matrix factorization model (GMF) with side information, called SGMF, to improve prediction performance. The collaborative filtering module of SGMF utilizes a computational method similar to the inner product of MF. SGMF utilizes the extracted user and item features to calculate the element-wise product of the user and item latent vectors, and outputs the calculated vectors to a fully connected neural layer. The element-wise products of the user and item latent vectors in the first neural CF layer are defined as follows:

$$\Psi_1(u_i, v_j) = u_i \odot v_j \tag{4}$$

Then, SGMF projects the vectors to the output layer:

$$\hat{R}_{ij} = a_{out}(h^\top \Psi(u_i, v_j)) = a_{out}(h^\top(u_i \odot v_j)) \tag{5}$$

where a_{out} and h denote the activation function and edge weights of the output layer, and $\hat{R}_{ij}$ represents the predicted rating. Intuitively, SGMF is equivalent to MF, as a_{out} is an identity function and h is a uniform vector of 1.

Under the NVCF framework, a_{out} can be a nonlinear activation function and h can be learned from training data, so SGMF has more powerful learning capability than MF. Unlike the original GMF only relying on implicit feedback, SGMF incorporates both user and item side information into latent representations learning, and employs VAE to obtain user and item latent vectors, which can lead to better performance.

3.4   Optimization

Generally, loss functions consist of the reconstruction error in feature extraction and the prediction error. The loss of feature extraction contains user and item features extraction, and the loss functions of VAEs for user and item feature extraction are equivalent to their ELBOs, respectively. For convenience, the ELBOs of for user and item network are denoted by $\mathcal{L}_u$ and $\mathcal{L}_v$, respectively.

Similar to NCF, the loss function of SGMF is defined as follows.

$$\mathcal{L}_{\text{SGMF}} = - \sum_{(i,j)\in\mathcal{T}\cup\mathcal{T}^-} R_{ij} \log \hat{R}_{ij} + (1 - R_{ij}) \log(1 - \hat{R}_{ij}) \tag{6}$$

where $\mathcal{T}$ denotes the set of observed instances and $\mathcal{T}^-$ denotes a set of negative instances, which can be sampled from unobserved user-item interactions. This objective function is the same as binary cross-entropy loss, which is appropriate for binary classification problems.

Thus, the general loss function for training NVCF is defined as,

$$\mathcal{L} = \mathcal{L}_{\text{SGMF}} + \alpha \cdot \mathcal{L}_u + \beta \cdot \mathcal{L}_v \tag{7}$$

where α and β denote the hyper parameters of the loss function.

3.5   Prediction

After the model training and parameters learning, we can predict the probability that user will rate an item for a user-item pair (u_i, v_i). Given a trained model, for a user-item pair (u_i, v_i) with no observed relation, the predicted rating can be written as:

$$\hat{R}_{ij} = a_{out}(h^\top(u_i \odot v_j)) = a_{out}(h^\top(\epsilon_i + Z_{u_i}) \odot (\epsilon_j + Z_{v_j})) \tag{8}$$

4   Experiments and Results

4.1   Experimental Settings

Datasets. In this section, two public datasets from GroupLens are collected to evaluate our model, which are MovieLens-100K and MovieLens-1M.

The MovieLens-100K (ML100K) and MovieLens-1M (ML1M)datasets have been widely utilized to evaluate CF-based recommendation algorithms. The former one contains 943 users and 1,682 movies with 100,000 ratings, while the latter one includes 6,040 users and 3,706 movies with 1,000,209 ratings. Each rating value is on a scale of 1 to 5, and each user has rated at least 20 movies. These two datasets are explicit feedback data, while our goal is to investigate the performance of learning from the implicit feedback. Thus, the MovieLens-100K and MovieLens-1M is transformed into implicit data, where each entry is marked as 1 if the corresponding rating is no less than 4, otherwise marked as 0. For side information, user demographics including age, occupation and gender are regarded as collaborative information, and movie descriptions (genre) are taken as auxiliary item information. Table 1 summarizes the characteristics of MovieLens datasets.

Table 1. Statistics of two MovieLens datasets

Dataset	Users	Items	Ratings	Sparsity	User features	Item features
ML100K	943	1,682	100,000	93.70%	Demographics	Genres
ML1M	6040	3,706	1,000,209	95.53%	Demographics	Genres

Baselines and Evaluation Metrics. To evaluate the proposed NVCF model, six representative CF models are selected as baselines.

BPR [28] optimizes the MF model with a pairwise ranking loss, which is tailored to learn from implicit feedback. It is a highly competitive baseline for item recommendation.

mDACF [8] employs SDAE to extract features from user and item auxiliary information and uses MF to determine user-item latent relations.

NeuMF [10] is a model proposed within the NCF framework, which combines hidden layer of GMF and MLP to learn the user-item interaction function.

NFM [12] generalizes factorization machines for CF, and combines the factorization machine and neural network to incorporate both feedback information and content information.

CVAE [19] is a Bayesian generative model that jointly models CTR and deep generative model to bridge auxiliary information together with deep architecture.

VAECF [21] is a state-of-the-art method that directly apply VAE to CF to for implicit feedback.

To evaluate the performance of our model, two common evaluation metrics for top-K recommendation are adopted: Hit Radio (HR) [10] and Normalized Discounted Cumulative Gain NDCG [30]. HR@k is a recall-based metric, measuring whether the testing item is in the top-K position. NDCG@k assigns the higher scores to the items within the top-K positions of the ranking list.

Parameter Settings. For the training set, we sampled four negative instances for each positive instance. We randomly initialized the model parameters using a Gaussian distribution with mean of 0 and standard deviation of 0.01. Similar to [29], we used a mini-batch Adam method to optimize the model and set the learning rate to 0.001 and the batch size to 128. In the feature extraction step, K is set to 128. The two generative networks both are two latent layers with Relu activation. The last layer of generative network is sigmod activation, and the parameter η is set to be 0.2 in order to achieve the best performance of VAE, according to [21]. The two prior networks are one latent layer. In neural collaborative step, the latent vector dimension is defined as the number of neurons in the last neural collaborative filtering layer of neural collaborative filtering.

4.2   Experimental Results

In our experiments, each dataset is split into two parts: training datasets and testing datasets. For the training set, experiments are carried out with a setting of 80% random sample of each user ratings, and the rest of user ratings (20%) are used for testing. Tables 2 and 3 list the top-K recommendation performance of all methods on the MovieLens datasets, in terms of HR@5/NDCG@5 and HR@10/NDCG@10, respectively.

Table 2. The performance comparison between all methods on HR@5 and NDCG@5

Dataset	Metrics	BPR	mDACF	NeuMF	NFM	CVAE	VAECF	NVCF
ML100K	HR@5	0.4789	0.4706	0.4944	0.5053	0.4720	0.5039	**0.5082**
	NDCG@5	0.3185	0.3268	0.3356	0.3392	0.3181	0.3407	**0.3489**
ML1M	HR@5	0.5305	0.5391	0.5489	0.5624	0.5390	0.5646	**0.5709**
	NDCG@5	0.3642	0.3688	0.3866	0.3887	0.3764	0.3924	**0.4003**

Table 3. The performance comparison between all methods on HR@10 and NDCG@10

Dataset	Metrics	BPR	mDACF	NeuMF	NFM	CVAE	VAECF	NVCF
ML100K	HR@10	0.6233	0.6227	0.6416	0.6554	0.6248	0.6542	**0.6610**
	NDCG@10	0.3502	0.3619	0.3851	0.3990	0.3596	0.4005	**0.4093**
ML1M	HR@10	0.6819	0.6962	0.6993	0.7095	0.6971	0.7178	**0.7239**
	NDCG@10	0.4117	0.4206	0.4368	0.4428	0.4234	0.4403	**0.4495**

Table 4. The performance comparison between selected methods in cold-start scenarios on NDCG@5

Dataset	Scenario	mDACF	NFM	CVAE	NVCF
ML100K	Cold-U	0.1651	0.1820	–	**0.2014**
	Cold-V	0.1439	0.1777	0.1382	**0.1906**
ML1M	Cold-U	0.1918	0.2269	–	**0.2412**
	Cold-V	0.1879	0.1951	0.1567	**0.2133**

From Tables 2 and 3, we can find that most neural network-based methods (NeuMF, NFM, VAECF and NVCF) outperform linear baselines, which demonstrates that deep neural network can help to achieve more subtle and better latent user and item representations. It is clear that the NVCF achieves the best performance on all datasets with two metrics, which indicates the effectiveness of NVCF to perform CF task. It is also found that the VAE-based method (VAECF and NVCF) achieve promising performance, which means the Bayesian nature and non-linearity of neural network can facilitate inferring better latent preferences of users and items. Although based on VAE, the NVCF outperforms VAE-CF and CVAE in terms of all datasets and all metrics, which shows the advantage of our VAE-boosted NCF framework.

To evaluate our model on different cold-start scenarios, we form evaluation sets in different cold ratios. For 30% cold users, we random choose 30% samples in the test sets and give each sample a specific user id only for the sample. We evaluate our model in 30% user cold (Cold-U) and 30% item cold (Cold-V) scenarios on all datasets with NDCG@5. BPR, NeuMF and VAECF only use feedback information and are not able to manage cold-start scenario well, so they are not compared with NVCF.

Table 4 shows the performance of NVCF and other hybrid methods in different cold-start scenarios. Because CVAE cannot handle cold user problem, we do not conduct experiments in cold user scenario. It is obvious that NVCF significantly outperforms other methods in the scenarios of both cold items and cold users, which indicates that using neural network to model interactions between users and items works better than those of simply using inner product.

5   Conclusion

In this paper, we proposed a new recommendation algorithm, NVCF, which is a unified deep generative model for hybrid collaborative filtering. The NVCF models both users' and items' generative process, which enables it to generate recommendation under cold-start scenario. Our method incorporates users' and items' side information through neural networks, to mitigate data sparsity and facilitate modeling users' and items' features. For inference, we proposed a SGVB algorithm to approximate posteriors of the latent variables related to users and items. Due to Bayesian nature and non-linearity, NVCF can learn better user and item latent factors and deal with the cold-start problem via a full Bayesian probabilistic view. Experimental results show that the NVCF achieve the best performance and can effectively handles cold-start problem. In future work, we plan to incorporate more auxiliary information to further improve the recommendation precision, such as adopting knowledge graph.

References

1. Shi, Y., Larson, M., Hanjalic, A.: Collaborative filtering beyond the user-item matrix: a survey of the state of the art and future challenges. ACM Comput. Surv. **47**(1), 3 (2014)
2. Mnih, A., Salakhutdinov, R.R.: Probabilistic matrix factorization. In: Advances in Neural Information Processing Systems, pp. 1257–1264 (2008)
3. Koren, Y., Bell, R., Volinsky, C.: Matrix factorization techniques for recommender systems. IEEE Comput. **42**(8), 30–37 (2009)
4. Zhong, J., Li, X.: Unified collaborative filtering model based on combination of latent features. Expert Syst. Appl. **37**(8), 5666–5672 (2010)
5. Wang, C., Blei, D.M.: Collaborative topic modeling for recommending scientific articles. In: Proceedings of the 17th ACM SIGKDD International Conference on Knowledge Discovery and Data Mining, pp. 448–456 (2011)
6. Wang, H., Wang, N., Yeung, D.Y: Collaborative deep learning for recommender systems. In: Proceedings of the 21st ACM SIGKDD International Conference on Knowledge Discovery and Data Mining, pp. 1235–1244 (2015)
7. Ying, H., Chen, L., Xiong, Y., Wu, J.: Collaborative deep ranking: a hybrid pairwise recommendation algorithm with implicit feedback. In: Proceedings of the 20th Pacific-Asia Conference on Knowledge Discovery and Data Mining, pp. 555–567 (2016)
8. Li, S., Kawale, J., Fu, Y.: Deep collaborative filtering via marginalized denoising auto-encoder. In: Proceedings of the 24th ACM International on Conference on Information and Knowledge Management, pp. 811–820 (2015)
9. Dong, X., Yu, L., Wu, Z., Sun, Y., Yuan, L., Zhang, F.: A hybrid collaborative filtering model with deep structure for recommender systems. In: Proceedings of 31st AAAI Conference on Artificial Intelligence, pp. 1309–1315 (2017)
10. He, X., Liao, L., Zhang, H., Nie, L., Hu, X., Chua, T.S.: Neural collaborative filtering. In: Proceedings of the 26th International Conference on World Wide Web, pp. 173–182 (2017)
11. Xue, H.J., Dai, X.Y., Zhang, J., Huang, S., Chen, J.: Deep matrix factorization models for recommender systems. In: Proceedings of the 26th International Joint Conference on Artificial Intelligence, pp. 3203–3209 (2017)

12. He, X., Chua, T.S.: Neural factorization machines for sparse predictive analytics. In: Proceedings of the 40th International ACM SIGIR Conference on Research and Development in Information Retrieval, pp. 355–364 (2017)
13. Guo, H., Tang, R., Ye, Y., Li, Z., He, X.: DeepFM: a factorization-machine based neural network for CTR prediction. In: Proceedings of the 26th International Joint Conference on Artificial Intelligence, pp. 1725–1731 (2017)
14. Zhang, Y., Ai, Q., Chen, X., Croft, W.B.: Joint representation learning for top-n recommendation with heterogeneous information sources. In: Proceedings of the 2017 ACM on Conference on Information and Knowledge Management, pp. 1449–1458 (2017)
15. Berg, R.V.D., Kipf, T.N., Welling, M.: Graph convolutional matrix completion. In: Proceedings of the 23rd ACM SIGKDD International Conference on Knowledge Discovery and Data Mining, pp. 1–7 (2018)
16. Wang, J., et al.: IRGAN: a minimax game for unifying generative and discriminative information retrieval models. In: Proceedings of the 40th International ACM SIGIR conference on Research and Development in Information Retrieval, pp. 515–524 (2017)
17. He, X., Du, X., Wang, X., Tian, F., Tang, J., Chua, T.S.: Outer product-based neural collaborative filtering. In: Proceedings of the 27th International Joint Conference on Artificial Intelligence, pp. 2227–2233 (2018)
18. Kingma, D.P., Welling, M.: Auto-encoding variational bayes. arXiv preprint arXiv:1312.6114 (2013)
19. Li, X., She, J.: Collaborative variational autoencoder for recommender systems. In: Proceedings of the 23rd ACM SIGKDD International Conference on Knowledge Discovery and Data Mining, pp. 305–314 (2017)
20. Lee, W., Song, K., Moon, I.C.: Augmented variational autoencoders for collaborative filtering with auxiliary information. In: Proceedings of the 2017 ACM on Conference on Information and Knowledge Management, pp. 1139–1148 (2017)
21. Liang, D., Krishnan, R.G., Hoffman, M.D., Jebara, T.: Variational autoencoders for collaborative filtering. In: Proceedings of the 2018 World Wide Web Conference, pp. 689–698 (2018)
22. Hoffman, M.D., Johnson, M.J.: ELBO surgery: yet another way to carve up the variational evidence lower bound. In: Workshop in Advances in Approximate Bayesian Inference, pp. 1–4 (2016)
23. Krizhevsky, A., Sutskever, I., Hinton G.E.: ImageNet classification with deep convolutional neural networks. In: Advances in Neural Information Processing Systems, pp. 1097–1105 (2012)
24. LeCun, Y., Bengio, Y., Hinton, G.: Deep learning. Nature 521(7553), 436 (2015)
25. Zhang, S., Yao, L., Sun, A.: Deep learning based recommender system: a survey and new perspectives. ACM Comput. Surv. (CSUR) 52(1), 5 (2019)
26. Rezende, D.J., Mohamed, S., Wierstra, D.: Stochastic backpropagation and approximate inference in deep generative models. In: Proceedings of the 31st International Conference on International Conference on Machine Learning, pp. 1278–1286 (2014)
27. Bowman, S.R., Vilnis, L., Vinyals, O., Dai, A.M., Jozefowicz, R., Bengio, S.: Generating sentences from a continuous space. In: Proceedings of The 20th SIGNLL Conference on Computational Natural Language Learning, pp. 10–21 (2016)
28. Rendle, S., Freudenthaler, C., Gantner, Z., Schmidt-Thieme, L.: BPR: Bayesian personalized ranking from implicit feedback. In: Proceedings of the 25th Conference on Uncertainty in Artificial Intelligence, pp. 452–461 (2009)

29. Shi, S., Zhang, M., Liu, Y., Ma, S.: Attention-based adaptive model to unify warm and cold starts recommendation. In: Proceedings of the 27th ACM International Conference on Information and Knowledge Management, pp. 127–136 (2018)
30. He, X., Chen, T., Kan, M.Y., Chen, X.: TriRank: review-aware explainable recommendation by modeling aspects. In: Proceedings of the 24th ACM International on Conference on Information and Knowledge Management, pp. 1661–1670 (2015)

Author Index